Statistical Physics

The Manchester Physics Series

General Editors

F. MANDL: R. J. ELLISON: D. J. SANDIFORD

*Physics Department, Faculty of Science,
University of Manchester*

Properties of Matter:	B.H. Flowers and E. Mendoza
Optics: *Second Edition*	F.G. Smith and J.H. Thomson
Statistical Physics: *Second Edition*	F. Mandl
Solid State Physics: *Second Edition*	J.R. Hook and H.E. Hall
Electromagnetism: *Second Edition*	I.S. Grant and W.R. Phillips
Statistics:	R.J. Barlow
Particle Physics:	B.R. Martin and G. Shaw
Quantum Mechanics:	F. Mandl

STATISTICAL PHYSICS

Second Edition

F. Mandl

*Department of Theoretical Physics,
University of Manchester*

John Wiley & Sons

CHICHESTER   NEW YORK   BRISBANE   TORONTO   SINGAPORE

First published 1971

Second edition 1988

Reprinted November 1989

Reprinted November 1991

Library of Congress Cataloging-in-Publication Data:

Mandl, F. (Franz), (date)
  Statistical physics.

  (The Manchester physics series)
  Bibliography: p.
  Includes index.
  1. Statistical physics.   I. Title.   II. Series.
  QC174.8.M27   1988   530.1'3   87-8283

ISBN 0 471 91532 7
ISBN 0 471 91533 5 (pbk.)

British Library Cataloguing in Publication Data:

Mandl, F.
  Statistical physics.—2nd ed.—
  (The Manchester physics series).
  1. Statistical mechanics
  I. Title   II. Series
  530.1'3   QC174.8

ISBN 0 471 91532 7
ISBN 0 471 91533 5 Pbk

Phototypeset by Dobbie Typesetting Service,
Plymouth, Devon
Printed and Bound in Great Britain by
Courier International Limited, East Kilbride, Scotland

Editors' Preface to the Manchester Physics Series

The first book in the Manchester Physics Series, *Properties of Matter*, was published in 1970. Since then, eight volumes have appeared in the series and we have been extremely encouraged by the response of readers, both colleagues and students. These books have been reprinted many times and are being used world-wide in the English language edition and in translations. All the same, both the editors and authors feel the time is right for a revision of some of the books in order to take into account the feedback received and to reflect the changing style and needs of undergraduate courses.

The Manchester Physics Series is a series of textbooks on physics at undergraduate level. It grew out of our experience at Manchester University Physics Department, widely shared elsewhere, that many textbooks contain much more material than can be accommodated in a typical undergraduate course and that this material is only rarely so arranged as to allow the definition of a shorter self-contained course. In planning these books, we have had two objectives. One was to produce short books: so that lecturers should find them attractive for undergraduate courses; so that students should not be frightened off by their encyclopaedic size or their price. To achieve this, we have been very selective in the choice of topics, with the emphasis on the basic physics together with some instructive, stimulating and useful applications. Our second aim was to produce books which allow courses of different length and difficulty to be selected, with emphasis on different applications. To achieve such flexibility we have encouraged authors to use flow diagrams showing the logical connections between different chapters and to put some topics in starred sections. These cover more advanced and alternative material which is not required for the understanding of later parts of each volume. Although these books were conceived as a series, each of them is self-contained and can be used independently of the others. Several of them are suitable for use outside physics courses; for example, *Electronics* is used by electronic engineers, *Properties of Matter* by chemists and metallurgists. Each author's preface gives details about the level, prerequisites, etc., of his volume.

We are extremely grateful to the many students and colleagues, at Manchester and elsewhere, whose helpful criticisms and stimulating comments have led to many improvements. Our particular thanks go to the authors for all the work they have done, for the many new ideas they have contributed, and for discussing patiently, and often accepting, our many suggestions and requests. We would also like to thank the publishers, John Wiley & Sons, who have been most helpful.

F. MANDL
R. J. ELLISON
D. J. SANDIFORD

January, 1987

Preface to the Second Edition

My motivation for producing this second edition is to introduce two changes which, I believe, are substantial improvements.

First, I have decided to give much greater prominence to the Gibbs distribution. The importance of this formulation of statistical mechanics is due to its generality, allowing applications to a wide range of systems. Furthermore, the introduction of the Gibbs distribution as the natural generalization of the Boltzmann distribution to systems with variable particle numbers brings out the simplicity of its interpretation and leads directly to the chemical potential and its significance. In spite of its generality, the mathematics of the Gibbs approach is often much simpler than that of other approaches. In the first edition, I avoided the Gibbs distribution as far as possible. Fermi–Dirac and Bose–Einstein statistics were derived within the framework of the Boltzmann distribution. In this second edition, they are obtained much more simply taking the Gibbs distribution as the starting point. (For readers solely interested in the Fermi–Dirac and Bose–Einstein distributions, an alternative derivation, which does not depend on the Gibbs distribution, is given in section 11.3.) The shift in emphasis to the Gibbs approach has also led me to expand the section on chemical reactions, both as regards details and scope.

Secondly, I have completely revised the treatment of magnetic work in section 1.4, with some of the subtler points discussed in the new Appendix C. As is well known, the thermodynamic discussion of magnetic systems easily leads to misleading or even wrong statements, and I fear the first edition was not free from these. My new account is based on the work of two colleagues of mine, Albert Hillel and Pat Buttle, and represents, I believe, an enlightening addition to the many existing treatments.

I have taken this opportunity to make some other minor changes: clarifying some arguments, updating some information and the bibliography, etc. Many of these points were brought to my attention by students, colleagues, correspondents and reviewers, and I would like to thank them all—too many to mention by name—for their help.

I would like to thank Henry Hall, Albert Hillel and Peter Lucas for reading the material on the Gibbs distribution, etc., and suggesting various improvements. I am most grateful to Albert Hillel and Pat Buttle for introducing me to their work on magnetic systems and for allowing me to use it, for many discussions and for helpful comments on my revised account. It is a pleasure to thank David Sandiford for his help throughout this revision, particularly for critically reading all new material, and Sandy Donnachie for encouraging me to carry out this work.

January 1987   FRANZ MANDL

Preface to First Edition

This book is intended for an undergraduate course in statistical physics. The laws of statistical mechanics and thermodynamics form one of the most fascinating branches of physics. This book will, I hope, impart some of this fascination to the reader. I have discarded the historical approach of treating thermodynamics and statistical mechanics as separate disciplines in favour of an integrated development. This has some decisive advantages. Firstly, it leads more directly to a deeper understanding since the statistical nature of the thermodynamic laws, which is their true explanation, is put in the forefront from the very beginning. Secondly, this approach emphasizes the atomic nature of matter which makes it more stimulating and, being the mode of thought of most working physicists, is a more useful training. Thirdly, this approach is more economical on time, an important factor in view of the rapid expansion of science.

It is a consequence of this growth in scientific knowledge that an undergraduate physics course can no longer teach the whole of physics. There are many ways of selecting material. I have tried to produce a book which allows maximum flexibility in its use: to enable readers to proceed by the quickest route to a particular topic; to enable teachers to select courses differing in length, difficulty and choice of applications. This flexibility is achieved by means of the flow diagram (on the inside front cover) which shows the logical connections of the chapters. In addition, some sections are marked with a star ★ and some material, insufficient to justify a separate section, is printed on a tinted background. Material distinguished in either of these ways may be omitted. It is not needed later except very occasionally in similarly marked parts, where explicit cross-references are always given.

My aim has been to explain critically the basic laws of statistical physics and to apply them to a wide range of interesting problems. A reader who has mastered this book should have no difficulties with one of the more advanced treatises or with tackling quite realistic problems. I have limited myself to systems in equilibrium, omitting irreversible thermodynamics, fluctuation phenomena and transport theory. This was partly for reasons of time and space, but largely because these topics are hardly appropriate for a fairly elementary account. For this reason also, I have not discussed the foundations of statistical physics but have based the theory on some simple intuitively plausible axioms. The ultimate justification of this approach lies in its success.

The development of statistical physics which I have given is self-contained, but the level of sophistication presupposes some previous acquaintance with the kinetic theory of gases, with the elementary descriptive ideas of atomic physics and with the rudiments of quantum theory. Fortunately, very little of the latter is required.

Over the past ten years I have given various undergraduate and post-graduate courses on statistical physics at Manchester University. In its present form, this book developed out of a course given to second-year undergraduates in physics, chemical physics and electronic engineering. This course of 26 lectures, of 50 minutes each, approximately covered the unstarred sections of the book, as well as the material of chapter 5 and of sections 7.5, 7.7, 11.4 and 11.5,* omitting all material printed on tinted background. In addition, students were expected to solve about 20 problems. The answers were corrected, returned together with sample solutions, and discussed in class.

The problems and hints for solving them form an important part of the book. Attempting the problems and *then* studying the hints will deepen the reader's understanding and develop his skill and self-confidence in facing new situations. The problems contain much interesting physics which might well have found its way into the main body of the text.

This book was preceded by a preliminary edition which was used in my lecture course and was also distributed fairly widely outside Manchester University. I have received many comments and suggestions for improvements and additions from readers. I also had many stimulating discussions with students and colleagues at Manchester. As a result the original text has been greatly improved. I would like to thank all these people most warmly; there are too many of them to thank them all by name. However, I would

---

*In the second edition, the numbers of these sections have become 11.5 and 11.6.

like to express my appreciation for their help to Professor Henry Hall and to Dr David Sandiford who read the whole manuscript and with whom I discussed difficult and obscure points until—temporarily at least—they seemed clear. Not only was this intellectual pursuit of great benefit to this book, but to me it was one of the joys of writing it.

May, 1970   F. MANDL

Contents

★ Starred sections may be omitted as they are not required later in the book.

11   SYSTEMS WITH VARIABLE PARTICLE NUMBERS

A   MATHEMATICAL RESULTS

B   THE DENSITY OF STATES

C   MAGNETIC SYSTEMS

D   HINTS FOR SOLVING PROBLEMS

*Sections 11.1–11.2 and section 11.3 are alternative treatments which can be read independently of each other. Either suffices for the applications in sections 11.4 to 11.6. Sections 11.7 to 11.9 depend on section 11.1 only.

The first law of thermodynamics

1.1 MACROSCOPIC PHYSICS

Statistical physics is devoted to the study of the physical properties of macroscopic systems, i.e. systems consisting of a very large number of atoms or molecules. A piece of copper weighing a few grams or a litre of air at atmospheric pressure and room temperature are examples of macroscopic systems. In general the number of particles in such a system will be of the order of magnitude of Avogadro's number $N_0 = 6 \times 10^{23}$. Even if one knows the law of interaction between the particles, the enormousness of Avogadro's number precludes handling a macroscopic system in the way in which one would treat a simple system—say planetary motion according to classical mechanics or the hydrogen molecule according to quantum mechanics. One can never obtain experimentally a complete microscopic* specification of such a system, i.e. a knowledge of some 10^{23} coordinates. Even if one were given this initial information, one would not be able to solve the equations of motion; some 10^{23} of them!

In spite of the enormous complexity of macroscopic bodies when viewed from an atomistic viewpoint, one knows from everyday experience as well as from precision experiments that macroscopic bodies obey quite definite

---

*'Microscopic' is here used in contrast to 'macroscopic'. It means a complete atomistic specification.

Gas

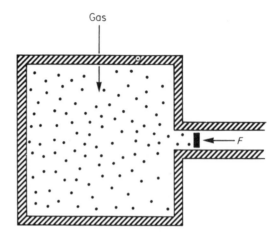

Fig. 1.1. Gas exerting pressure on movable piston, balanced by
external applied force F.

laws. Thus when a hot and a cold body are put into thermal contact
temperature equalization occurs; water at standard atmospheric pressure
always boils at the same temperature (by definition called 100 °C); the pressure
exerted by a dilute gas on a containing wall is given by the ideal gas laws.
These examples illustrate that the laws of macroscopic bodies are quite
different from those of mechanics or electromagnetic theory. They do not
afford a complete microscopic description of a system (e.g. the position of
each molecule of a gas at each instant of time). They provide certain
macroscopic observable quantities, such as pressure or temperature. These
represent averages over microscopic properties. Thus the macroscopic laws
are of a statistical nature. But because of the enormous number of particles
involved, the fluctuations which are an essential feature of a statistical theory
turn out to be extremely small. In practice they can only be observed under
very special conditions. In general they will be utterly negligible, and the
statistical laws will in practice lead to statements of complete certainty.

To illustrate these ideas consider the pressure exerted by a gas on the walls
of a containing vessel. We measure the pressure by means of a gauge attached
to the vessel. We can think of this gauge as a freely movable piston to which
a variable force F is applied, for example by means of a spring (Fig. 1.1).
When the piston is at rest in equilibrium the force F balances the pressure
P of the gas: $P = F/A$ where A is the area of the piston.

In contrast to this macroscopic determination of pressure consider how
the pressure actually comes about.* According to the kinetic theory the

---

*For a detailed derivation of the pressure of a perfect gas from kinetic theory, see Flowers
and Mendoza,[26] section 5.1.2, R. Becker,[2] section 24, or Present,[11] Chapter 2.

molecules of the gas are undergoing elastic collisions with the walls. The pressure due to these collisions is certainly not a strictly constant time-independent quantity. On the contrary the instantaneous force acting on the piston is a rapidly fluctuating quantity. By the pressure of the gas we mean the average of this fluctuating force over a time interval sufficiently long for many collisions to have occurred in this time. We may then use the steady-state velocity distribution of the molecules to calculate the momentum transfer per unit area per unit time from the molecules to the wall, i.e. the pressure. The applied force F acting on the piston can of course only approximately balance these irregular impulses due to molecular collisions. On average the piston is at rest but it will perform small irregular vibrations about its equilibrium position as a consequence of the individual molecular collisions. These small irregular movements are known as Brownian motion (Flowers and Mendoza,[26] section 4.4.2). In the case of our piston, and generally, these minute movements are totally unobservable. It is only with very small macroscopic bodies (such as tiny particles suspended in a liquid) or very sensitive apparatus (such as the very delicate suspension of a galvanometer — see section 7.9.1) that Brownian motion can be observed. It represents one of the ultimate limitations on the accuracy of measurements that can be achieved.

There are two approaches to the study of macroscopic physics. Historically the oldest approach, developed mainly in the first half of the 19th century by such men as Carnot, Clausius, William Thomson (the later Lord Kelvin), Robert Mayer and Joule, is that of classical thermodynamics. This is based on a small number of basic principles — the laws of thermodynamics — which are deductions from and generalizations of a large body of experiments on macroscopic systems. They are phenomenological laws, justified by their success in describing macroscopic phenomena. They are not derived from a microscopic picture but avoid all atomic concepts and operate exclusively with macroscopic variables, such as pressure, volume, temperature, describing the properties of systems in terms of these. Of course, the avoidance of atomic concepts severely limits the information that thermodynamics can provide about a system. In particular, the equation of state (e.g. for an ideal gas: $PV = RT$) which relates the macroscopic variables and which distinguishes one system from another must be derived from experiment. But there are many situations where a microscopic description is not necessary or not practicable and where thermodynamics proves its power to make far-reaching deductions of great generality.*

The second approach to macroscopic physics is that of statistical mechanics. This starts from the atomic constitution of matter and endeavours to derive

*For a superb if not easy account of classical thermodynamics, showing its aesthetic appeal, logical structure and power, see the book by Pippard.[9]

the laws of macroscopic bodies from the atomic properties. This line of approach originated in Maxwell's kinetic theory of gases which led to the profound works of Boltzmann and of Gibbs. There are two aspects to statistical mechanics. One aim is to *derive* the thermodynamic laws of macroscopic bodies from the laws governing their atomic behaviour. This is a fascinating but very difficult field. Nowadays one has a fairly general understanding of the underlying physics but most physicists working in the field would probably agree that no real proofs exist. In this book we shall not consider these aspects of statistical mechanics and shall only give arguments which make the thermodynamic laws plausible from the microscopic viewpoint.

The second objective of statistical mechanics is to derive the properties of a macroscopic system — for example, its equation of state — from its microscopic properties. Essentially this is done by averaging over unobservable microscopic coordinates leaving only macroscopic coordinates such as the volume of a body, as well as other macroscopic variables, such as temperature or specific heat, which have no counterpart in mechanics and which represent averages over unobservable microscopic coordinates.

This division of macroscopic physics into thermodynamics and statistical mechanics is largely of historical origin. We shall not follow this development. Instead we shall emphasize the unity of the subject, showing how the two aspects illuminate each other, and we shall use whichever is more appropriate.

1.2 SOME THERMAL CONCEPTS

Some of the variables which were introduced in the last section to describe a macroscopic system, such as its volume or pressure, have a direct meaning in terms of mechanical concepts, e.g. one can measure the pressure of gas in a container by means of a mercury manometer. However, some of the concepts are quite foreign to mechanics. Of these the one most basic to the whole of statistical thermodynamics is that of temperature. Originally temperature is related to the sensations of 'hot' and 'cold'. The most remarkable feature of temperature is its tendency to equalization: i.e. if a hot and a cold body are put into thermal contact, the hot body cools down and the cold body warms up until both bodies are at the same temperature. This equalization is due to a net flow of energy from the hotter to the colder body. Such a flow of energy is called a flow of heat. When this flow of heat ceases, the two bodies are in thermal equilibrium. The basic fact of experience which enables one to compare the temperatures of two bodies by means of a third body is that if two bodies are each in thermal equilibrium with a third body they are also in thermal equilibrium with each other. This statement is sometimes referred to as the zeroth law of thermodynamics. To measure temperature, one can utilize any convenient property of matter which depends on its degree of hotness, such as the electric resistance of a platinum wire, the volume (i.e. length in a glass capillary) of a mass of mercury, the pressure of a given mass of gas contained in a fixed volume. For each of these thermometers one can then define a Celsius (centigrade) scale by calling the temperatures of the ice and steam points* 0 °C and 100 °C and interpolating linearly for other temperatures. It turns out that these different temperature scales do not agree exactly (except at the fixed points, of course). They depend on the particular thermometer used. We shall see presently that this arbitrariness is removed by the second law of thermodynamics which enables one to define an *absolute temperature scale*, i.e. one which is independent of the experimental arrangement used for measuring the temperature. The physical meaning of the absolute temperature is revealed by statistical mechanics. It turns out to be a measure of the energy associated with the molecular, macroscopically unobserved, motions of a system.

Above we considered temperature equilibrium. More generally, let us consider an isolated system. This system may be in a state containing all sorts of pressure differences, temperature gradients, inhomogeneities of density, concentrations, etc. A system in such a state is of course not in equilibrium. It will change with time as such processes as pressure equalization, thermal conduction, diffusion, etc., occur. Left to itself, the system eventually reaches a state in which all these pressure gradients, etc., have disappeared and the system undergoes no further macroscopically observable changes. We call such a state an *equilibrium state*. Of course, this is not static equilibrium. Sufficiently refined experiments will show up the thermal motions, a typical example being Brownian motion. The time that a system requires to reach equilibrium depends on the processes involved. In general there will be several mechanisms, as we have seen; each will possess its own characteristic relaxation time. After a time long compared to all relaxation times the system will be in equilibrium.

On the other hand there are frequently situations where the relaxation time for a particular process is very long compared with the time for which a system is observed. One can then ignore this process altogether. It occurs too slowly to be of any consequence. In many cases the relaxation time is for practical purposes infinite. Consider a binary alloy, for example β-brass which consists of Cu and Zn atoms in equal numbers. At sufficiently low temperatures, the stable equilibrium configuration of the atoms is one where they are ordered in a regular mosaic-like pattern in the crystal lattice. No such ordering occurs at high temperatures. The two situations are schematically illustrated for a two-dimensional model lattice in Figs. 1.2(a) and (b). If such an

*The ice point is defined as the temperature at which pure ice coexists in equilibrium with air-saturated water at a pressure of one standard atmosphere. The steam point is the temperature at which pure water and pure steam coexist in equilibrium at a pressure of one standard atmosphere.

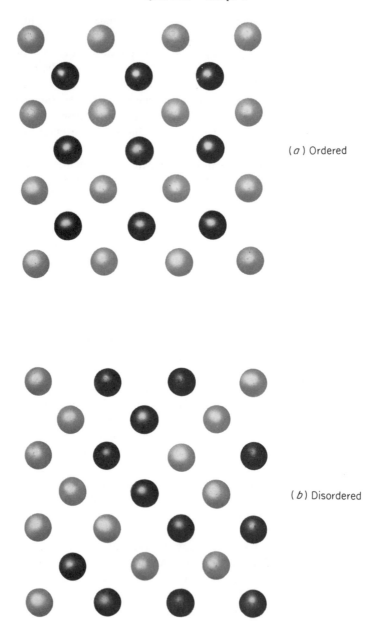

Fig. 1.2. Schematic two-dimensional model of a binary alloy: (*a*) in ordered state, (*b*) in disordered state.

alloy is rapidly cooled from a high to a low temperature, the atoms get 'frozen' into their instantaneous disordered pattern. This is a metastable state but the rate of migration of the atoms at the low temperature is so small that for practical purposes the disorder will persist for all times.

In β-brass the Cu and Zn atoms each form a simple cubic lattice, the two lattices being interlocked so that each Cu atom is at the centre of a cube formed by 8 Zn atoms, and vice versa. There is an attractive force between the Cu and Zn atoms. At low temperatures this attraction dominates over the comparatively feeble thermal motion resulting in an ordered state, but at high temperatures the thermal agitation wins. The ordering shows up as extra diffraction lines in x-ray diffraction, since the two types of atom will scatter x-rays differently.

We have discussed relaxation times in order to explain what is meant by equilibrium. The calculation of how long it takes for equilibrium to establish itself, and of non-equilibrium processes generally, is extremely difficult. We shall not consider such questions in this book but shall exclusively study the properties of systems in equilibrium without inquiring how they reached equilibrium. But we shall of course require a criterion for characterizing an equilibrium state. The second law of thermodynamics provides just such a criterion.

The description of a system is particularly simple for equilibrium states. Thus for a fluid not in equilibrium it may be necessary to specify its density at every point in space as a function of time, whereas for equilibrium the density is uniform and constant in time. The equilibrium state of a system is fully determined by a few macroscopic variables. These variables then determine all other macroscopic properties of the system. Such properties which depend only on the state of a system are called *functions of state*. The state of a homogeneous fluid is fully determined by its mass M, volume V, and pressure P. Its temperature T is then a function of state determined by these, i.e.

$$T = f(P, V, M) \ . \tag{1.1}$$

Eq. (1.1) is called the equation of state of the fluid. Of course, we could have chosen other independent variables to specify the state of the fluid, for example M, V and T, and found P from Eq. (1.1).

In our discussion of a fluid we tacitly assumed the characteristic property of a fluid: that its thermodynamic properties are independent of its shape. This makes a fluid a very simple system to discuss. More complicated systems require a larger number of parameters to determine a unique state and lead to a more complicated equation of state. This mode of description of a system breaks down if its state depends not only on the instantaneous values of certain parameters but also on its previous history, i.e. in the case of

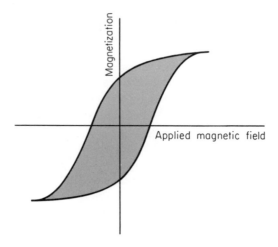

Fig. 1.3. Hysteresis in a ferromagnetic material.

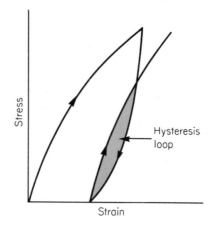

Fig. 1.4. Stress-strain relationship in a solid showing
the hysteresis loop.

hysteresis effects such as occur in ferromagnetic materials or the plastic deformation of solids. In the former example the magnetization is not a unique function of the applied magnetic field (Fig. 1.3); in the latter, the strain is not a unique function of the applied stress (Fig. 1.4).*

In general the equation of state of a substance is very complicated. It must be found from experiment and does not allow a simple analytic representation. The perfect (or ideal) gas is an exception. For real gases, at sufficiently

*The elastic example is discussed by Flowers and Mendoza,[26] section 9.1.1, and more fully by A. H. Cottrell,[25] section 4.4.

low pressures, the pressure and volume of a fixed mass of gas are very nearly related by

$$PV = \text{const.} \tag{1.2}$$

at a given temperature. An equation such as (1.2), relating different states of a system, all at the *same* temperature, is called an *isotherm*. A perfect gas is defined to be a fluid for which the relation (1.2) holds *exactly* for an isotherm, i.e. a perfect gas represents an extrapolation to zero pressure from real gases. We can use this to define a (perfect) gas temperature scale T by the relation

$$T \propto \lim_{P \to 0} PV \ . \tag{1.3}$$

The gas temperature scale is then completely determined if we fix *one* point on it *by definition*. This point is taken as the triple point of water, i.e. the temperature at which ice, water and water vapour coexist in equilibrium. The reason for this choice is that the triple point corresponds to a unique temperature and pressure of the system (see section 8.3). The triple point temperature T_{tr} was chosen so that the size of the degree on the gas scale equals as nearly as possible the degree Celsius, i.e. according to the best available measurements there should be a temperature difference of 100 degrees between the steam and ice points. This criterion led to

$$T_{tr} \equiv 273.16 \, \text{K} \tag{1.4}$$

being internationally adopted in 1954 as the *definition* of the triple point. (Very accurate measurements, in the future, of the steam and ice points on this temperature scale may result in their temperature difference being not *exactly* 100 degrees.) In Eq. (1.4) we have written K (kelvin) in anticipation of the fact that the gas scale will turn out to be identical with the absolute thermodynamic temperature scale. (Older notations for K are deg. or °K.) Any other absolute temperature is then, in principle, determined from Eqs. (1.3) and (1.4). The temperature of the ice point becomes 273.15 K.

The constant of proportionality, still missing in Eq. (1.3), is determined from accurate measurements with gas thermometers. For one mole (we shall always use the gram-mole) of gas one finds that

$$PV = RT \tag{1.5}$$

with the gas constant R having the value

$$R = 8.31 \, \text{J} \, \text{mol}^{-1} \, \text{K}^{-1} \ . \tag{1.6}$$

From Avogadro's number

$$N_0 = 6.02 \times 10^{23} \text{ molecules/mole} , \tag{1.7}$$

we can calculate Boltzmann's constant k, i.e. the gas constant per molecule

$$k \equiv R/N_0 = 1.38 \times 10^{-23} \text{ J/K} . \tag{1.8}$$

The equation of state of a perfect gas consisting of N molecules can then be written

$$PV = NkT . \tag{1.9}$$

The physically significant quantity in this equation is the energy kT. Under classical conditions, i.e. when the theorem of equipartition of energy holds (see, for example, section 7.9.1 below, or Flowers and Mendoza,[26] sections 5.3 and 5.4.4), kT is of the order of the energy of one molecule in a macroscopic body at temperature T. By contrast, Boltzmann's constant is merely a measure of the size of the degree Celsius. At $T = 290$ K (room temperature)

$$kT = 4.0 \times 10^{-21} \text{ J} = \tfrac{1}{40} \text{ eV} \tag{1.10}$$

where we introduced the electron-volt (eV):

$$\begin{aligned} 1 \text{ eV} &= 1.60 \times 10^{-19} \text{ J} \\ &= 1.60 \times 10^{-12} \text{ erg} . \end{aligned} \tag{1.11}$$

The electron-volt is a reasonably-sized unit of energy on the atomic scale. For example, the ionization energy of atoms varies from about 4 eV to about 24 eV; the cohesive energy of solids varies from about 0.1 eV to about 10 eV per molecule, depending on the type of binding force.

1.3   THE FIRST LAW

We shall now consider the application of the universally valid principle of conservation of energy to macroscopic bodies. The new feature, which makes this different from merely a very complicated problem in mechanics, is that we do not want to describe the system on the microscopic scale, i.e. in terms of the individual molecular motions. This is of course impossibly complicated. Instead we want to describe the motion associated with these internal degrees of freedom in terms of macroscopic parameters.

Consider a system enclosed in walls impervious to heat transmission. Such walls are called adiabatic walls. (In practice one uses a dewar flask to obtain these conditions.) We can change the state of such a thermally isolated system by doing work on it. There is overwhelming experimental evidence that for a change from a definite state 1 to another definite state 2 of the system the same amount of work W is required irrespective of the mechanism used to perform the work or the intermediate states through which the system passes. Historically the earliest precise evidence comes from Joule's work, published in 1843, on the mechanical equivalent of heat. He produced given changes of state in a thermally isolated liquid in different ways. These included vigorously stirring the liquid with a paddle-wheel driven by weights (Fig. 1.5) and supplying electrical work by inserting a resistor carrying a current in the liquid (Fig. 1.6). The work done on the system—known in the first case from the motion of the weights, in the second from the current through the resistor and the potential drop across it—is the same in both cases.

We can hence define a *function of state E*, such that for a change from a state 1 to a state 2 of a *thermally isolated system* the work done on the system equals the change in E:

$$W = \Delta E \equiv E_2 - E_1 \ . \tag{1.12}$$

E is called the energy of the system. Except for an arbitrary choice of the zero of the energy scale (i.e. of the energy of a standard reference state) Eq. (1.12) determines the energy of any other state.

Suppose we now consider changes of state of the system no longer thermally isolated. It turns out that we can in general still effect the same change from state 1 to state 2 of the system but in general the work W done on the system does not equal the increase in energy ΔE of the system. We define the deficit

$$Q = \Delta E - W \tag{1.13}$$

as the heat supplied *to* the system. Eq. (1.13) is the general statement of the first law of thermodynamics. It is the law of conservation of energy applied to processes involving macroscopic bodies. The concept of heat, as introduced here, has all the properties associated with it from calorimetry experiments, etc. These are processes in which no work is done, the temperature changes being entirely due to heat transfer.

Let us consider how the energy E of a given state of a macroscopic system subdivides. (For definiteness you might think of the system as a gas or a crystal.) According to the laws of mechanics, the energy E is the sum of two contributions: (i) the energy of the macroscopic mass motion of the system, (ii) the internal energy of the system.

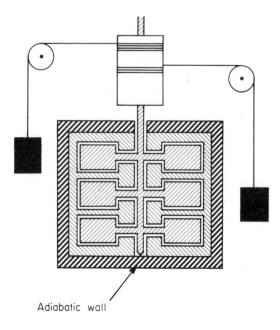

Adiabatic wall

Fig. 1.5. Schematic picture of Joule's paddle-wheel experiment. A system for doing mechanical work on the liquid in the calorimeter.

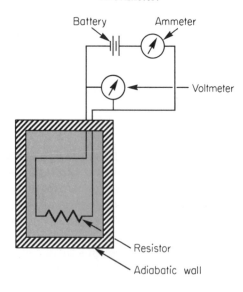

Fig. 1.6. A system for doing electrical work on the liquid in the calorimeter.

The energy of the mass motion consists of the kinetic energy of the motion of the centre of mass of the system, plus any potential energy which the system might possess due to the presence of an external field of force. For example, the system might be in a gravitational field. In statistical physics one is usually interested in the internal properties of systems, not in their macroscopic mass motion. Usually we shall be considering systems at rest and the potential energy of any external fields will be unimportant so that we shall not distinguish between the energy and the internal energy of a system.

The internal energy of a system is the energy associated with its internal degrees of freedom. It is the kinetic energy of the molecular motion (in a frame of reference in which the system is at rest) plus the potential energy of interaction of the molecules with each other. In an ideal gas at rest the internal energy is the sum of the kinetic energies of the translational motions of the molecules plus the internal energies of the molecules due to their rotations, etc. In a crystal the internal energy consists of the kinetic and potential energies of the atoms vibrating about their equilibrium positions in the crystal lattice. Thus the internal energy is the energy associated with the 'random' molecular motion of the system. We shall see later that the temperature of a system is a measure of its internal energy, which is therefore also called the thermal energy of the system.

The internal energy of a system is a function of state. For a fluid we could write $E = E(P, T)$ or $E = E(V, T)$, depending on which independent variables we choose to specify the state of the fluid. (We have suppressed the dependence on the mass of the fluid in these expressions for E as we shall usually be considering a constant mass, i.e. size of system, and are only interested in the variation of the other variables. In most cases the dependence on the size is trivial.) Thus for the change of a system from a state 1 to a state 2, ΔE in Eq. (1.13) is the difference of two energies, E_1 and E_2, for these two states as given by Eq. (1.12). By contrast Q and W are *not* changes in functions of state. There exists *no* function of state 'heat of a system' such that the system has a definite 'heat' in state 1 and a definite 'heat' in state 2, with Q the difference of these 'heats'. Similarly there exists *no* function of state 'work of a system' such that the system has a definite 'work' in state 1 and a definite 'work' in state 2, with W the difference of these 'works'. It follows that there is no conservation of 'heat' by itself, nor conservation of 'work' by itself. We only have conservation of energy, given by Eq. (1.13). *Work and heat flow are different forms of energy transfer.* The physical distinction between these two modes is that work is energy transfer via the macroscopically observable degrees of freedom of a system, whereas heat flow is the direct energy transfer between microscopic, i.e. internal, degrees of freedom. For examples of these two modes of energy transfer we again consider a gas. If the gas is contained in a thermally isolated cylinder, closed off at one end by a movable piston (Fig. 1.7), then work can be done on

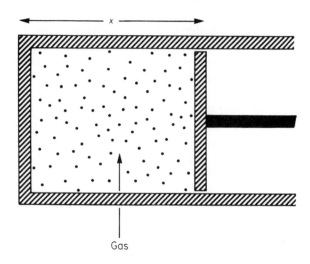

Fig. 1.7. Adiabatic compression of a gas.

the gas by compressing it. The macroscopic degree of freedom here corresponds to the position coordinate x of the piston. During the compression the gas is warmed up. From the molecular standpoint this warming up comes about because in elastic collisions with the moving piston the molecules gain energy which, as a result of subsequent collisions between molecules, is shared by all of them. Next assume that the gas is contained in a vessel with fixed walls and that there exists a temperature gradient in the gas. If we consider an element of area normal to this gradient, then a net transport of energy occurs across this area. This is the process of thermal conduction in the gas. Its explanation on the molecular scale is that molecules traversing this element of area from opposite sides possess different kinetic energies on average, corresponding to the different temperatures which exist in the regions from which those molecules came (for details, see Flowers and Mendoza,[26] Chapter 6, or Present,[11] Chapter 3).

Eq. (1.13) expresses the conservation of energy for finite changes. For infinitesimal changes we correspondingly write

$$dE = đQ + đW \ . \tag{1.14}$$

Here dE is the infinitesimal change in the energy of the system, brought about by an infinitesimal amount of work $đW$ and an infinitesimal heat transfer $đQ$. We write $đW$ and $đQ$ (*not* dW and dQ) to emphasize that, as discussed, these infinitesimal quantities are *not* changes in functions of state.

For a change from a definite state 1 to a definite state 2, ΔE is determined and hence, from Eq. (1.13), so is $(Q + W)$; but not Q and W separately.

Q and *W* depend on *how* the change from state 1 to state 2 takes place, i.e. *on the particular path taken by the process.* (Corresponding statements hold for infinitesimal changes.) Of course, for adiabatic changes, $Q = 0$, the work is determined by initial and final states only, as we saw in Eq. (1.12). Similarly for a change involving no work ($W = 0$), the heat transfer Q is determined. But these are the exceptions.

Of particular importance are *reversible changes*. For a process to be reversible it must be possible to reverse its direction by an infinitesimal change in the applied conditions. For a process to be reversible two conditions must be satisfied: (i) it must be a quasistatic process; (ii) there must be no hysteresis effects.

A *quasistatic* process is defined as a succession of equilibrium states of the system. Thus it represents an idealization from reality. For, to produce actual changes one must always have pressure differences, temperature differences, etc. But by making these sufficiently small one can ensure that a system is arbitrarily close to equilibrium at any instant. Strictly speaking, processes must occur infinitely slowly under these conditions. But in practice a process need only be slow compared to the relevant relaxation times in order that it may be considered quasistatic.

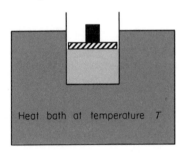

Fig. 1.8. Isothermal compression of a gas.

The importance of reversible processes is that for these the work performed on the system is well defined by the properties of the system. Consider the isothermal compression of a gas contained in a cylinder closed off by a piston (Fig. 1.8). To ensure isothermal compression (i.e. at constant temperature) the cylinder is placed in thermal contact with a *heat bath* at temperature *T*. By a heat bath we mean a body whose heat capacity is *very large* compared to that of the system it serves. Because of its large heat capacity, the temperature of the heat bath stays constant in spite of heat exchange with the system. The system is then also at the same constant temperature when in thermal equilibrium with the heat bath. To perform the compression quasistatically, the weight on the piston must be increased in a large number of very small increments. After each step we must wait for thermal and

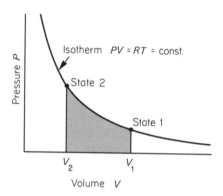

Fig. 1.9. The isotherm of an ideal gas. The shaded area
is the work W done on the gas in compressing it
isothermally from volume V_1 to V_2, Eq. (1.16).

mechanical equilibrium to establish itself. At any instant the pressure of the
gas is then given from the equation of state in terms of the volume V,
temperature T and mass M of gas. Let us consider one mole of an ideal gas.
The locus of equilibrium states along which the quasistatic compression occurs
is then given by the perfect gas law, Eq. (1.5), with T the temperature of
the heat bath. This isotherm is plotted on the (P, V) diagram in Fig. 1.9.
The work done on the gas in compressing it from V to $V + dV$ is

$$\text{d}W = -P\,\text{d}V \; ; \tag{1.15a}$$

for compression one has $dV < 0$, making $\text{d}W > 0$. In a finite change from
volume V_1 to V_2 the work done on the system is

$$W = -\int_{V_1}^{V_2} P\,\text{d}V = RT\ln\frac{V_1}{V_2} \, . \tag{1.16}$$

The significance of carrying changes out slowly becomes clear if we consider
a fast change, such as a sudden compression of the gas in Fig. 1.8. We may
imagine the piston initially clamped in a fixed position with a weight resting
on it which exerts a pressure P_0 on the piston, which exceeds the gas pressure
P. If the piston is unclamped and the volume of gas changes by $dV (<0)$,
then the work done by the weight on the system is $\text{d}W = -P_0\,\text{d}V$; from
$P_0 > P$ it follows that (remember $dV < 0$!)

$$\text{d}W > -P\,\text{d}V \, . \tag{1.15b}$$

This inequality also holds for a sudden expansion of the gas, for example
by suddenly lifting the piston. This produces a rarefaction of gas near the

piston initially, and the work done by the gas $(-\text{d}W)$ during this expansion is less than the work which would be done by the gas during a slow expansion for which the gas pressure would be uniform: $(-\text{d}W) < P\text{d}V$, in agreement with Eq. (1.15b). An extreme example of this kind of process is the expansion of a gas into a vacuum (see the end of this section). In this case no work is done by the gas $(-\text{d}W = 0)$, but $P\text{d}V$ is of course positive. In these sudden volume changes of the gas, pressure gradients are produced. The equalization of such gradients through mass flow of the gas is of course an irreversible process. (This point is discussed further in section 2.1.)

In order that the work done on the system (Fig. 1.8) in compressing the gas from volume V to $V + \text{d}V$ be given by Eq. (1.15a), there must be no frictional forces between cylinder and piston. Only in this way is there equality between the applied pressure P_0 due to the weight and the pressure P of the gas: $P_0 = P$. Only in this case is the work done on the system by the applied pressure $(\text{d}W = -P_0\text{d}V)$ equal to $-P\text{d}V$. If there is friction then in compressing the gas the applied pressure P_0 must be greater than the gas pressure P, in order to overcome the frictional forces, and the work performed $\text{d}W$ exceeds $-P\text{d}V$, i.e. we have again obtained the inequality (1.15b). This inequality also holds for the expansion of the gas when there is friction. This relation between the applied pressure P_0 and the pressure P of the gas is illustrated in Fig. 1.10. The dashed curve shows the gas pressure P, the continuous curves the applied pressure P_0 for compression and for expansion. When there is no friction $(P_0 = P)$ the two continuous curves coalesce on top of the dashed curve. If we carry out the cycle ABCDA by first

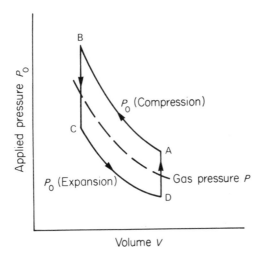

Fig. 1.10. Relation between applied pressure P_0 and volume V of a gas.

compressing and then expanding the gas, then the applied force has done a net amount of work on the system which is positive and is given by the area of the cycle ABCDA. This work is converted into thermal energy due to the friction, i.e. it produces a temperature rise. With the arrangement shown in Fig. 1.8 this thermal energy is of course passed on to the heat bath. The path ABCDA has the features of a hysteresis curve, similar to the cases illustrated in Figs. 1.3 and 1.4. All such hysteretic processes are irreversible. On reversing the external conditions, the system won't traverse the original path in the reverse direction but will follow a new path. It is only for a reversible process that the work done on the system can be expressed in terms of the parameters of state of the system; in the above example in terms of P and V.

We can sum up our conclusions about work in reversible and irreversible processes in the relations

$$\mathrm{d}W \geqslant -P\mathrm{d}V \tag{1.15c}$$

where $\mathrm{d}W$ is the work done on the system by external agencies while P and V are parameters of state of the system. The equality sign holds for reversible changes, the 'greater than' sign for irreversible changes.

Although the work done in a reversible change is well-defined, it does depend on the path, i.e. the intermediate states of the process. Thus we could join the states 1 and 2 in Fig. 1.9 by many paths other than the isotherm shown. Two such paths are shown in Fig. 1.11(a). The work done in going from state 1 to state 2 via paths A or B is given by the areas under the two curves respectively. For curve A

$$W_{\mathrm{A}} = -\int_{V_1}^{V_2} P\mathrm{d}V$$
$$\text{(path A)}$$

and similarly for path B. The work depends on the path. This result is put differently in Fig. 1.11(b). Consider the cyclic process: from state 1 via path A to state 2 and then via path B′ (which is the reverse of path B of Fig. 1.11(a)) back to state 1. The work done on the system in this cycle is represented by the shaded area of Fig. 1.11(b) and is given by

$$-\oint P\mathrm{d}V = W_{\mathrm{A}} - W_{\mathrm{B}} \neq 0 \ .$$

Thus the work around a complete cycle does not vanish. In contrast the change in internal energy around any complete cycle does vanish,

$$\oint \mathrm{d}E = 0 \ ,$$

because at *any* point on the path (e.g. at 1 in Fig. 1.11(b)) E has a definite value. *This distinction is characteristic of a function of state.*

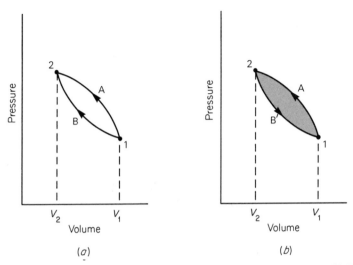

Fig. 1.11. (a) The work done in going from 1 to 2 along different paths. (b) The net work done in the cyclic process shown is given by the shaded area.

Having considered the work term in Eq. (1.14) we shall next look at the heat term in that equation. Combining it with Eq. (1.15a)—i.e. we are still considering a fluid—the first law, for *a reversible change*, becomes

$$\text{d}Q = \text{d}E + P\text{d}V \; . \tag{1.17}$$

From this equation we get the heat capacities. (We shall prefer the term heat capacity to the term specific heat which will be reserved for heat capacity per unit mass.) We shall consider one mole of fluid and so get molar heat capacities. Since $\text{d}Q$ does not define a state function, the heat capacity depends on the mode of heating the system.

The heat capacity at constant volume is given by

$$C_V = \left(\frac{\text{d}Q}{\partial T} \right)_V = \left(\frac{\partial E}{\partial T} \right)_V \; , \tag{1.18a}$$

i.e. it is the ratio of the quantity of heat $\text{d}Q$ to the temperature change $\text{d}T$ which this produces, the volume V of the system being kept constant. In the last expression in Eq. (1.18a), we must think of E as a function of V and T: $E = E(V, T)$.

The heat capacity at constant pressure is similarly defined by

$$C_P = \left(\frac{\text{d}Q}{\partial T} \right)_P = \left(\frac{\partial E}{\partial T} \right)_P + P \left(\frac{\partial V}{\partial T} \right)_P \tag{1.18b}$$

where we now think of both E and V as functions of P and T: these are the appropriate independent variables for *this* problem.

To conclude this section we apply Eqs. (1.18) to a perfect gas. For this purpose we must use the fact that for a perfect gas the internal energy E depends only on the temperature:

$$E = E(T) .\qquad(1.19)$$

The explanation of this is that in a perfect gas the volume occupied by the molecules and their mutual interactions are negligible, so the mean spacing between molecules cannot matter.

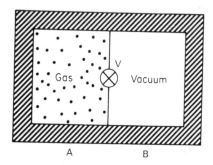

Fig. 1.12. Free expansion of a completely isolated gas
from compartment A to compartment B via valve V.

The experimental evidence for Eq. (1.19) comes from Joule's experiment. Consider the expansion of a gas into a vacuum. This is schematically illustrated in Fig. 1.12. Initially compartment A contains the gas, compartment B vacuum. The entire system is totally isolated. The valve V is opened and the gas expands into compartment B. Eventually the gas is again in a state of equilibrium. Under the conditions of this experiment no work is done on the system and no heat enters it: $W = Q = 0$. Hence from the first law, the internal energy of the gas remains unchanged: $\Delta E = 0$. For a real gas this free expansion leads to a small temperature drop due to the work done against the cohesive forces in the gas. At low pressures (i.e. under ideal gas conditions) this temperature change becomes negligible. Hence an ideal gas is defined by two conditions: (i) it satisfies the equation of state (1.9): $PV = NkT$; (ii) its internal energy is independent of P and V, and depends on T only, Eq. (1.19): $E = E(T)$.

It follows from Eqs. (1.18) and (1.19) that for a perfect gas

$$C_P - C_V = P\left(\frac{\partial V}{\partial T}\right)_P = R ,\qquad(1.20)$$

where the last step follows from the equation of state (1.5).

We have so far considered work done by a hydrostatic pressure on a fluid. In general other forms of work may occur. In writing down the first law (1.14) one must allow for *all* these forms of work. Correspondingly one then defines other heat capacities with the appropriate parameters held constant.

★ 1.4   MAGNETIC WORK

For the study of the magnetic properties of matter, one requires the expression for the work of magnetizing a material. We shall now derive this. We shall consider a process in which an initially unmagnetized sample of material is magnetized by applying a magnetic field to it. This can be done in various ways. For example, we might place the sample inside a solenoid in which a current is gradually switched on, or we might move the sample into the field of the solenoid in which a constant current is maintained. In either case we end up with the magnetized sample *situated in the external applied field*. We shall consider these two procedures in turn, starting with the stationary sample inside the solenoid.

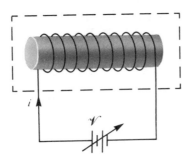

Fig. 1.13. Magnetization of a material inside a solenoid.

Consider a solenoid of length L, cross-sectional area A, and n turns per unit length. We shall assume the magnetic field inside the solenoid uniform* and that the coil has zero resistance. A current i in the solenoid produces a uniform magnetic intensity $\mathcal{H}$ parallel to the axis of the solenoid given by

$$\mathcal{H} = ni \ . \tag{1.21}$$

(We are using SI units.) It follows from Ampère's law that, with the specimen inside the solenoid (Fig. 1.13), the magnetic intensity for a given current i

---

*Pippard[9] gives an ingenious proof which avoids this assumption. It leads to the same result as we shall obtain.

is still given by Eq. (1.21).* Since the magnetic field $\mathscr{B}$ is parallel to the axis of the solenoid, the total magnetic flux through the solenoid is given by

$$\Phi = A\,(nL)\mathscr{B} \ . \tag{1.22}$$

A change of flux induces an E.M.F. of magnitude $\mathscr{V}$ given by

$$\mathscr{V} = \frac{d\Phi}{dt} = A\,(nL)\frac{d\mathscr{B}}{dt} \tag{1.23}$$

in the circuit. Correspondingly during a time interval dt the battery does work dW_1 on the system in driving the current against this induced E.M.F., given by

$$dW_1 = \mathscr{V} i \; dt \ . \tag{1.24}$$

The system here consists of the solenoid and the specimen. In Fig. 1.13 the system is marked off by the dashed lines. When compressing a gas (Fig. 1.7) the analogously defined system — on which the external agency does work — consists of cylinder, piston and gas. We rewrite Eq. (1.24) by substituting for i and $\mathscr{V}$ from Eqs. (1.21) and (1.23), and using the fact that $\mathscr{B}$ and $\mathscr{H}$ are parallel:

$$dW_1 = V\mathscr{H} \cdot d\mathscr{B} \tag{1.25}$$

where $V = AL$. We now write

$$\mathscr{B} = \mu_0(\mathscr{H} + \mathscr{I}) \ , \tag{1.26}$$

where $\mathscr{H}$ is the applied magnetic field intensity (1.21) and $\mathscr{I}$ is the magnetization (magnetic moment per unit volume) of the specimen. Eq. (1.25) then becomes

$$dW_1 = d(V\tfrac{1}{2}\mu_0\mathscr{H}^2) + V\mu_0\mathscr{H} \cdot d\mathscr{I} \ , \tag{1.27}$$

which can also be written

$$dW_1 = d\left(\int \tfrac{1}{2}\mu_0\mathscr{H}^2 \, dV \right) + \int (\mu_0\mathscr{H} \cdot d\mathscr{I})dV \ . \tag{1.28}$$

---

*The electromagnetic theory which is used in this section will be found in most standard textbooks on the subject; for example, in I. S. Grant and W. R. Phillips, *Electromagnetism* (Manchester Physics Series), Wiley, Chichester, England, 1975.

Pippard has shown that Eq. (1.28) also holds if the fields depend on position. If the specimen is situated in a homogeneous magnetic field $\mathscr{H}$ and possesses a magnetic moment

$$\mathscr{M} = \int \mathscr{I} \, dV \, , \tag{1.29}$$

we can rewrite Eq. (1.28) as

$$\mathrm{d}W_1 = \mathrm{d}\left(\int \tfrac{1}{2} \mu_0 \mathscr{H}^2 \, \mathrm{d}V \right) + \mu_0 \mathscr{H} \cdot \mathrm{d}\mathscr{M} \, . \tag{1.30}$$

Eq. (1.30) is our final result. It gives the work done by the battery in an infinitesimal change of the solenoid current. As this current is increased, the battery does work to increase the vacuum field energy

$$E_{\mathrm{vac}} = \int \tfrac{1}{2} \mu_0 \mathscr{H}^2 \, \mathrm{d}V \tag{1.31}$$

i.e. E_{vac} is the energy stored in the magnetic field of the solenoid in the absence of the sample. The first term in Eq. (1.30) represents this work. The second term in (1.30) therefore represents *all* other changes in energy of the system due to work having been done on it. We shall analyse this further later in this section.

Before doing so, we shall consider a second way of magnetizing a sample, namely by moving it from far away into the field of the solenoid in which a constant current i is maintained. In the above analysis of the stationary sample, we were able to treat the magnetic field as uniform. For a large-sized sample, it may not be justified to treat the field as uniform throughout the motion of the sample. To overcome this difficulty, we shall consider a small sample. (For a large sample, one would first consider individual volume elements of the sample and then integrate over the volume of the sample.) For a small sample, the magnetic field due to the solenoid is approximately uniform over the sample. We shall move the sample along the solenoid axis,

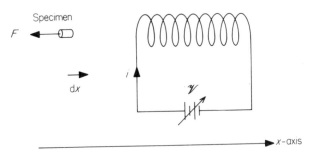

Fig. 1.14. Magnetizing a specimen by bringing it up to a current-carrying solenoid.

taken as the x-axis, through the distance dx towards the solenoid (Fig. 1.14). A small sample, positioned at x, will experience an attractive force

$$\mu_0 \mathcal{M}(x) \cdot \frac{d\mathcal{H}(x)}{dx}$$

towards the solenoid. Here $\mathcal{M}(x)$ is the magnetic moment of the sample when situated at x, and $\mathcal{H}(x)$ is the magnetic field intensity at x due to the solenoid. In order to carry out the process quasistatically, we apply a balancing force

$$F = -\mu_0 \mathcal{M}(x) \cdot \frac{d\mathcal{H}(x)}{dx}$$

to the sample (see Fig. 1.14). For example, the sample might lift a weight as it is pulled into the field of the solenoid. In moving the sample through dx, the applied force F does work

$$đW = Fdx = -\mu_0 \mathcal{M}(x) \cdot d\mathcal{H}(x) \ . \tag{1.32}$$

This expression does not represent the total work done in this process. Moving the magnetized sample produces a change of magnetic flux through the solenoid. Thus an E.M.F. $\mathcal{V}$ is induced in the solenoid, and the battery must do work against this E.M.F. to maintain a *constant* current i. This work is given by

$$đW' = i\mathcal{V}\,dt = id\Phi$$

where $d\Phi$ is the change in flux through the solenoid due to the sample being moved through dx. Since the solenoid current is kept constant in this process the self-flux of the solenoid is constant, and we shall take $\Phi = \Phi(x)$ as the flux through the solenoid due to the sample, situated at x. Also, since i is constant, we can write the last expression

$$đW' = d(i\Phi) \ . \tag{1.33}$$

Now the flux Φ through the solenoid due to the sample does not depend on the detailed structure of the sample but only on its magnetic moment $\mathcal{M}$. We shall assume that the sample has the shape of a cylinder of cross-sectional area A_m, with a magnetization current i_m circulating around its curved surface. The magnetic moment of the sample is then given by

$$\mathcal{M} = \mathbf{A}_m i_m \ , \tag{1.34}$$

the direction of the vector area $\mathbf{A}_m$ being defined by the right-hand screw sense of the circulating current i_m.

We can express ($i\Phi$) in Eq. (1.33) in terms of the magnetic moment $\mathcal{M}$ in the following way. Let M be the mutual inductance of the solenoid and the equivalent elementary current loop by which we represented the sample. In terms of M, the flux through the solenoid due to the sample is given by

$$\Phi = M i_m \tag{1.35}$$

and the flux through the current loop due to the solenoid by

$$\mu_0 \mathcal{H} \cdot \mathbf{A}_m = M i \ . \tag{1.36}$$

It follows from the last equation and Eq. (1.34) that

$$M i i_m = \mu_0 \mathcal{H} \cdot \mathcal{M} \ , \tag{1.37}$$

and hence from (1.35) that

$$i\Phi = \mu_0 \mathcal{H} \cdot \mathcal{M} \ . \tag{1.38}$$

Note that in deriving Eq. (1.38) we did not have to assume that the field due to the sample is uniform over the volume of the solenoid.

Substituting Eq. (1.38) in Eq. (1.33) gives

$$đW' = d(\mu_0 \mathcal{H} \cdot \mathcal{M}) \ . \tag{1.39}$$

Adding Eqs. (1.32) and (1.39), we obtain for the total work done in moving the sample through dx

$$đW_2 = đW + đW' = \mu_0 \mathcal{H}(x) \cdot d\mathcal{M}(x) \ . \tag{1.40}$$

Eq. (1.40) is our final result for the work done in this second way of magnetizing the sample. Comparison with the corresponding result (1.30) for the first method shows that these results are consistent as in the second method the solenoid current, and hence the vacuum field, are held constant, so that E_{vac} does not change.

In analysing magnetic work further, we shall for simplicity restrict ourselves to the case of uniform fields and a uniformly magnetized sample, i.e. to the magnetization of the stationary sample filling the solenoid, which we considered above.* The work done by the battery in this magnetization process is obtained by integrating Eq. (1.30):

---

*This analysis can be extended to non-uniform fields, and the results which are derived below can be shown to hold in the more general case. The general derivation, on which our more specialized treatment is based, has been given by A. J. Hillel and P. J. Buttle (private communication). See also J. R. Waldram,[19] chapter 15, for a treatment based on the same point of view.

$$W_1 = \int \tfrac{1}{2}\mu_0 \mathscr{H}^2 \, dV + \int \mu_0 \mathscr{H} \cdot d\mathscr{M} \tag{1.41a}$$

or, on integrating the second integral by parts,

$$W_1 = \int \tfrac{1}{2}\mu_0 \mathscr{H}^2 \, dV + \mu_0 \mathscr{H} \cdot \mathscr{M} - \int \mathscr{M} \cdot \mu_0 d\mathscr{H} . \tag{1.41b}$$

To interpret the expressions (1.41), we go back to the general expression for the energy stored in a magnetic field $\mathscr{B}$:

$$E_{\text{mag}} = \frac{1}{2\mu_0} \int \mathscr{B}^2 dV . \tag{1.42}$$

With

$$\mathscr{B} = \mu_0(\mathscr{H} + \mathscr{I}) , \tag{1.26}$$

the integral in Eq. (1.42) breaks up into three terms

$$E_{\text{mag}} = \int \tfrac{1}{2}\mu_0 \mathscr{H}^2 dV + \int \tfrac{1}{2}\mu_0 \mathscr{I}^2 dV + \int \mu_0 \mathscr{H} \cdot \mathscr{I} dV . \tag{1.43}$$

These three terms admit a simple interpretation. The first term is just the vacuum field energy (1.31). The second term does not depend on the applied field $\mathscr{H}$ but only on the magnetization $\mathscr{I}$ of the sample, i.e. it represents a magnetic self-energy of the sample. To interpret the third term, we remind the reader that just as $\mathscr{H}$ is the magnetic field intensity originating from the free currents (in our case the solenoid current), so $\mathscr{I}$ is the magnetic field intensity due to the magnetization currents.* Thus the third term in Eq. (1.43) represents a *mutual field energy* due to the superposition of the fields originating from the solenoid and from the magnetized sample. For the case of a uniform applied field, which we are considering, we have from Eq. (1.29) that this mutual field energy reduces to

$$\int \mu_0 \mathscr{H} \cdot \mathscr{I} dV = \mu_0 \mathscr{H} \cdot \mathscr{M} . \tag{1.44}$$

This result can also be understood if we consider a single dipole of the sample, which we again represent by an elementary current loop, placed in the field of the solenoid. The magnetic energy can be written in the usual way as

$$E_{\text{mag}} = \tfrac{1}{2}L i^2 + \tfrac{1}{2}L_m i_m^2 + M i i_m$$

---

*It is these statements which hold for uniform fields only, necessitating a more indirect treatment for non-uniform fields.

where L and L_m are the self-inductances of the solenoid and of the current loop, M is the mutual inductance of the solenoid and the loop, and i and i_m are the solenoid and loop currents. The three terms in the last equation correspond to the vacuum field energy, the magnetic self-energy and the mutual field energy in Eq. (1.43). We see from Eq. (1.37) that the mutual field energy Mii_m reduces to $\mu_0 \mathscr{H} \cdot \mathscr{M}$, in agreement with Eq. (1.44).

We now use Eq. (1.43) to interpret Eq. (1.41b) for the work done by the battery in magnetizing the sample and establishing the fields. The first term on the right-hand side of Eq. (1.41b) is just the work done in establishing the vacuum field. From Eq. (1.44), we recognize the second term in Eq. (1.41b) as the work done in generating the mutual field energy. Hence the third term in (1.41b) is the work done in magnetizing the sample, *exclusive of establishing the mutual field energy*.

In studying a magnetic system, we usually subtract off the vacuum field energy, i.e. we do not count the vacuum field energy as part of the system. This still leaves two possibilities, depending on whether we include the mutual field energy (1.44) as part of the system or not. If it is *excluded*, we see from Eq. (1.41b) that the magnetic work in an infinitesimal change is given by

$$\text{d}W = -\mathscr{M} \cdot \mu_0 \text{d}\mathscr{H} . \tag{1.45}$$

If we *include* it as part of the system, we obtain from Eq. (1.41a) for the magnetic work in an infinitesimal change

$$\text{d}W_1 = \mu_0 \mathscr{H} \cdot \text{d}\mathscr{M} . \tag{1.46}$$

We next consider the first law of thermodynamics, i.e. we take into account that we can change the energy of a system by doing work on it or by heat transfer. We note first of all that the expressions (1.45) and (1.46) for the infinitesimal work are generally valid. However, we want to use these in the first law (1.14) and we want the infinitesimal work to be defined by the parameters of state of the system. We know from section 1.3 that for this to be the case we must restrict ourselves to reversible processes, i.e. to processes which are quasistatic and which do not involve hysteresis effects. This means that we must exclude ferromagnetic materials from our discussion. We require that the magnetization is a single-valued function of the applied magnetic field.

Above we had two different expressions for the magnetic work and correspondingly we obtain two different forms for the first law (1.14). If the mutual field energy as well as the vacuum field energy are *not* counted as part of the system, the magnetic work is given by Eq. (1.45) and the first law becomes

$$\text{d}E = \text{d}Q - \mathscr{M} \cdot \mu_0 \text{d}\mathscr{H} . \tag{1.47}$$

Alternatively, if E' is the energy of the system includng the mutual field energy, but still excluding the vacuum field energy, then

$$E' = E + \mu_0 \mathcal{H} \cdot \mathcal{M} , \tag{1.48}$$

whence Eq. (1.47) becomes

$$dE' = dQ + \mu_0 \mathcal{H} \cdot d\mathcal{M} . \tag{1.49}$$

This is in agreement with the expression (1.46) for the work in this case. We shall see later that in statistical physics, i.e. from the microscopic point of view, it is more natural and useful not to count the mutual field energy as part of the system and to employ Eqs. (1.45) and (1.47).

Strictly speaking, statements of the first law must contain terms corresponding to *all* types of work of which the system is capable. Thus the right-hand sides of Eqs. (1.47) and (1.49) should be augmented by a term $-P\,dV$ corresponding to volume changes, as in Eq. (1.17). Frequently one particular work term dominates so that the others can be omitted. For example, in studying magnetic properties of solids, volume changes are generally unimportant.

Analogously to the definitions of heat capacities at constant volume and at constant pressure from Eq. (1.17), we can define heat capacities at constant magnetization, $C_{\mathcal{M}}$, and at constant magnetic field strength, $C_{\mathcal{H}}$,

$$C_{\mathcal{M}} = \left(\frac{dQ}{\partial T} \right)_{\mathcal{M}} , \tag{1.50}$$

$$C_{\mathcal{H}} = \left(\frac{dQ}{\partial T} \right)_{\mathcal{H}} . \tag{1.51}$$

From the expressions of the first law of a magnetic system, these heat capacities can be expressed in various ways. For example, Eqs. (1.51) and (1.47) lead to

$$C_{\mathcal{H}} = \left(\frac{\partial E}{\partial T} \right)_{\mathcal{H}} . \tag{1.52}$$

SUMMARY

In this summary we collect together those definitions and equations from this chapter with which the reader may not have been familiar and which will be used frequently later on.

Quasistatic (p. 15)   A quasistatic process is defined as a succession of equilibrium states.

Reversible (p. 15)   A process is reversible if it is possible to reverse its direction by an infinitesimal change in the applied conditions. For a process to be reversible it must be quasistatic and there must be no hysteresis effects.

The first law
  (i) Generally:

$$\Delta E = Q + W \ .$$ (1.13)

  (ii) For infinitesimal changes:

$$dE = dQ + dW \ .$$ (1.14)

(In these equations, only the energy E is a function of state.)

Work done on a fluid in an infinitesimal change:

$$dW \geqslant -PdV$$ (1.15c)

where the $=$ and $>$ signs apply to reversible and irreversible changes respectively.

Heat bath (p. 15)   is a system at a definite temperature whose heat capacity is so large compared with that of another system that when the two systems are put into thermal contact the temperature of the heat bath remains essentially constant.

PROBLEMS 1

1.1 Show that for a quasistatic adiabatic process in a perfect gas, with constant specific heats,

$$PV^\gamma = \text{const.}$$

$(\gamma \equiv C_P/C_V)$.

1.2 The molar energy of a monatomic gas which obeys van der Waals' equation is given by

$$E = \frac{3}{2}RT - \frac{a}{V} \ ,$$

where V is the molar volume at temperature T, and a is a constant. Initially one mole of the gas is at the temperature T_1 and occupies a volume V_1. The gas is allowed to expand adiabatically into a vacuum so that it occupies a total volume V_2. What is the final temperature of the gas?

1.3 Calculate the work done on 1 mole of a perfect gas in an adiabatic quasistatic compression from volume V_1 to V_2.

1.4 The enthalpy H is defined by $H \equiv E + PV$. Express the heat capacity at constant pressure in terms of H.

1.5 One mole of a perfect gas performs a quasistatic cycle which consists of the following four successive stages: (i) from the state (P_1, V_1) at constant pressure to the state (P_1, V_2), (ii) at constant volume to the state (P_2, V_2), (iii) at constant pressure to the state (P_2, V_1), (iv) at constant volume back to the initial state (P_1, V_1). Find the work done on the gas in the cycle and the heat absorbed by the gas in the cycle.

1.6 The same as problem 1.5 with the four stages of the cycle: (i) at constant volume from (T_1, V_1) to (T_2, V_1), (ii) isothermally to (T_2, V_2), (iii) at constant volume to (T_1, V_2), (iv) isothermally back to (T_1, V_1).

1.7 Calculate the change in internal energy when 1 mole of liquid water at 1 atm and 100 °C is evaporated to water vapour at the same pressure and temperature, given that the molar volumes of the liquid and the vapour under these conditions are 18.8 cm³/mol and 3.02×10^4 cm³/mol, and that the latent heat of evaporation is 4.06×10^4 J/mol.

CHAPTER

2

The second law of thermodynamics I

2.1 THE DIRECTION OF NATURAL PROCESSES

The first law of thermodynamics deals with the energy balance in processes, the second with their direction. Left to itself, a system which initially is not in equilibrium (for example, it may contain gradients of temperature, pressure or concentrations) always changes in a definite direction, towards equilibrium, although a change in the opposite sense would equally conserve energy. Reversible changes, discussed in section 1.3, are a limiting idealization of real processes, only approximately attainable with special precautions; but in general real processes are irreversible. For example *heat, by itself, cannot pass from a colder to a hotter body*. ('By itself' is essential. Heat can flow in the 'wrong' direction—for example in a refrigerator—if there are accompanying changes which in some sense compensate for this wrong heat flow. The exact meaning of these compensating changes will become clear presently.) This is the form of the second law due to Clausius (1850). Any other irreversible change could be used as the basis. Joule's paddle-wheel experiment leads to the second law as put by Kelvin (1851): *a process whose only effect is the complete conversion of heat into work cannot occur.* (Again, the word 'only' is an essential safeguard against compensating changes. Complete conversion of heat into work does occur; for example, in the isothermal expansion of a perfect gas. But this is not the only change: after the expansion the gas in the cylinder, Fig. 1.8, occupies a bigger volume.)

The general formulation of the second law, without reference to specific types of processes, was achieved by Clausius (1854, 1865). He introduced the new concept of entropy which provides a quantitative criterion for the direction in which processes occur and, consequently, for the equilibrium of a system.

It was the superb achievement of Boltzmann, in the 1870s, to relate entropy, which is a macroscopic concept, to the molecular properties of a system. The basic idea is that the macroscopic specification of a system is very imperfect. A system in a given macroscopic state can still be in any one of an enormously large number of microscopic states: the coarse macroscopic description cannot distinguish between these. The microscopic state of a system changes all the time; for example, in a gas it changes due to collisions between molecules. But the number of microscopic states which correspond to macroscopic equilibrium is overwhelmingly large compared with all other microscopic states. Hence the probability of appreciable deviations from equilibrium occurring is utterly negligible.

Consider an enclosure containing gas (Fig. 2.1(a)). The number of gas molecules is, say, 10^{20}. In equilibrium the density will be uniform. In particular each half of the enclosure (labelled A and B in Fig. 2.1(a)) will contain half the molecules at any instant. Of course, it will not be exactly half. There will be fluctuations but these will be small compared to the total number of molecules. (We shall see later — see section 11.7.1 — that typical fluctuations for 10^{20} molecules are of the order of 10^{10}.) But it is utterly improbable that the gas, starting from a state of uniform density (Fig. 2.1(a)), should spontaneously change to a state where all the molecules are in one half (A) of the enclosure (Fig. 2.1(b)). Of course one can especially prepare the system to be in this state; for example by compressing the gas into half the space and inserting a partition (Fig. 2.1(c)). On removing the partition (if we imagine that this can be done sufficiently quickly), the gas will in the first instance be in the state depicted in Fig. 2.1(b), but the gas will very rapidly expand to fill the whole available space, i.e. return to the uniform state of Fig. 2.1(a). Thereafter fluctuations will be very small. One would have to wait for times enormously long compared with the age of the universe for a fluctuation to occur which is large on the macroscopic scale and then it would only last a very small fraction of a second. Thus one may safely ignore such large fluctuations altogether.

The expansion of a gas into a vacuum is a typical example of an *irreversible* process. Left to itself, a gas initially in the state (b), Fig. 2.1, will end up in the state (a), but not vice versa. The basic distinction between the initial and final states in such an irreversible process is that in the final state we have a less complete knowledge of the state of the system, in the sense that initially we are certain that all the molecules are in the same half (A) of the container whereas finally each molecule can be in either half, A or B. We have less information about the position coordinates of the molecules.

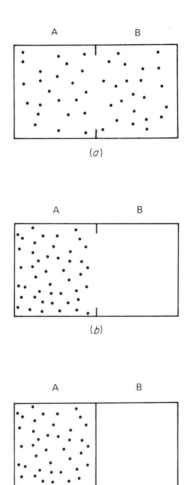

Fig. 2.1. Schematic picture of gas molecules in an enclosure. (*a*) Uniform distribution throughout the whole volume (A + B). (*b*) Large density fluctuation, with all molecules in half the volume (A). (*c*) The gas compressed into half the volume (A) and kept there by means of a partition.

We call the final state a less ordered state or a more random state of the system.

The entropy of the state of a system (or its statistical weight which is closely related to the entropy) is a quantitative measure of the degree of disorder of that state: the more disordered the state is, the larger its entropy. For example, if a monatomic crystal is sublimated into a vapour, the initial state is highly ordered with the atoms performing small vibrations about the regularly spaced equilibrium positions of the crystal lattice while in the final state the atoms of the vapour are distributed almost completely at random throughout the volume occupied by the vapour.*

Since the entropy is a measure of the degree of disorder of the state of a system, it is a function of state, as defined in section 1.3. For a change in a system between two definite states, the entropy change is independent of how the change in the system is brought about. We may, for example, change the temperature of 1 kg of water at standard atmospheric pressure from 20 °C to 21 °C by supplying heat to it (putting it into thermal contact with a heat bath at 21 °C) or by doing mechanical work on it (as in Joule's paddle-wheel experiment). In both cases the same increase in disorder, associated with the thermal agitation of the molecules, occurs, i.e. the same change in entropy. In this example there is also a change in energy of the system. This is not essential. For example in the adiabatic expansion of a perfect gas into a vacuum the energy of the gas stays the same, since no heat enters the system and no work is being done, but as we saw above the entropy of the gas increases. (A quantitative discussion of this case is given in section 4.3.4.) Entropy is not concerned with the energy balance of processes (that is the domain of the first law). Entropy provides a quantitative criterion for the direction of processes in nature. Because entropy is a function of state, its usefulness extends much further and it is of fundamental importance in discussing the properties of macroscopic systems.

After this qualitative discussion of these basic ideas we must now formulate them precisely. In the next section we shall derive a quantitative measure for the disorder of a system, and in section 2.3 we shall obtain a criterion for the equilibrium of a macroscopic system.

2.2   THE STATISTICAL WEIGHT OF A MACROSTATE

In this section we shall consider the relation between the description of macroscopic and of microscopic states. For brevity these will be referred to as *macrostates* and *microstates*.

---

*See Flowers and Mendoza,[26] section 2.2 and Fig. 7.6. For a more quantitative discussion of the spatial ordering of the atoms in gases, liquids and solids, which in particular mentions how these orderings show up in x-ray diffraction experiments, see T. L. Hill,[28] section 6.3. (See also the 'Hints for solving problems', problem 7.8.)

For simplicity we shall treat a system consisting of one component only, i.e. all molecules are of the same type. Let there be N of them. A macrostate of such a system can be specified in various ways, depending on the constraints to which the system is subject. The volume of gas within a cylinder with perfectly rigid walls is constrained to have a definite constant value. If instead, one end of the cylinder is closed off by a freely movable piston to which a constant external pressure is applied, the gas, if in a state of equilibrium, is constrained to be at the same pressure. Generally, any conditions imposed on a system forcing certain variables to assume definite values will be called constraints. In this section we shall assume the system totally isolated. In this case its energy E, volume V and number of molecules N are fixed, and (E, V, N) fully determine a macrostate of the system in equilibrium. For a non-equilibrium state other macroscopic quantities must be specified. For example, in the case of a gas one may have to specify the particle density $\rho(\mathbf{r}, t)$ at every point $\mathbf{r}$ at every instant t or, rather more realistically, the particle density averaged over finite intervals of volume and time, depending on the degree of refinement of the observations. In the equilibrium state these additional variables have definite values. For a gas in equilibrium the particle density is given by $\rho = N/V$. Generally we shall label these additional variables α (which may stand for many of them: $\alpha_1, \alpha_2, \ldots$). A macrostate is then specified by (E, V, N, α). We see that for a macrostate not in equilibrium what is meant by specification of the state depends on the refinements of observation and that is how it should be.

The complete description of a microstate is very complicated; whether one uses classical or quantum mechanics, some 10^{23} parameters are involved. The saving grace is that for equilibrium statistical mechanics one does not require a detailed knowledge of the microstates. All one needs to know is how many microstates correspond to a given macrostate. The general idea is simple. Given a gas in a box, of volume V, completely isolated, there are still enormously many ways of assigning the position coordinates $\mathbf{r}_1, \ldots \mathbf{r}_N$ and the momentum coordinates $\mathbf{p}_1, \ldots \mathbf{p}_N$ of the molecules to correspond to a given macrostate; for example in the state of equilibrium with energy E. The position coordinate $\mathbf{r}_1$, for example, can assume a 'continuous infinity' of values (like points on a line) and the same is true of the other position and momentum coordinates. In this classical description, these different microstates form a continuum and 'counting' them contains an element of arbitrariness. In the hands of Boltzmann and Gibbs this approach was outstandingly successful.

The problem of counting states gains great clarity if one adopts a quantum-mechanical viewpoint. According to quantum mechanics the microstates do not form a continuum but a discrete set. There is an integral number of such states; they can be counted. Thus each spectral line of an atom corresponds to the emission of light quanta (photons) of definite energy, the atoms at the same time making transitions between two sharply defined discrete states, each of definite energy. Other examples of this 'quantization' will occur later in this book.* There appear to be exceptions to this discreteness of states. For example, an atom when ionized (e.g. in the photoelectric effect where an electron is knocked out of the atom by a very energetic photon (ultraviolet light)) consists of the positive ion and an electron, and the free electron appears to be able to have any energy. This is correct. But suppose we enclose our system, consisting of positive ion plus electron, in a box of finite volume. One can then show that the states become discrete again and this happens however large the box, provided its volume is *finite*.

A simple example which may be familiar to the reader is given by waves on a string. For a string of infinite length any wavelength is possible and the most general waveform results from a superposition of waves of all wavelengths and appropriate amplitudes and phases. Mathematically such a wave is represented by a Fourier *integral*. On the other hand for a string of finite length the most general waveform is represented by a superposition of a fundamental wave and its harmonics, i.e. by a Fourier *series*. These fundamental and harmonic modes for a string of finite length correspond to the discrete states of a system enclosed in a box of finite volume, however large. (The similarity of these two situations is due to the wave nature of matter. This is discussed further in Appendix B.)

A sufficiently large box (say 10 light-years across) will clearly not affect the properties of our system, ion plus electron sitting somewhere near the middle. This result is quite general. We can always force a system to have discrete states by enclosing it in a sufficiently large box. In view of this, we shall generally assume that *every macrostate of a system comprises a perfectly definite number of microstates of the system*. This formulation was introduced by Planck at the beginning of this century.

We shall denote by $\Omega(E, V, N, \alpha)$ the number of microstates comprising a macrostate specified by V, N, α, and having energy in the small interval E to $E + \delta E$.[†] $\Omega(E, V, N, \alpha)$ is called the statistical weight or the thermodynamic probability (not a good name since Ω is not a probability in the ordinary sense of the word, e.g. usually $\Omega \gg 1$) of this macrostate.

---

*For the atomic physics background used in this section see, for example, Eisberg,[39] Fano and Fano,[40] French and Taylor[42] or Willmott.[48]

†There are several reasons for defining a macrostate as having an imprecise energy, lying in an interval of width δE. It corresponds to the fact that experimentally the energy is only determined up to some finite accuracy. In studying a system theoretically we can choose δE to suit our convenience. If the energy interval δE is chosen sufficiently large so that it contains many microstates, then $\Omega(E, V, N, \alpha)$ will be a smoothly varying function of E. If δE is chosen very small, Ω will be wildly fluctuating, being different from zero whenever E coincides with the energy of one of the discrete microstates and being zero for all other values of E. Of course altering δE changes $\Omega(E, V, N, \alpha)$, but for macroscopic systems these changes in Ω are insignificantly small. This point will be discussed further, following Eq. (2.3).

Correspondingly the statistical weight of an equilibrium state will be denoted by $\Omega(E, V, N)$.

To illustrate these ideas we shall consider a simple model of a paramagnetic solid when placed in a magnetic field. This model consists of molecules, each possessing a permanent magnetic moment μ, like little bar magnets, arranged on the sites of a crystal lattice. We shall neglect the vibrational motion of the molecules and shall only take into account the interactions of the magnetic dipoles with the applied magnetic field but neglect all other interactions, such as those between the magnetic dipoles. We shall return to these questions in Chapter 3. Here we only wish to consider the problem of counting states.

In the absence of a magnetic field, the magnetic dipoles of our solid will be oriented at random. When the field $\mathscr{B}$ is applied, each dipole acquires an interaction energy $-\mu \cdot \mathscr{B}$ (Fig. 2.2) which tends to orient the dipole along the field.* This tendency to align is opposed by the disorganizing thermal motion. Which of these effects wins depends on the strength of the magnetic field and on the temperature. Strong fields and low temperatures favour alignment; reducing the field and raising the temperature increasingly favours random orientation.

Fig. 2.2. A magnetic dipole, magnetic moment μ, placed in a magnetic field $\mathscr{B}$. Interaction energy $= -\mu \cdot \mathscr{B}$.

Classically any relative orientation of the magnetic moment μ and the applied field $\mathscr{B}$ can occur. According to quantum mechanics only certain specific orientations are permitted. These depend on the angular momentum properties of the magnetic dipoles. The simplest situation, to which we shall restrict ourselves, occurs if the dipoles possess angular momentum $\hbar/2$, usually referred to as 'spin $\frac{1}{2}$'. ($\hbar$ is Planck's constant divided by 2π.) In this case only two orientations occur: parallel and antiparallel to the field, with interaction energies $\mp \mu \mathscr{B}$ respectively, as shown in Fig. 2.3. Thus for a single dipole there are two possible states with different energies, the energy splitting being $2\mu \mathscr{B}$ as shown in the energy level scheme in Fig. 2.3.

*This interaction energy $-\mu \cdot \mathscr{B}$ is derived in Appendix C.

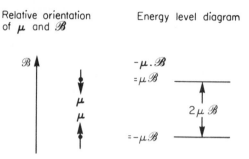

Fig. 2.3. Interaction energy $(-\mu\cdot\mathscr{B})$ and energy level diagram of a spin $\frac{1}{2}$ magnetic dipole (magnetic moment μ) in a magnetic field $\mathscr{B}$.

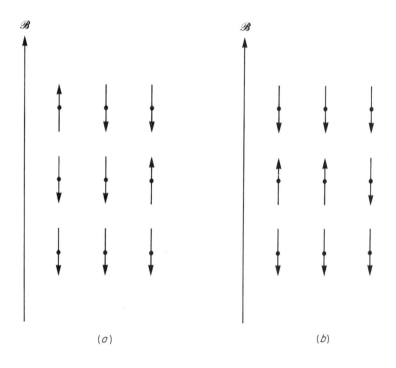

Fig. 2.4. Array of 9 spin $\frac{1}{2}$ dipoles in a magnetic field $\mathscr{B}$. Figures (a) and (b) show two *different* microscopic states, both corresponding to the *same* macrostate specified by $n=2$.

Consider now a system of N dipoles placed in a magnetic field $\mathcal{B}$. If n dipoles are oriented parallel to the field, and the remaining $(N-n)$ antiparallel, the energy of the system is given by

$$E(n) = n(-\mu\mathcal{B}) + (N-n)(\mu\mathcal{B}) = (N-2n)\mu\mathcal{B} \ . \qquad (2.1)$$

We have written $E(n)$ rather than $E(\mathcal{B}, V, N, n)$ since we are going to keep N and $\mathcal{B}$ constant, and in the present problem the energy is clearly independent of the volume. n corresponds to the parameter which was called α in the general case.

We see from (2.1) that, in the present problem, n determines the energy, and vice versa. However, n does not determine the microstate of the system. Fig. 2.4 shows, for a two-dimensional model and $N=9$, two different microstates both of which correspond to the same macrostate $n=2$. The two states shown are distinct, physically different states because the two dipoles with 'spin up', i.e. with μ parallel to $\mathcal{B}$, are located at different lattice sites in the two cases. The distinctness of these states is not due to different dipoles being in any sense distinguishable. All the dipoles are *identical*; they do *not* possess any property which we can use to label them A, B, ... H, I (for $N=9$) and then permute dipoles with different labels. For example, exchanging the two 'spin up' dipoles in Fig. 2.4(a) with each other does not lead to a new microstate: *we cannot distinguish these two possibilities.**

Altogether our system of N dipoles possesses 2^N microstates: two possible orientations for the dipole at each of the N sites. The number of microstates with energy $E(n)$ is given by the number of ways in which n sites can be selected, out of the total of N sites, at which 'up-spins' are placed, all other sites having 'down-spins'. This number is

$$\Omega(n) = \binom{N}{n} \equiv \frac{N!}{n!(N-n)!} \ . \qquad (2.2)$$

$\Omega(n)$, short for $\Omega(E(n), \mathcal{B}, N)$, is the number of microstates with n up-spins. We choose the energy interval δE, which enters the definition of the statistical weight of a macrostate, such that $\delta E < 2\mu\mathcal{B}$. Then the interval δE contains only one energy level and $\Omega(n)$ is the statistical weight of the macrostate with energy $E(n)$.

Eq. (2.2) is our final result. For $n=N$ all dipoles are aligned parallel to the magnetic field and the resultant magnetic moment of the system attains its maximum magnitude μN. Correspondingly the energy is $E(N) = -\mu\mathcal{B}N$.

---

*We shall frequently encounter this type of problem. But two electrons or two hydrogen atoms or two dipoles, in our problem, are *always identical*. The only thing which can distinguish them is if, in some sense, they are in different states. In our problem the *location* of the 'spin up' dipoles varies.

This is the state of lowest energy, i.e. the ground state of the system. From Eq. (2.2), $\Omega(N) = 1$; there exists only one such microstate, as is obvious! On the other hand for $n = N/2$, the resultant magnetic moment vanishes, there is no net orientation of dipoles; $E(N/2) = 0$.* The system behaves as in the case of zero field where the dipoles are oriented at random. The statistical weight $\Omega(n)$ attains its maximum value for $n = N/2$ and is monotonically decreasing as $|n - \frac{1}{2}N|$ increases, as the reader should easily be able to verify from Eq. (2.2). For $\frac{1}{2}N < n < N$ one deals with an intermediate situation in which the dipoles are partially aligned. We shall return to this problem in the next chapter where we shall consider how the resultant alignment of the dipoles depends on the strength of the magnetic field and on the temperature of the system. We see that the statistical weight Ω is a measure of the order or disorder of the system. The minimum value of Ω ($= 1$) corresponds to the system being in a unique microstate, i.e. perfect order. In the macrostate of maximum Ω (with equally many spins pointing either way) the microstate of the system is least well defined: it is the macrostate of maximum disorder.

2.3 EQUILIBRIUM OF AN ISOLATED SYSTEM

We return now to the study of the isolated system considered at the beginning of section 2.2. We want to find an equilibrium condition for this system. In equilibrium it is fully specified by (E, V, N); for nonequilibrium macrostates additional variables α must be specified. To each choice of α corresponds a statistical weight $\Omega(E, V, N, \alpha)$, the number of microstates comprising that macrostate. Due to collisions, etc., in a gas for example, the microstate of the system and hence α and $\Omega(E, V, N, \alpha)$ change continually. In the course of time, the system will pass through all microstates consistent with the fixed values of E, V and N. We now introduce the postulate of equal *a priori* probabilities: *for an isolated system all microstates compatible with the given constraints of the system (in our case E, V and N) are equally likely to occur, i.e. have equal a priori probabilities.* As a result of this postulate, the probability that the system is in the macrostate specified by (E, V, N, α) is proportional to $\Omega(E, V, N, \alpha)$.

The equilibrium state of the system corresponds to a particular value of α. We now make a second postulate: *equilibrium corresponds to that value of α for which $\Omega(E, V, N, \alpha)$ attains its maximum value, with (E, V, N) fixed.* The meaning of this equilibrium postulate is thus that the equilibrium state is the state of maximum probability: it corresponds to the maximum statistical weight. The explanation is that of all the states (E, V, N, α), specified by different α, the *overwhelming majority* correspond to the equilibrium

---

*There is no need to distinguish the cases of even and odd values of N since N is of the order of 10^{23} in practice.

macrostate. Hence a system especially prepared to be in a state different from equilibrium (see the discussion in section 2.1) rapidly approaches equilibrium. Even very small fluctuations away from equilibrium will occur only extremely rarely. Substantial deviations will, for practical purposes, never occur unless the system is disturbed from outside.

We have tried to make these statements appear plausible. We have not proved them. We shall take as their justification the universal successes of the theory, based on these postulates, which we shall develop.

One justification for the postulate of equal *a priori* probabilities is that if all we know about a system is its macro-specification (E, V, N, α) then we must supplement this by some probabilistic assumption about which microstate the system is in, and since we have no *a priori* reason to prefer one microstate to another we must assign equal *a priori* probabilities to them. This is the viewpoint taken by Planck[10] (p. 221 *et seq.*). But a proper derivation of the laws of statistical mechanics should start from the laws of mechanics which govern the microscopic behaviour of the system. The difficulty of this approach is that the concept of probability is quite foreign to that of mechanics. The fundamental work which still forms the basis of much present-day research is due to Boltzmann. He showed that for an isolated gas the entropy (which will be defined below) increases with time as a result of molecular collisions and reaches its maximum value when the gas is in equilibrium. In his proof, Boltzmann had to make a probabilistic assumption about the nature of molecular collisions, and it does not seem consistent to introduce such an assumption in deriving statistical mechanics from mechanics. These problems are not yet resolved.

Instead of the statistical weight Ω we shall now introduce the entropy as a measure of the disorder of a system in a given macrostate. The entropy of a system in the macrostate (E, V, N, α) is defined by the equation

$$S(E, V, N, \alpha) = k \ln \Omega(E, V, N, \alpha) \; , \tag{2.3}$$

where k is Boltzmann's constant, Eq. (1.8). The reason for introducing the entropy rather than working with the statistical weight will be discussed below, following Eq. (2.6).

The entropy of a macrostate, like its statistical weight, apparently depends on the energy interval δE used in defining the macrostate (E, V, N, α). In practice this dependence is totally negligible. The reason for this is that for a macroscopic system the number of microstates is an extremely rapidly increasing function of the energy. More precisely, let $\Phi(E)$ be the number of microstates of the system with energy less than or equal to E. If the system possesses ν degrees of freedom, then Φ is generally of the form

$$\Phi(E) = CE^{\nu}$$

where C is a constant. (For the particular case of a perfect gas this result is derived in section 7.7, Eq. (7.71a).) For a macroscopic system ν is of the order of 10^{24}. It follows that

$$\Omega(E) = \Phi(E + \delta E) - \Phi(E) = \frac{d\Phi}{dE}\,\delta E = \nu C E^{\nu-1}\delta E \ ,$$

whence

$$\ln\Omega(E) = \ln\Phi(E) + \ln\nu + \ln\left(\frac{\delta E}{E}\right) \ .$$

Now for all E, except $E \approx 0$, $\ln\Phi(E)$ is of the order of ν; (for the ground state of the system, $E = 0$, we have $\Phi = 1$ and $\ln\Phi = 0$). We shall show that for any reasonable interval δE, $\ln(\delta E/E)$ is of the order of $\ln\nu$ and is therefore utterly negligible compared to the term $\ln\Phi$. (Remember that $\ln\nu \ll \nu$ for ν very large; e.g. for $\nu = 10^{24}$, $\ln\nu = 55.3$.) In other words, we shall show that the term $\delta E/E$ in the expression for $\ln\Omega$ is completely negligible, whatever δE, as long as δE has a reasonable value at all. If δE lies in the range

$$1 \geqslant \frac{\delta E}{E} \geqslant \nu^{-p} \ ,$$

where p is a positive number, then $\ln(\delta E/E)$ lies between 0 and $(-p\ln\nu)$. Hence unless p is of the order of ν, $\ln(\delta E/E)$ is at most of order $\ln\nu$. But a reasonable value of p is of order unity, since E/ν is the average energy per degree of freedom of the system, and with $\nu \sim 10^{24}$, E/ν^2 is already an extremely small energy for δE. Hence we conclude that $\ln(\delta E/E)$ is at most of order $\ln\nu$ and is completely negligible compared with $\ln\Phi$.

In our development of statistical physics Eq. (2.3) appears merely as the definition of entropy. Nevertheless it is one of the most fundamental equations of physics. We shall see that the entropy, independently of its atomic interpretation through Eq. (2.3), has a macroscopic meaning of its own in terms of thermal measurements. In fact, it was first introduced in this purely macroscopic way by Clausius. The importance of Eq. (2.3) lies in the fact that it relates this macroscopically observable entropy to the microscopic properties of a system. To establish this connection was the great achievement of Boltzmann. But Eq. (2.3) was first written down by Planck in studying black-body radiation. It differs in two important respects from that of Boltzmann. Firstly, Boltzmann used classical mechanics. As discussed in section 2.2 this implies that Ω is not fully determined, because of the arbitrary element in counting states. In fact Boltzmann's Ω is only determined up to an arbitrary constant factor, leading to an arbitrary additive constant in the entropy definition (2.3). Secondly, Boltzmann operated with moles rather than individual molecules, so his equation did not contain Boltzmann's constant k as a constant of proportionality.

On account of Eq. (2.3), we can now translate our original statements about Ω into statements about the entropy S:

During real (as distinct from idealized reversible) processes, the entropy of an isolated system always increases. In the state of equilibrium the entropy attains its maximum value.

This statement of the second law of thermodynamics, due to Clausius, is called the principle of the increase of entropy (entropy principle, for short) and we shall take it as the basic form of the second law. Clausius had defined entropy purely macroscopically, and Boltzmann produced the above connection with the microscopic viewpoint. Our presentation, in which entropy is defined by Eq. (2.3), hides the audacity of this step.

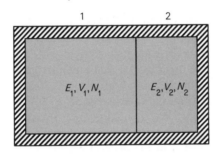

Fig. 2.5. Isolated system partitioned into two sub-systems 1 and 2.

The equilibrium postulate which was introduced at the beginning of this section is extremely powerful. From it we can derive the concepts of temperature and pressure, and their mathematical definitions in terms of statistical weights or, equivalently, in terms of the entropy. To see this, let us apply these results to our isolated system. We imagine it divided into two parts by means of a partition as shown in Fig. 2.5. The two parts are labelled 1 and 2 with corresponding labels for the energy, volume and number of particles in each subsystem. We have

$$E_1 + E_2 = E \ , \tag{2.4a}$$

$$V_1 + V_2 = V \ , \tag{2.4b}$$

$$N_1 + N_2 = N \ , \tag{2.4c}$$

E, V and N being fixed. We will suppose that this particular division of energy, etc., between 1 and 2 does not correspond to equilibrium. For equilibrium to establish itself the two subsystems must be able to interact suitably. In particular it is essential that exchange of energy can occur. This means that the energy of the combined system cannot be written in the form (2.4a), as the sum of two 'private' energies for the two subsystems. There must be an interaction energy ε, depending on both subsystems 1 and 2. We shall assume that the systems interact only very weakly, that is $\varepsilon \ll E_1$ and $\varepsilon \ll E_2$. In this case we can, with sufficient accuracy, write the total energy in the form (2.4a). The price we pay is that for such weak interactions equilibrium takes a very long time to establish itself. This is of no consequence since we are not attempting to study rates of processes but are merely considering devices for studying systems in equilibrium.

For each division of E, V and N between the two subsystems, according to Eqs. (2.4), we have

$$\Omega(E, V, N, E_1, V_1, N_1) = \Omega_1(E_1, V_1, N_1)\Omega_2(E_2, V_2, N_2) \ . \tag{2.5}$$

Here Ω_1 is the statistical weight of subsystem 1 in the macrostate specified by (E_1, V_1, N_1), and similarly for Ω_2. Ω, the left-hand side of Eq. (2.5), is the statistical weight of the composite system with this division of energy, etc. Eq. (2.5) states the obvious fact that any microstate of subsystem 1 can be combined with any microstate of subsystem 2. Because of conditions (2.4) which act as constraints on the variables E_1, E_2, V_1, V_2, N_1 and N_2 we cannot choose them all freely. We are taking E_1, V_1 and N_1 as independent variables, as is indicated on the left-hand side of (2.5). They correspond to the quantities we called α in the general discussion.

On account of Eq. (2.3), we can rewrite Eq. (2.5) as

$$S(E, V, N, E_1, V_1, N_1) = S_1(E_1, V_1, N_1) + S_2(E_2, V_2, N_2) \tag{2.6}$$

where S_1, S_2 and S are the entropies of the subsystems and the combined system.

Eq. (2.6) for the entropy is of the same form as Eqs. (2.4) for E, V and N. These equations state the additivity of these quantities: they are proportional to the size of the system, i.e. if we double the size (e.g. consider 100 g of copper instead of 50 g of copper) all these quantities are doubled. Quantities which are proportional to the size of the system are called *extensive* variables. In order that the energy of a system be proportional to its size, it is necessary that surface energies should be negligible. This is usually the case. A similar proviso with regard to surface effects applies to other extensive variables. The introduction into thermodynamics of the entropy (rather than of the statistical weight) is probably due to its extensive nature, since the experimental study of macroscopic systems leads rather naturally to quantities proportional to the size of the system.

In contrast to *extensive* variables, one defines *intensive* variables as ones which are independent of the size of a system, e.g. the temperature or

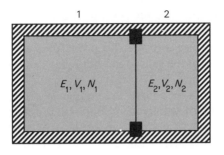

Fig. 2.6. Isolated system partitioned into two subsystems 1 and 2 by means of a fixed diathermal wall.

pressure. Any extensive quantity can be converted into an intensive one by dividing it by the size of the system; e.g. the mass and the density of a body are related in this way.

We shall now consider different kinds of partition separating off the two parts of the system. At first we shall assume the partition fixed but permeable to heat, a so-called diathermal wall (Fig. 2.6). V_1, V_2, N_1 and N_2 are now fixed, but the two subsystems are in thermal contact and heat transfer can occur. We have one independent variable E_1 (E_2 is given by Eq. (2.4a)). From Clausius's entropy principle, we obtain the equilibrium condition for the system by maximizing the entropy. From Eq. (2.6)

$$\left(\frac{\partial S}{\partial E_1}\right)_{E,V,N,V_1,N_1} = \left(\frac{\partial S_1}{\partial E_1}\right)_{V_1,N_1} + \left(\frac{\partial S_2}{\partial E_2}\right)_{V_2,N_2} \frac{dE_2}{dE_1} = 0 \tag{2.7}$$

and, on account of (2.4a),

$$\left(\frac{\partial S_1}{\partial E_1}\right)_{V_1,N_1} = \left(\frac{\partial S_2}{\partial E_2}\right)_{V_2,N_2}. \tag{2.8}$$

Eq. (2.8) is the condition for thermal equilibrium, i.e. no heat transfer between the two subsystems. In other words, the two subsystems must be at the same temperature; $(\partial S_i/\partial E_i)_{V_i,N_i}$ must be a measure of the temperature of subsystem $i(i=1$ or $2)$. We can use this to define an absolute temperature $T_i(i=1,2)$ for each subsystem by

$$\left(\frac{\partial S_i}{\partial E_i}\right)_{V_i,N_i} = \frac{1}{T_i}, \qquad i=1,2. \tag{2.9}$$

Eq. (2.8) then tells us that $T_1 = T_2$, as required in thermal equilibrium. This definition of temperature is quite general. It depends only on whether heat transfer does or does not occur between two systems when put into thermal contact. We did not use any specific properties, e.g. particular internal molecular properties, of the subsystems but relied completely on Eqs. (2.4a), (2.6) and the second law.

The definition (2.9) is, of course, not unique. We could have replaced $1/T$ by any other function of temperature. But with the choice (2.9), T becomes identical with the perfect gas scale (see section 1.2), as will be shown later.* That temperature as defined by Eq. (2.9) has the correct qualitative attribute can be seen by calculating the rate of change with time of the entropy of a system not in equilibrium. Analogously to Eq. (2.7) one finds, using Eq. (2.9), that

$$\frac{dS}{dt} = \left(\frac{1}{T_1} - \frac{1}{T_2} \right) \frac{dE_1}{dt} > 0 \;, \tag{2.10}$$

where the inequality sign expresses Clausius's principle of increase of entropy. It follows that if $T_1 < T_2$, we must have $dE_1/dt > 0$, i.e. heat flows from the subsystem at the higher temperature to that at lower temperature, in agreement with the usual meaning of 'hotter' and 'colder'.

We note a point regarding the derivation of Eq. (2.9) which sometimes causes confusion. Above we maximized the entropy S of the composite system with respect to the energy E_1 of one subsystem. We cannot maximize S with respect to the total energy E of the composite system. Unlike E_1 which is a variable parameter for the composite system, the total energy E is a constant for the isolated system. Hence the entropy principle does *not* lead to an equilibrium condition of the type $\partial S/\partial E = 0$. The latter equation is clearly nonsense since, from the temperature definition $\partial S/\partial E = 1/T$, it would imply that in equilibrium one necessarily has $T = \infty$.

Secondly, we see from the temperature definition (2.9) that the absolute temperature T is positive.[†] For the statistical weight $\Omega(E)$ is a rapidly increasing function of the energy E for a macroscopic system. This follows from the fact that the energy E, as it increases, can be distributed in many more ways over the microscopic degrees of freedom of the system, resulting in a rapid increase in the number of microstates.

---

*In section 5.2 this identity will be established using a Carnot engine with a perfect gas as working substance. In Chapter 7 we shall give a different proof using statistical rather than thermodynamic methods. We shall there derive the equation of state of a perfect gas in terms of the temperature parameter $\beta = 1/kT$, introduced statistically; we shall obtain the equation $PV = NkT$, showing that T is the usual gas temperature. It will be convenient to anticipate these proofs and assume the identity of the two temperature scales already now.

†The possibility of negative temperatures, in a certain restricted sense, is discussed in section 3.4.

We next consider a movable diathermal partition (Fig. 2.7). N_1 and N_2 are still fixed, but the energies and the volumes of the subsystems can adjust themselves. In equilibrium we expect temperature and pressure equalization to have occurred. On account of Eqs. (2.4a) and (2.4b), we choose E_1 and V_1 as independent variables, and we maximize S, Eq. (2.6), with respect to them. We omit all details which are left as an exercise to the reader. One again obtains the condition (2.8) and in addition, as second condition,

$$\left(\frac{\partial S_1}{\partial V_1}\right)_{E_1, N_1} = \left(\frac{\partial S_2}{\partial V_2}\right)_{E_2, N_2} \quad , \tag{2.11}$$

which must be interpreted as implying equal pressures in the two subsystems. We define the pressure for each subsystem by

$$P_i = T_i \left(\frac{\partial S_i}{\partial V_i}\right)_{E_i, N_i} , \qquad i = 1, 2 \ . \tag{2.12}$$

This definition is, as we shall see presently, identical with the conventional one. For example, applied to an ideal gas Eq. (2.12) reduces to $PV = NkT$. Eq. (2.11) thus implies pressure equilibrium: $P_1 = P_2$. It is left to the reader to verify that the Clausius principle $dS/dt > 0$, applied to our system (Fig. 2.7) when not in pressure equilibrium, implies that the subsystem at the higher pressure expands, that at the lower pressure contracts.

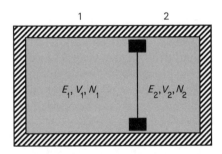

Fig. 2.7. Isolated system partitioned into two subsystems 1 and 2 by means of a movable diathermal wall.

Fig. 2.8 shows our system partitioned into two subsystems, each of fixed volume. The partition allows particles to pass through as well as heat transfer. We may think of it as a porous membrane but we may equally consider it a theoretical construct which divides the system into two definite regions of volume V_1 and V_2 respectively. Maximizing S with respect to N_1 leads to a new condition

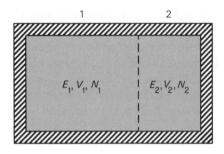

Fig. 2.8. Isolated system partitioned into two subsystems 1 and 2, of fixed volumes V_1 and V_2 respectively, by a fixed partition permeable to heat and particles.

$$\left(\frac{\partial S_1}{\partial N_1}\right)_{E_1, V_1} = \left(\frac{\partial S_2}{\partial N_2}\right)_{E_2, V_2}. \tag{2.13}$$

This expresses the equilibrium with respect to particle transfer between the two subsystems. Such conditions occur, for example, in studying the equilibria of systems containing several phases, e.g. water in contact with water vapour (see Chapter 8).

2.4 THE SCHOTTKY DEFECT*

At the absolute zero of temperature, the atoms (or ions or molecules) of a solid are ordered completely regularly on a crystal lattice. As the temperature is increased thermal agitation introduces several kinds of irregularities. Firstly it causes vibrations of the atoms about their equilibrium sites in the lattice. (Due to quantum-mechanical effects, small vibrations occur even at $T = 0$ K; cf. section 6.2.) In addition, atoms may be altogether displaced from their lattice sites, migrating elsewhere and leaving vacant lattice sites. These kinds of crystal defects are called point defects.[†] There exist different kinds of point defects. In this section we shall consider *Schottky defects*: in these the displaced atoms, causing the vacancies, migrate to the surface of the crystal. Fig. 2.9 illustrates a Schottky defect. Fig. 2.9(a) shows a perfect lattice. In Fig. 2.9(b) a vacancy has been created by the migration of an atom to the crystal surface.

---

*In this section we shall illustrate the theory, which has so far been developed, for a simple physical problem. Readers who prefer an uninterrupted development of the general formalism may omit this section and read it only later when it is referred to.

[†]For a fuller treatment of point defects, which in particular discusses the properties of crystals which depend on point defects, see Flowers and Mendoza,[26] Chapter 9, and Kittel,[29] Chapter 18. See also Problem 2.1.

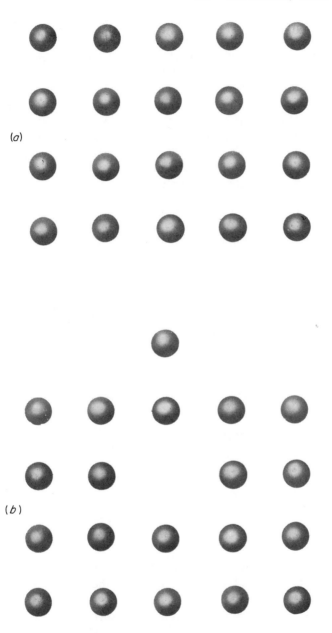

Fig. 2.9. Schottky defect. (*a*) shows a two-dimensional perfect lattice. (*b*) shows the same lattice containing one Schottky defect, i.e. one vacancy, the displaced atom having migrated to the surface of the crystal.

We now want to find out how, for a crystal in thermal equilibrium, the number of Schottky defects depends on the temperature of the crystal. At $T = 0\,K$, we know there will be no defects. As the temperature increases, the number of defects will gradually increase. Atoms at a lattice site inside the crystal have a lower energy than atoms at the surface, since interior atoms are bonded to a larger number of surrounding atoms. Let ε be the energy of formation of a Schottky defect, i.e. the energy of an atom at the surface measured relative to the energy of an interior atom (i.e. the energy of an interior atom is taken as the zero of the energy scale). Let us consider a crystal consisting of N atoms and possessing n Schottky defects. The energy associated with these defects is then given by

$$E = n\varepsilon \ .$$

This expression assumes that the vacant sites are well separated so that each vacancy is surrounded by a well-ordered crystal lattice. Only in this case can we attribute a definite energy ε to each vacancy. Hence we shall assume $n \ll N$, i.e. that the number of defects is small compared with the number of atoms. We shall see below that this condition is well satisfied in practice.

We next require the statistical weight $\Omega(n)$ of the macrostate of a crystal consisting of N atoms and possessing n Schottky defects. To create n Schottky defects in a perfect crystal of N atoms we must move n of these atoms to surface sites. This can be done in $\binom{N}{n}$ ways, since there are $\binom{N}{n}$ ways of choosing n sites from which atoms are removed. Hence we have

$$\Omega(n) = \binom{N}{n} \equiv \frac{N!}{n!(N-n)!} \ .$$

For the entropy associated with the disorder resulting from n Schottky defects we obtain, from Eq. (2.3),

$$S(n) = k\ln\Omega(n) = k\ln\frac{N!}{n!(N-n)!} \ .$$

In writing down this expression we have neglected surface effects. These will be small compared with bulk properties, since the number of sites at the surface of the crystal is of order $N^{2/3}$, and this is very small compared with N. In particular, we have omitted from $S(n)$ a surface entropy term which corresponds to the different ways of arranging the n displaced atoms at surface sites of the crystal. (We also neglect volume changes due to the formation of defects which is justified for $n \ll N$.)

We have considered the energy and the entropy associated with the Schottky defects only. This requires an explanation, since the crystal possesses other *aspects*, such as the vibrational motion. Each of these aspects possesses its own energy and entropy. The point is that these different aspects interact very weakly. Hence each such aspect can be considered separately in calculating the energy and the entropy, as if it existed by itself. Of course there is weak interaction between these different aspects, and this interaction leads to thermal equilibrium. This is exactly the philosophy underlying the treatment of section 2.3; we divided the system into two weakly interacting subsystems (labelled 1 and 2). Now we think of the different weakly interacting aspects as these subsystems; say the vibrational motion and the Schottky defects. We could repeat the treatment of section 2.3 thinking of the energy and entropy of subsystems 1 and 2 as the corresponding quantities associated with these two aspects. This will again lead to Eqs. (2.8) and (2.9). Eq. (2.9) defines a temperature associated with each aspect, and Eq. (2.8) is the condition for thermal equilibrium between different aspects. For a crystal in thermal equilibrium the temperature T is then given by Eq. (2.9), i.e.

$$\frac{1}{T} = \frac{\partial S}{\partial E} = \frac{dS(n)}{dn}\frac{dn}{dE} = \frac{1}{\varepsilon}\frac{dS(n)}{dn} \; ,$$

where the last step follows from $E = n\varepsilon$.

To calculate $dS(n)/dn$ from the expression for $S(n)$, we use Stirling's formula for the factorial function (Appendix A1, Eq. (A.2))

$$\ln n! = n\ln n - n$$

which holds for $n \gg 1$. Hence we obtain

$$S(n) = k[N\ln N - n\ln n - (N-n)\ln(N-n)]$$

whence

$$\frac{dS(n)}{dn} = k[-\ln n + \ln(N-n)]$$

and

$$\frac{1}{T} = \frac{k}{\varepsilon}\ln\frac{N-n}{n} \; .$$

Taking exponentials and solving for n leads to

$$\frac{n}{N} = \frac{1}{\exp(\varepsilon/kT)+1}$$

which for $n \ll N$ (i.e. $\varepsilon \gg kT$) can be written

$$n = N\exp(-\varepsilon/kT) \; .$$

This equation expresses our final result. In this form it suggests an obvious reinterpretation: instead of expressing the temperature T of a crystal in thermal equilibrium in terms of the number of Schottky defects (and of N), we can think of it as giving the concentration n/N of Schottky defects in a crystal which is in thermal equilibrium at the temperature T. We ensure that the crystal is at temperature T by placing it in a heat bath at this temperature, as discussed in section 1.3. This modified viewpoint, of a system in a heat bath instead of completely isolated, will be developed systematically in section 2.5. It is extremely useful and will be used frequently later on.

Our result for the concentration n/N of Schottky defects has the expected form. At $T = 0$ K, one has $n = 0$: there are no vacant sites. This is the state of lowest energy of the system, the ground state: $E = 0$. The statistical weight of the ground state is $\Omega(n = 0) = 1$; it is the state of complete order, with no defects. As the temperature increases the number of defects increases. It is of course only for $n \ll N$, i.e. a low concentration of defects, that the system displays the properties of a crystal. At ordinary temperatures, this condition ($n \ll N$) is satisfied. The excitation energy ε is typically about 1 eV. At room temperature ($T = 290$ K) $kT = 1/40$ eV (see Eq. (1.10)), so that $\varepsilon/kT = 40$ and $n'/N = e^{-40} \approx 10^{-17}$. But on account of the exponential form, the concentration of defects is extremely sensitive to temperature changes. Thus at $T = 1000$ K, one obtains $n/N \approx 10^{-6}$ (for $\varepsilon = 1$ eV).

2.5   EQUILIBRIUM OF A SYSTEM IN A HEAT BATH

So far we have considered the equilibrium of an isolated system. We shall now treat the same system when placed in a heat bath at temperature T (Fig. 2.10). As in section 2.3 we are again dealing with a composite system but the two subsystems, our system and the heat bath, are now of very different size. There is energy transfer between the two subsystems but the temperature of the heat bath, because of its size, remains constant. The temperature of our system when in thermal equilibrium with the heat bath is then also T but the energy of the system is not fixed. We now want to find out in what sense we can ascribe a value to this energy.

The combined system is totally isolated as shown in Fig. 2.10. We assume that the particle number N and the volume V of the system are fixed. The macrostate of the system in equilibrium is then specified by T, V and N.

The system will possess a discrete set of microstates which will be labelled $1, 2, \ldots, r, \ldots$, and in these states the system will have the energy $E_1, E_2, \ldots, E_r, \ldots$. The energy by itself will in general not suffice to specify a unique microstate. Many microstates will possess the same energy and they will differ in some other characteristics. We had an example of this in our model of a paramagnetic solid (section 2.2) where the same energy could result from a large number of different configurations of dipoles. Thus a given energy

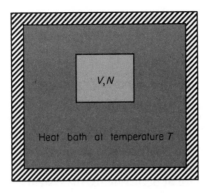

Fig. 2.10. System of fixed volume V, containing
N particles, in a heat bath at temperature T.

will occur repeatedly in the above sequence which we assume ordered so that

$$E_1 \leqslant E_2 \leqslant \ldots \leqslant E_r \leqslant \ldots . \qquad (2.14)$$

We now choose an energy interval δE which is smaller than the minimum spacing of the energy levels (2.14). A given interval δE will then contain one energy level at most, but this will usually correspond to several different states of the system.

The system plus heat bath, Fig. 2.10, is identical with the composite system of Fig. 2.5, if we think of subsystem 1 as our system and subsystem 2 as the heat bath. We can therefore straightaway take over all results from section 2.3. In particular, the probability p_r that our system will be in a definite state r (with energy E_r) will be proportional to the number of states of the heat bath compatible with this, given that the total energy of the combined system (i.e. system plus heat bath) has a constant value E_0. These states of the heat bath must have an energy lying in the interval $E_0 - E_r$ to $E_0 - E_r + \delta E$. There are $\Omega_2(E_0 - E_r)$ such states, so that

$$p_r = \text{const}.\Omega_2(E_0 - E_r) \ . \qquad (2.15)$$

In Eq. (2.15) and similar expressions below we only write down the energy arguments but omit volumes and particle numbers which are constant throughout. The correctly normalized probability is given by

$$p_r = \frac{\Omega_2(E_0 - E_r)}{\sum\limits_r \Omega_2(E_0 - E_r)} \qquad (2.16)$$

where the summation in the denominator (indicated by $\sum$) is over all *states* of the system. Using Eq. (2.3), we can express (2.15) in terms of the entropy of the heat bath

$$p_r = \text{const.} \exp\left[S_2(E_0 - E_r)/k \right]. \qquad (2.17)$$

Our discussion so far is generally valid. Now we introduce the specific assumptions which make subsystem 2 a heat bath: its average energy is very large compared with that of our system, i.e. $E_r \ll E_0$. This inequality cannot hold for all possible states r of the system, but it will hold for all states which have a reasonable chance of occurring for the system in thermal contact with the heat bath, in particular for the states near thermal equilibrium. It follows that the temperature of the heat bath is constant, since the fluctuations in energy are too small to affect it.

We expand $S_2(E_0 - E_r)$ into a Taylor series

$$\frac{1}{k} S_2(E_0 - E_r) = \frac{1}{k} S_2(E_0) - \frac{E_r}{k} \frac{\partial S_2(E_0)}{\partial E_0} + \frac{1}{2} \frac{E_r^2}{k} \frac{\partial^2 S_2(E_0)}{\partial E_0^2} + \cdots . \qquad (2.18)$$

It follows from (2.9) that

$$\frac{\partial S_2(E_0)}{\partial E_0} = \frac{1}{T} \qquad (2.19)$$

where T is the temperature of the heat bath. (The partial derivatives in the last two equations are again taken at constant volume and constant particle number for the system.) The third term on the right-hand side of Eq. (2.18) thus involves the change in temperature of the heat bath due to heat exchange with the system. By the very definition of a heat bath (i.e. its temperature does not change) this term is negligible.

One can easily see this explicitly. The ratio of the third to the second term on the right-hand side of Eq. (2.18) is $\frac{1}{2}(E_r/T)(\partial T/\partial E_0)$. To estimate this expression we use the classical result (section 7.9.1) that the energy of a system of N molecules at temperature T is of the order of NkT. Examples of this result, which the reader may have met, are the specific heat of a perfect gas and Dulong and Petit's law of specific heats. If the system and heat bath have N_1 and N_2 particles respectively with $N_1 \ll N_2$ then $E_r \sim N_1 kT$, $E_0 \sim N_2 kT$, and the above ratio of terms is of the order of $N_1/N_2 \ll 1$.

The Taylor expansion (2.18) raises the question whether we could have expanded $\Omega_2(E_0 - E_r)$ in Eq. (2.16) *directly*, instead of its logarithm. The answer is: *no*. The expansion of $\Omega_2(E_0 - E_r)$ does not converge sufficiently rapidly. We shall illustrate this briefly. The heat bath is a macroscopic system with a *very large* number of degrees of freedom v. For such a system one typically has*

$$\Omega_2(E_0 - E_r) = C(E_0 - E_r)^v$$

where C is a constant. Hence Ω_2 has the Taylor series

$$\Omega_2(E_0 - E_r) = \Omega_2(E_0) \left\{ 1 - v\frac{E_r}{E_0} + \tfrac{1}{2}v(v-1) \left(\frac{E_r}{E_0}\right)^2 + \cdots \right\}.$$

For the term quadratic in (E_r/E_0) to be small compared with the linear term in this series, we require $E_r \ll 2E_0/(v-1)$, i.e. E_r must be very small compared to the energy of *one* degree of freedom, which is of the order of kT. In general this will not be the case.

On the other hand we have

$$\begin{aligned}
\ln\Omega_2(E_0 - E_r) &= \ln\left[C(E_0 - E_r)^v \right] \\
&= \ln\left[CE_0^v \right] + v\ln\left(1 - \frac{E_r}{E_0} \right) \\
&= \ln\left[CE_0^v \right] - v\left\{ \frac{E_r}{E_0} + \frac{1}{2}\left(\frac{E_r}{E_0}\right)^2 + \cdots \right\},
\end{aligned}$$

and now the quadratic term is very small compared to the linear term if only $\tfrac{1}{2}E_r \ll E_0$, and this certainly is the case by our assumptions about the heat bath. $E_r \ll E_0$ is, of course, just the condition which was derived above for Eq. (2.18), from a slightly different viewpoint.

Retaining only the term linear in E_r in Eq. (2.18) we obtain

$$\frac{1}{k} S_2(E_0 - E_r) = \frac{1}{k} S_2(E_0) - \beta E_r \tag{2.20}$$

where we introduced the 'temperature parameter'

$$\beta \equiv \frac{1}{kT}. \tag{2.21}$$

We shall see that it is β rather than T which occurs naturally, as has already

---

*See the discussions following Eq. (2.3) and Eq. (2.32) below. For a perfect gas this result is derived in section 7.7, Eq. (7.71a).

been mentioned in section 1.2. On account of Eqs. (2.20) and (2.21) we can write (2.17) in the form

$$p_r = \frac{1}{Z}\, e^{-\beta E_r} \qquad (2.22)$$

where the normalization constant Z is given by

$$Z = \sum_r e^{-\beta E_r} \qquad (2.23)$$

so as to ensure correct normalization of probabilities ($\sum_r p_r = 1$).

Eq. (2.22) is our final result. It is called the *Boltzmann distribution*. It gives the *probability that a system when placed in a heat bath at temperature T (i.e. with a temperature parameter $\beta = 1/kT$) should be in a particular state.* We see that this probability depends on the energy of the state. The *only* property of the heat bath on which it depends is the *temperature* of the heat bath. The quantity Z, defined by (2.23), is known as the *partition function* of the system. We shall see that it plays a central role in studying systems at a fixed temperature.

Eq. (2.23) is a sum over all microstates of the system. Since the energies of these different states are not necessarily different, we can rewrite Eq. (2.23). Instead of repeating equal terms by summing over *all states*, we can group the terms in Eq. (2.23) as follows:

$$Z = \sum_{E_r} g(E_r)e^{-\beta E_r} \; . \qquad (2.24)$$

Here the summation $\sum_{E_r}$ is only over *all different energies* E_r, and $g(E_r)$ is the number of states possessing the energy E_r. (A reader who is not clear how Eqs. (2.23) and (2.24) are related should consider a particular case; for example, $E_1 < E_2 = E_3 = E_4 < E_5 = E_6 < E_7 < \ldots$.) In quantum mechanics E_r is called the energy eigenvalue of the state r, and $g(E_r)$ is known as the degeneracy of the energy eigenvalue E_r. The probability $p(E_r)$ that the system be in a state with the energy E_r is then given, from Eq. (2.22), by

$$p(E_r) = \frac{1}{Z}\, g(E_r)e^{-\beta E_r} \; , \qquad (2.25)$$

since there are just $g(E_r)$ different states with this energy, and each of these states has the same probability (2.22) of occurring.*

---

*It might appear desirable to label energy eigenvalues differently, according to whether degenerate eigenvalues are explicitly written down repeatedly, as in (2.23), or only once and multiplied by their degeneracy $g(E_r)$, as in (2.24). As is customary, we shall use the same notation for both cases; it will always be clear from the context which convention is being used.

From the Boltzmann distribution, we can at once obtain the mean energy $\bar{E}$ of the system in its heat bath:

$$\bar{E} \equiv \sum_r p_r E_r = - \frac{\partial \ln Z}{\partial \beta} \qquad (2.26)$$

as follows from Eqs. (2.22) and (2.23). (Remember that $d(\ln x)/dx = 1/x$.) The *partial* derivative in (2.26) signifies that in differentiating (2.23) the energy eigenvalues E_r are held constant. These energy levels do not depend on the temperature. They depend purely on the microscopic structure of the system. Hence they depend on those parameters — usually called *external parameters* — which determine the structure of the system. In the case of the paramagnetic solid, treated in section 2.2, the energy levels depend on the applied magnetic field. For a perfect gas in an enclosure the energy levels depend on the volume of the enclosure, as we shall see later. Thus it is such other quantities as magnetic field, volume, etc., which are kept fixed in (2.26). We shall return to this point in more detail later. In order to make the discussion less abstract, we shall in this section definitely be considering the volume as the variable external parameter.

The energy (2.26) is only the *mean* energy of the system. Because of the heat bath its *actual* energy will fluctuate and we now want to see how large these fluctuations are. The magnitude of these fluctuations is measured by the standard deviation ΔE. Defining the mean value $\overline{f(E)}$ of a quantity $f(E_r)$ with respect to a probability distribution p_r by

$$\overline{f(E)} = \sum_r p_r f(E_r) \ , \qquad (2.27)$$

the standard deviation ΔE is defined by

$$(\Delta E)^2 \equiv \overline{(E - \bar{E})^2} = \overline{E^2} - \bar{E}^2 \ . \qquad (2.28)$$

Differentiating $\ln Z$ twice with respect to β, using Eqs. (2.26), (2.22) and (2.23), and comparing with Eq. (2.28), one obtains

$$(\Delta E)^2 = \frac{\partial^2 \ln Z}{\partial \beta^2} = - \frac{\partial \bar{E}}{\partial \beta} = - \frac{dT}{d\beta} \frac{\partial \bar{E}}{\partial T} = kT^2 C \ , \qquad (2.29)$$

where $C \equiv \partial \bar{E}/\partial T$ is the heat capacity of the system at constant external parameters, i.e. at constant volume. The quantity of interest is the relative fluctuation:

$$\frac{\Delta E}{\bar{E}} = \frac{(kT^2 C)^{1/2}}{\bar{E}} \ . \qquad (2.30)$$

In this equation C and $\bar{E}$ are extensive quantities, i.e. they are proportional to the number of molecules N in the system, whereas kT^2 is independent of N. Hence the dependence of $\Delta E/\bar{E}$ on the size of the system (i.e. on N) is, from Eq. (2.30), given by

$$\frac{\Delta E}{\bar{E}} \sim \frac{1}{\sqrt{N}} . \tag{2.31}$$

This equation tells us that the relative fluctuation of the energy of a system in a heat bath is inversely proportional to the square-root of its size. *For a macroscopic system where $N \sim 10^{23}$, $\Delta E/\bar{E} \sim 10^{-11}$: the fluctuations are extremely small: the energy of a macroscopic body in a heat bath is, for practical purposes, completely determined.* Eq. (2.31) is an extremely important and very general type of result. *It contains the justification why statistical physics can make definite quantitative statements about macroscopic systems; similar results—always containing a $1/\sqrt{N}$—apply to the relative fluctuations of other properties of macroscopic systems.*

However, there are exceptional situations where large relative fluctuations occur in macroscopic systems. Consider the system shown in Fig. 2.11: a liquid in contact with its saturated vapour. The system is immersed in a heat bath at temperature T, and the pressure P applied to the freely movable piston is adjusted to equal the saturation vapour pressure corresponding to the temperature T. (We are appealing to the fact—to be discussed in section 8.3—that for a liquid–vapour system in equilibrium the temperature determines the vapour pressure, and vice versa.) For this system in equilibrium, the temperature does not determine the energy of the system. As energy is transferred between the liquid–vapour system and the heat bath, liquid will evaporate or vapour condense, but the temperature of the system will remain constant. Correspondingly the energy of the system can vary widely between the extreme situations: 'only liquid present' and 'only vapour present', i.e. the energy of the system is not well determined: it possesses large fluctuations. Similarly there are large fluctuations in the volume of this system corresponding to the two extreme situations mentioned above. Another example of large fluctuations is given by the fluctuations near the critical point of a substance. (See section 8.6 for the definition of the critical point, etc.) A full discussion of these critical point phenomena is very difficult. For a partial discussion of some aspects see Flowers and Mendoza,[26] section 7.7, and Present,[11] section 9.2.

The fact that usually $\Delta E/\bar{E}$ is so small for macroscopic systems means that the probability distribution (2.25) has an extremely sharp maximum at energy $\bar{E}$, as illustrated schematically in Fig. 2.12. One can understand this sharp maximum as follows. For a macroscopic body the density of energy levels E_r, i.e. the number of such states per unit energy interval, is very large.

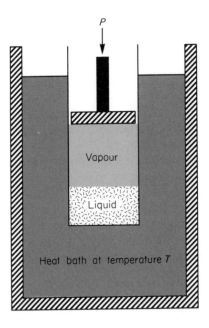

Fig. 2.11. Liquid–vapour system in a heat bath.

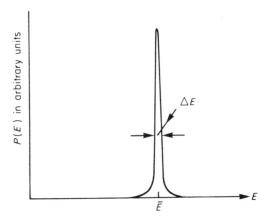

Fig. 2.12. Schematic picture of the energy distribution $P(E)$ for a macroscopic body. The width of the peak ΔE is the standard deviation.

They practically form a continuum. So it is convenient to introduce $f(E)\,dE$, the number of microscopic states of the system with energy eigenvalues in the energy interval $(E, E + dE)$. The probability that the system should have energy in this interval is then, from Eq. (2.25), given by

$$P(E)\,dE = \frac{1}{Z} f(E) e^{-\beta E}\,dE \ . \tag{2.32}$$

For a macroscopic system the density of states $f(E)$ is a very rapidly increasing function of E. (For a perfect gas, for example, we shall later see that $f(E) \propto E^{3N/2-1}$. See section 7.7, Eq. (7.71).) The Boltzmann factor $\exp(-\beta E)$, on the other hand, being an exponential is rapidly decreasing with energy. The product of these two factors has a very sharp maximum, as shown in Fig. 2.12. (See problem 2.8.)

Our result allows generalization. For most purposes a macroscopic system behaves the same way whether we think of it as isolated (E, V and N specified), or as in a heat bath (T, V and N specified) with average energy $\bar{E} = E$. The same conclusion holds for other choices of independent variables, e.g. for T, P and N. Which of these choices one adopts is largely a matter of convenience. Theoretically, it is usually easiest to take T and V as independent variables.* We shall frequently write E instead of $\bar{E}$, etc., for macroscopic quantities whose fluctuations are negligible.

The situation is quite different for a small system consisting of a few molecules or only one molecule! Now the relative fluctuation (2.31) is very large. The reader may wonder what we mean by one molecule in a heat bath. But consider a gas and, in particular, one specific molecule of the gas. Then the rest of the gas will act as a heat bath for this one molecule, whose energy will indeed fluctuate violently with each collision.

In the same way, we can think of, say, one cubic centimetre of copper inside a block of copper weighing several kilograms as a system plus its heat bath. But now the fluctuations will be small. On the other hand, our whole treatment of system plus heat bath presupposed only weak thermal interactions between them, and if one considers some part of a macroscopic system it is not clear that this assumption of weak interactions will always hold. One can adopt a different point of view in studying a macroscopic system. One actually imagines a very large number ν of identical replicas of this system and puts all these systems into weak thermal contact with each other. We can call any one of these our system and all the others the heat bath. We can make ν as large as we like and so get rid of unwanted fluctuations. No problem arises over counting the systems (we shall see later that because of quantum-mechanical effects great care is required when dealing with identical atoms, etc.); they are macroscopic systems; we can label them, e.g. write numbers on them: $1, 2, 3, \ldots$, with a piece of chalk. This ensemble of systems in thermal contact with each other is known as a *canonical ensemble*. The usefulness of such ensembles was particularly emphasized by Gibbs.

---

*To keep P rather than V constant, means that the energy levels E_r of the system which depend on volume vary, and this makes the analysis much harder.

Previously we defined the entropy for the special case of an *isolated* system. We now give a *general* definition which will then be applied to the particular case of a system in a heat bath.

Let us, quite generally, consider a macroscopic system. Let us label its microstates $1, 2, \ldots, r, \ldots$, and let p_r be the probability that the system is in the state r. We are not stating what the conditions or constraints of the system are. Hence we do not know the values of these probabilities p_r, except for the normalization condition

$$\sum_r p_r = 1 \; .$$

One can show that the entropy of the system is, quite generally, given by

$$S = -k \sum_r p_r \ln p_r \; . \tag{2.33}$$

To derive Eq. (2.33), let us consider not a single system but an ensemble of a very large number ν of identical replicas of our system. (This ensemble is essentially more general than the canonical ensemble defined above, since the constraints have not been specified.) In particular, each system in this ensemble of replicas will have the *same* probability distribution $p_1, p_2, \ldots, p_r, \ldots$ of being in the microstate $1, 2, \ldots, r, \ldots$ respectively. It follows that for ν sufficiently large, the number of systems of the ensemble in the state r is given by

$$\nu_r = \nu p_r \; .$$

Consider the statistical weight Ω_ν of the ensemble when ν_1 of its systems are in the state 1, ν_2 of its systems in the state 2, and so on. Ω_ν is just the number of ways in which this particular distribution can be realized, i.e.

$$\Omega_\nu = \frac{\nu!}{\nu_1! \, \nu_2! \, \ldots \, \nu_r! \, \ldots} \; .$$

It follows from Boltzmann's entropy definition (2.3) that the entropy S_ν of our ensemble is given by

$$S_\nu = k \ln \Omega_\nu$$

$$= k \ln \frac{\nu!}{\nu_1! \, \nu_2! \, \ldots \, \nu_r! \, \ldots}$$

$$= k [\nu \ln \nu - \sum_r \nu_r \ln \nu_r] \; .$$

The last line followed from Stirling's formula

$$\ln v! = v\ln v - v$$

(cf. Appendix A1, Eq. (A.2)) which is applicable since for sufficiently large values of v the numbers $v_r (= vp_r)$ also become very large. In the expression for S_v we substitute $v_r = vp_r$. On account of $\sum p_r = 1$ we finally obtain for the entropy of the ensemble

$$S_v = -vk\sum_r p_r \ln p_r \ .$$

But we saw in section 2.3 that the entropy is an extensive quantity. Hence the entropy S of a *single system* is related to the entropy S_v of the *ensemble* by

$$S = \frac{1}{v} S_v = -k \sum_r p_r \ln p_r$$

which is the basic result stated in Eq. (2.33) above.

We shall first of all show that for an isolated system in equilibrium this general definition (2.33) of the entropy reduces to our original definition (2.3), i.e. these two definitions are consistent. For an isolated system with energy in the interval $(E, E + \delta E)$, there are $\Omega(E, V, N)$ different microstates with energy E_r in this interval. Hence the probability p_r of finding the system in one of these states is $1/\Omega(E, V, N)$, while the probability of finding the system in a state with energy outside this interval δE is zero. We have here used the postulate of equal *a priori* probabilities and the normalization of probability: $\sum p_r = 1$. If we use the fact that there are $\Omega(E, V, N)$ non-zero terms in the sum of Eq. (2.33), and that for these $p_r = 1/\Omega(E, V, N)$, Eq. (2.33) at once reduces to

$$S(E, V, N) = k\ln\Omega(E, V, N) \ , \tag{2.34}$$

in agreement with (2.3).

In analogy to the canonical ensemble, discussed above, the *microcanonical* ensemble is defined as an ensemble of non-interacting systems with the probability distribution: $p_r = 1/\Omega(E, V, N)$ if E_r lies in the interval $(E, E + \delta E)$, and $p_r = 0$ otherwise. The analogous definition of the canonical ensemble is then an ensemble of systems whose probability distribution is given by Eq. (2.22).

To obtain the entropy of a system in a heat bath at temperature T, we substitute the Boltzmann distribution (2.22) into the general definition of the entropy, Eq. (2.33), giving

$$S(T, V, N) = k\ln Z + \bar{E}/T \ . \tag{2.35}$$

In Eq. (2.35) Z and $\bar{E}$, and consequently the entropy S, are defined as functions of T, V and N. This is in contrast to the definition (2.34) of the entropy of an *isolated* system which is a function of E, V and N. But we have seen that for a macroscopic system at temperature T the energy fluctuations are negligible, i.e. the energy is well defined and is equal to the mean energy $\bar{E}$. Hence the entropy of a macroscopic body in a heat bath at temperature T is also sharply defined and is equal to the entropy of an isolated body with energy E equal to the mean energy $\bar{E}$ of the system at temperature T:

$$S(T,V,N) = k\ln\Omega(\bar{E},V,N) \ . \tag{2.34a}$$

In discussing an isolated system, the basic statistical and thermodynamic quantities were the statistical weight $\Omega(E, V, N)$ and the entropy $S(E,V,N)$. For a system in a heat bath the basic statistical quantity is the partition function $Z(T, V, N)$, Eq. (2.23). (In specifying the independent variables, we have assumed that the volume is the appropriate external parameter.) The corresponding thermodynamic quantity is the Helmholtz free energy

$$F(T,V,N) = -kT\ln Z(T,V,N) \ . \tag{2.36}$$

Eliminating Z from Eqs. (2.35) and (2.36), we obtain the purely thermodynamic relation

$$F = E - TS \ , \tag{2.37}$$

where we have written E instead of $\bar{E}$ for the well-defined energy of a macroscopic system in a heat bath. We shall see that for a system at constant volume in a heat bath the Helmholtz free energy plays a role analogous to that of the entropy for an isolated system. In particular, Eqs. (2.34) and (2.36) show that the choice of (E, V, N) and (T, V, N) as independent variables leads naturally to the entropy S and the Helmholtz free energy F respectively. Furthermore, just as the entropy S is a maximum in equilibrium for an isolated system (i.e. E, V and N constant) so the Helmholtz free energy F is a minimum in equilibrium for a system of constant volume in a heat bath (i.e. T, V and N constant). This last result will be derived in sections 4.2 and 4.4 (see Eqs. (4.29) and (4.41)).

In our development of statistical physics, we have defined two thermodynamic functions: the entropy S and the Helmholtz free energy F. Which of these one uses depends on the conditions of the system. For an isolated system one deals with S; for a system in a heat bath with F. This correctly suggests that yet other conditions will lead to other thermodynamic functions (see sections 4.4 and 4.5).*

In this chapter most of the important ideas of statistical physics were introduced. In particular, we defined the thermodynamic functions of entropy and Helmholtz free energy. We shall see that from these functions one can derive all macroscopic properties (such as the temperature or pressure) of a system, as well as many laws relating different macroscopic properties. These results are of quite general validity. They can be applied to any system. For example, knowing the entropy of an isolated system as a function of the energy and other relevant parameters, the temperature is always given by an equation like Eq. (2.9). This is equally true for a gas or a paramagnetic solid! It is their generality which makes these arguments so important and useful.

Our approach in this chapter has been from the statistical, i.e. atomic viewpoint, and this approach is, of course, necessary if one wants to calculate the properties of a system completely from first principles. But there is an alternative less fundamental approach, that of classical thermodynamics. This is based on the fact that, as we shall see in Chapter 4, the entropy and other thermodynamic functions can be defined in terms of macroscopic, directly measurable quantities. Hence one can use relations involving thermodynamic functions without recourse to their statistical interpretation but using, instead, experimental data. This approach of classical thermodynamics is extremely powerful since it may be applied to systems which are too complex to admit a complete statistical treatment.

SUMMARY

The following definitions and equations which were given in this chapter are fundamental to the whole of statistical physics and will frequently be needed in later chapters.

Microstate (section 2.2)   The state of a system defined 'microscopically', i.e., a *complete* description on the atomic scale.

Macrostate (section 2.2)   The state of a system of macroscopic dimensions specified by (a few) macroscopically observable quantities only.

Statistical weight Ω of a macrostate   is the number of microstates comprising this macrostate (p. 36).

---

*The reader should be warned that neither the notation nor the names of thermodynamic quantities are fully standardized. The Helmholtz free energy F is sometimes called the Helmholtz function or the work function, etc., and is also denoted by A. The partition function is sometimes denoted by F, the internal energy by U; *and so on*!

Postulate of equal *a priori* probabilities   states that for an isolated system in a definite macrostate, the Ω microstates comprising this macrostate occur with equal probability (section 2.3).

Equilibrium postulate (section 2.3)   For an isolated macroscopic system, defined by E, V, N (which are fixed) and *variable* parameters α, equilibrium corresponds to those values of α for which the statistical weight $\Omega(E, V, N, \alpha)$ attains its maximum.

Boltzmann's definition of entropy:

$$S(E, V, N, \alpha) = k \ln \Omega(E, V, N, \alpha) \ . \tag{2.3}$$

The second law (section 2.3)   During real processes, the entropy of an isolated system always increases. It attains its maximum value in the state of equilibrium.

External parameter (p. 57)   A macroscopic variable (such as the volume of a system, or an applied electric or magnetic field) whose value affects the microstates (and energy levels $E_1, E_2, \ldots$) of a system.

Temperature definition

$$\frac{1}{T} = \left(\frac{\partial S(E, V, N)}{\partial E} \right)_{V, N} \ . \tag{2.9}$$

Pressure definition

$$P = T \left(\frac{\partial S(E, V, N)}{\partial V} \right)_{E, N} \ . \tag{2.12}$$

Heat bath   See definition at end of Chapter 1.

Partition function

$$Z = \sum_r \exp(-\beta E_r) \ , \tag{2.23}$$

where the summation is over all microstates of the system.

Temperature parameter

$$\beta \equiv \frac{1}{kT} \ . \tag{2.21}$$

Boltzmann distribution

$$p_r = \frac{1}{Z} \exp(-\beta E_r) \ . \tag{2.22}$$

(p_r is the probability that a system at temperature T is in the state r with energy E_r.)

Mean energy $(\bar{E})$ **of a system at temperature** T

$$\bar{E} = -\frac{\partial \ln Z}{\partial \beta} \ . \tag{2.26}$$

General entropy definition

$$S = -k\sum_r p_r \ln p_r \ . \tag{2.33}$$

For an isolated system in equilibrium this reduces to

Boltzmann's entropy definition for equilibrium

$$S(E, V, N) = k \ln \Omega (E, V, N) \ . \tag{2.34}$$

Helmholtz free energy F

$$F(T, V, N) = -kT \ln Z(T, V, N) \tag{2.36}$$

$$F = E - TS. \tag{2.37}$$

PROBLEMS 2

2.1 In a monatomic crystalline solid each atom can occupy either a regular lattice site or an interstitial site. The energy of an atom at an interstitial site exceeds the energy of an atom at a lattice site by an amount ε. Assume that the number of interstitial sites equals the number of lattice sites, and also equals the number of atoms N.

Calculate the entropy of the crystal in the state where exactly n of the N atoms are at interstitial sites. What is the temperature of the crystal in this state, if the crystal is in thermal equilibrium?

If $\varepsilon = 1$ eV and the temperature of the crystal is 300 K, what is the fraction of atoms at interstitial sites?

2.2 A system consists of N weakly interacting subsystems. Each subsystem possesses only two energy levels E_1 and E_2, each of them non-degenerate.

(i) Draw rough sketches (i.e. from common sense, not from exact mathematics) of the temperature dependence of the mean energy and of the heat capacity of the system.

(ii) Obtain an exact expression for the heat capacity of the system.

2.3 A system possesses three energy levels $E_1 = \varepsilon$, $E_2 = 2\varepsilon$ and $E_3 = 3\varepsilon$, with degeneracies $g(E_1) = g(E_3) = 1$, $g(E_2) = 2$. Find the heat capacity of the system.

2.4 Consider the problem of the Schottky defects studied in section 2.4. For a system consisting of N atoms and possessing n such defects write down the Helmholtz free energy $F(n)$. For a system in thermal equilibrium in a heat bath at temperature T the Helmholtz free energy is a minimum. Use this property to determine the temperature dependence of the concentration n/N of Schottky defects.

2.5 One knows from spectroscopy that the nitrogen molecule possesses a sequence of vibrationally excited states with energies $E_r = \hbar\omega(r + \frac{1}{2})$, $r = 0, 1, 2, \ldots$. If the level spacing $\hbar\omega$ is 0.3 eV, what are the relative populations of the first excited state $(r = 1)$ and the ground state $(r = 0)$, if the gas is in thermal equilibrium at 1000 K?

2.6 The first excited state of the helium atom lies at an energy 19.82 eV above the ground state. If this excited state is three-fold degenerate while the ground state is non-degenerate, find the relative populations of the first excited and the ground states for helium gas in thermal equilibrium at 10,000 K.

2.7 The following may be considered a very simple one-dimensional model of rubber. A chain consists of $n (\geqslant 1)$ links each of length a. The links all lie along the x-axis but they may double back on themselves. The ends of the chain are distance L apart and the system is isolated. Find the tension on the chain if both orientations of the links have the same energy. Show that for $L \ll na$ one obtains Hooke's law.

2.8 It follows from Eq. (2.32) that the partition function of a macroscopic system is given by

$$Z = \int f(E) e^{-\beta E} dE .$$ (2.38)

Show that for a macroscopic system one can approximate $\ln Z$ by

$$\ln Z = f(\tilde{E}) - \beta \tilde{E}$$ (2.39)

where $\tilde{E}$ is the value of the energy at which the integrand $f(E) \exp(-\beta E)$ has its maximum.

(Hint: expand $\ln[f(E) \exp(-\beta E)]$ in a Taylor series about $E = \tilde{E}$ up to the quadratic term.)

Paramagnetism

3.1 A PARAMAGNETIC SOLID IN A HEAT BATH

We shall now apply the results of the last chapter to the simple model of a paramagnetic solid, considered in section 2.2. This model consists of a system of magnetic dipoles placed in an external applied magnetic field (due, for example, to an electromagnet). Other aspects of the solid (such as the lattice vibrations) are neglected as they interact only weakly with the magnetic aspect. (See section 2.4 for discussion of this very important idea of different, more or less independent, aspects of a system.) They merely serve to produce thermal equilibrium.

The interaction energy of a dipole with magnetic moment μ placed in a magnetic field is $-\mu \cdot \mathscr{B}$. Here $\mathscr{B}$ is the *local* field which acts on the dipole; it is the field which is present at the location of the dipole if the dipole is removed, assuming that this removal in no way disturbs the surroundings.*
This local field is the resultant of the external applied field (due to electromagnets, etc.) and the field of the other dipoles in the specimen which produce a magnetization $\mathscr{I}$. In most cases the effects of these other dipoles are very small. (Ferromagnetic materials are an obvious exception.) Hence, although the interaction $-\mu \cdot \mathscr{B}$ and our subsequent expressions will strictly speaking involve the local field, in practice the local and the external applied fields will be the same, and we shall not have to distinguish between them.

*For a discussion of the local field concept see Kittel,[29] Chapter 13.

In this section we shall follow the simplest approach and consider a single magnetic dipole in a heat bath. We can limit ourselves to a single dipole because of the additivity of energy, Eq. (2.1), which we assumed, neglecting all interactions except those of each dipole with the magnetic field $\mathscr{B}$. We can then, if we wish, think of the rest of the solid as acting as a heat bath to the one dipole we are looking at. As in section 2.2 we shall restrict ourselves to spin $\frac{1}{2}$ dipoles.

A single spin $\frac{1}{2}$ dipole, with magnetic moment μ, placed in a magnetic field $\mathscr{B}$, can exist in two states, with energy $\mp\mu\mathscr{B}$. Hence the partition function for one dipole is (cf. Eq. (2.23))

$$Z_1 \equiv Z(T,\mathscr{B},1) = e^x + e^{-x} = 2\cosh x \qquad (3.1)$$

where

$$x \equiv \mu\mathscr{B}\beta \equiv \mu\mathscr{B}/kT \ . \qquad (3.2)$$

The probabilities p_+ and p_-, that the dipole should be in the states with μ parallel and antiparallel to $\mathscr{B}$ respectively, are given from Eq. (2.22) by

$$p_\pm = \frac{1}{Z_1} e^{\pm x} \ . \qquad (3.3)$$

These probabilities are plotted in Fig. 3.1. For $x \ll 1$ (i.e. very weak field, $\mathscr{B} \to 0$, high temperatures, $T \to \infty$) $p_+ = p_- = 0.5$: one has random orientation, as expected. For $x \gg 1$ (i.e. strong fields, $\mathscr{B} \to \infty$, low temperatures, $T \to 0$), $p_+ = 1$ and $p_- = 0$, i.e. complete alignment. The figure illustrates the *opposing tendencies of the magnetic field to bring about a state of order (alignment) and of the thermal motion to cause disorder (random orientation)*. This type of conflict is typical of thermal problems.

From Eq. (3.3) we obtain the mean magnetic moment $\bar\mu$ of the dipole in the direction of the field

$$\bar\mu = p_+(\mu) + p_-(-\mu) = \mu\tanh x \ , \qquad (3.4)$$

and the mean energy $\bar E_1$ of the dipole in the field

$$\bar E_1 = p_+(-\mu\mathscr{B}) + p_-(\mu\mathscr{B}) = -\mu\mathscr{B}\tanh x \ . \qquad (3.5)$$

These quantities can also be written in the forms

$$\bar\mu = \frac{1}{\beta}\left(\frac{\partial \ln Z_1}{\partial \mathscr{B}}\right)_\beta , \qquad (3.4a)$$

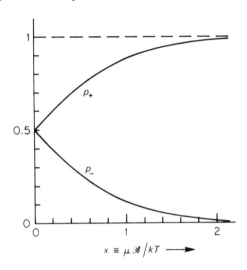

Fig. 3.1. Probabilities $p_\pm$ that a spin $\frac{1}{2}$ magnetic dipole μ placed in a magnetic field $\mathscr{B}$, should be orientated parallel/antiparallel to the field respectively.

and

$$\bar{E}_1 = - \left(\frac{\partial \ln Z_1}{\partial \beta} \right)_{\mathscr{B}} . \qquad (3.5a)$$

Because we are neglecting the interactions between the dipoles we can at once write down the corresponding results for a solid of N dipoles. The energy E is given by

$$E = N\bar{E}_1 = - N\mu\mathscr{B} \tanh x , \qquad (3.5b)$$

and the mean magnetic moment of the specimen (acting in the direction of the field) is given by

$$\mathscr{M} = N \bar{\mu} = N\mu \tanh x . \qquad (3.4b)$$

From the last two equations one obtains

$$E = - \mathscr{M}\mathscr{B} . \qquad (3.6)$$

The magnetization $\mathscr{I}$, i.e. the magnetic moment per unit volume, is given by

$$\mathscr{I} = \frac{\mathscr{M}}{V} = \frac{N}{V} \mu \tanh x . \qquad (3.7)$$

Eq. (3.7) is our basic result. It gives the dependences of the magnetization on the temperature and on the magnetic field. The magnetization (3.7) is plotted against x in Fig. 3.2.

In the limit of weak field and high temperature, i.e. $x \ll 1$, one has $\tanh x \approx x$ and Eq. (3.7) becomes

$$\mathscr{I} = \frac{N}{V}\mu x = \frac{N\mu^2}{VkT}\,\mathscr{B} \ , \tag{3.8}$$

i.e. the magnetization becomes proportional to the local field $\mathscr{B}$. If, as is usually the case in paramagnetic salts, the magnetization $\mathscr{I}$ is very small compared to the local field $\mathscr{B}/\mu_0$ (μ_0 = permeability of the vacuum), then we can interpret $\mathscr{B}$ in Eq. (3.8) as the applied magnetic field. Hence we obtain for the magnetic susceptibility χ:

$$\chi = \frac{\mathscr{I}}{\mathscr{H}} = \frac{N\mu^2\mu_0}{VkT} \tag{3.9}$$

where $\mathscr{H} = \mathscr{B}/\mu_0$ is the applied magnetic field strength. This temperature dependence of the susceptibility, $\chi \propto 1/T$, is known as Curie's law. For many salts it holds very accurately at quite low temperatures, down to about 1 K, so that it is used for temperature calibration in this

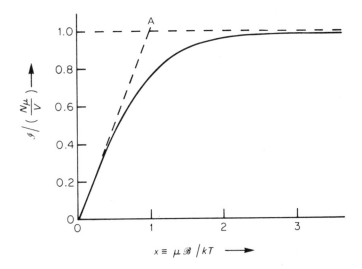

Fig. 3.2. Magnetization $\mathscr{I}$ (in units of $N\mu/V$) versus $x = \mu\mathscr{B}/kT$. The straight line OA marks the initial slope of the curve, corresponding to Eq. (3.8).

region.* (Cerium magnesium nitrate obeys Curie's law down to 0.01 K which makes it particularly useful as a magnetic thermometer.) Fig. 3.3 shows this $1/T$ dependence for a copper salt.

In the limit of strong field and low temperature, i.e. $x \gg 1$, one has $\tanh x \approx 1$, and Eq. (3.7) becomes

$$\mathscr{I} = \frac{N}{V} \mu , \tag{3.10}$$

i.e. the dipoles are completely aligned with the field, the specimen saturates.

We see that the magnitude of x, Eq. (3.2), determines the physics of the situation completely. Let us consider what determines x in a paramagnetic salt and let us estimate x in terms of T and $\mathscr{B}$. Paramagnetic salts consist of molecules possessing a permanent magnetic moment μ. Usually it is one particular ion which possesses the magnetic moment. Examples of paramagnetic salts are ferric ammonium alum and potassium chromium alum, the magnetic moment being carried by the iron and chromium ions respectively. The magnetic moment results from the intrinsic magnetic moments of the electrons and from their orbital motion. The value of μ depends on the atomic properties and the crystal structure of the solid, but it is of the order of magnitude of the Bohr magneton

$$\mu_B = \frac{e\hbar}{2m_e} = 9.3 \times 10^{-24} \text{ A.m}^2 .^\dagger \tag{3.11}$$

Thus x is of the order of magnitude of $\mu_B \mathscr{B}/kT$. With $\mathscr{B} = 1$ tesla ($= 10^4$ gauss), one has $\mu_B \mathscr{B} \approx 10^{-23}$ J. This must be compared with $kT \approx 4 \times 10^{-21}$ J at $T = 300$ K, and $kT = 1.4 \times 10^{-23}$ J at 1 K, showing that in the former case (room temperature) $\mu_B \mathscr{B}/kT \approx 2 \times 10^{-3}$ so that, according to Eq. (3.8), the magnetization will be proportional to $\mathscr{B}/T$, while in the latter case $\mu_B \mathscr{B}/kT \approx 1$, resulting in a substantial degree of alignment as can be seen from Fig. 3.2.

We have restricted ourselves to spin $\frac{1}{2}$ dipoles. The theory can be extended straightforwardly to more complex situations,‡ and excellent agreement is obtained with experiments, as shown in Fig. 3.4. The points are measurements at several temperatures, from 1.30 K to 4.21 K as labelled. The continuous curves are the theoretical values obtained from the appropriate generalizations of Eq. (3.4) for the three cases shown.

---

*See, for example, the books by Adkins,[1] Pippard,[9] or Zemansky and Dittman.[22]

†In Gaussian units $\mu_B = e\hbar/2m_e c = 0.93 \times 10^{-20}$ erg/gauss. Since at present solid-state physicists use both SI and Gaussian units we shall, in the following, state some quantities in both systems of units.

‡See, for example, the books by Blakemore,[24] Kittel[29] or Wannier.[20]

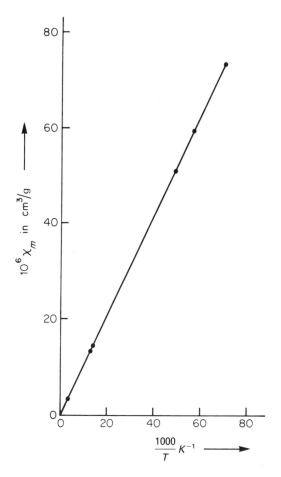

Fig. 3.3. Magnetic mass susceptibility

$$\chi_m = \frac{\chi}{\text{density}}$$

versus $1/T$ for powdered $CuSO_4 \cdot K_2SO_4 \cdot 6H_2O$. (After J. C. Hupse, *Physica*, **9**, 633 (1942).)

The excellent agreement between theory and experiment shows that the basic approximation of our simple model, namely the neglect of the interactions of the dipoles with each other, is well justified. The reason for this is that the paramagnetic salts used in practice have very large molecules (e.g. cerium magnesium nitrate is $Ce_2Mg_3(NO_3)_{12} \cdot 24H_2O$) so that the ions carrying the magnetic moments are spaced far apart and only interact very weakly with each other. If necessary, one can place the paramagnetic ions in a non-magnetic matrix and in this way ensure sufficient separation of the dipoles.

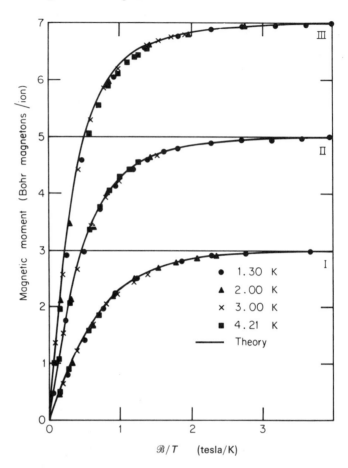

Fig. 3.4. Magnetic moment per ion versus $\mathscr{B}/T$ for spherical samples of (I) potassium chromium alum; (II) iron ammonium alum; (III) gadolinium sulphate. (After W. E. Henry, *Phys. Rev.*, **88**, 559 (1952).)

Large molecules, i.e. large dipole–dipole separations, imply a small value for N/V in Eq. (3.9), and hence for the magnetic susceptibility χ. We estimate χ from Eq. (3.9) for a paramagnetic salt. Typical values for the gram–ionic weight and the density of a paramagnetic salt are 400 g and 2 g/cm³. Hence $N/V \sim (6 \times 10^{23}/2 \times 10^{-4})\,\text{m}^{-3}$. We take $\mu \sim \mu_B$. Eq. (3.9) then gives $\chi \sim 0.03/T$ which is indeed very small down to quite low temperatures. Down to these temperatures the dipole–dipole interaction is negligible, and the local and the applied fields are very nearly the same.

★ 3.2 THE HEAT CAPACITY AND THE ENTROPY

Eqs. (3.5) and (3.6) show that the magnetic energy E of our dipole system is temperature dependent. Hence the system possesses a magnetic heat capacity. Experimentally one usually changes the temperature of the specimen keeping the applied field intensity $\mathcal{H}$ constant, i.e. one measures the heat capacity at constant applied field strength:

$$C_{\mathcal{H}} = (đQ/\partial T)_{\mathcal{H}} \ .$$

In section 1.4, we stated two forms of the first law for magnetic systems, depending on whether the mutual field energy is included as part of the system or not. The magnetic energy

$$E = -\mu_0 \mathcal{H} \mathcal{M} \tag{3.6}$$

does not include the mutual field energy, as is shown in Appendix C, i.e. we are not including the mutual field energy as part of the system. Hence the appropriate form of the first law is

$$dE = đQ - \mathcal{M}\mu_0 d\mathcal{H} \ , \tag{1.47}$$

and the magnetic contribution to the heat capacity of the system, at constant field $\mathcal{H}$, becomes

$$C_{\mathcal{H}} = \left(\frac{đQ}{\partial T}\right)_{\mathcal{H}} = \left(\frac{\partial E}{\partial T}\right)_{\mathcal{H}} = -\mu_0 \mathcal{H} \left(\frac{\partial \mathcal{M}}{\partial T}\right)_{\mathcal{H}} \ .$$

If the applied magnetic field $\mu_0 \mathcal{H}$ and the local field $\mathcal{B}$ are the same (and we have seen that this is usually so for paramagnetic substances), then differentiation at constant applied field intensity $\mathcal{H}$ or at constant local field $\mathcal{B}$ give the same result. The last equation, together with Eqs. (3.4b) and (3.2), then gives for the magnetic heat capacity of N dipoles at constant applied field intensity

$$C_{\mathcal{H}} = Nkx^2 \operatorname{sech}^2 x \ . \tag{3.12}$$

Rewriting Eq. (1.47) in the form

$$d(E + \mu_0 \mathcal{H} \mathcal{M}) = đQ + \mu_0 \mathcal{H} \, d\mathcal{M},$$

we obtain the heat capacity at constant magnetization

$$C_{\mathcal{M}} = \left(\frac{đQ}{\partial T}\right)_{\mathcal{M}} = \left(\frac{\partial (E + \mu_0 \mathcal{H} \mathcal{M})}{\partial T}\right)_{\mathcal{M}} \ .$$

For our paramagnetic system it follows from Eq. (3.6) that $C_{\mathscr{M}}=0$: the magnetic heat capacity at constant magnetization vanishes. This result holds in general, provided the magnetic moment $\mathscr{M}$ is a function of $\mathscr{H}/T$ only. (This property defines a so-called ideal paramagnetic substance.) This is shown in problem 5.11.

The heat capacity (3.12) is plotted against $kT/\mu\mathscr{B}$ in Fig. 3.5. This figure shows that for a given field the transition from complete order to random disorder occurs over essentially a finite temperature range corresponding to the hump of the heat capacity curve. (For a real solid, as distinct from our model, there is also a contribution to the heat capacity from the vibrational motion of the molecules.) The total energy ΔE required for this change is given by the area under the curve, Fig. 3.5:

$$\Delta E = \int_0^\infty C_{\mathscr{H}}\,dT = \mu\mathscr{B}N \ . \tag{3.13}$$

The entropy of the system can be obtained from the Helmholtz free energy $F=E-TS$, Eq. (2.37). We know E from Eqs. (3.5), and we have

$$F \equiv F(T,\mathscr{B},N) = NF(T,\mathscr{B},1) = -NkT\ln Z_1 \ , \tag{3.14}$$

with Z_1 given by Eq. (3.1). Combining these results leads to

$$S = Nk\,[\ln 2 + \ln(\cosh x) - x\tanh x] \ . \tag{3.15}$$

In the weak field/high temperature limit, when $x \ll 1$, we have

$$\tanh x \approx x \ , \qquad \ln(\cosh x) \approx \ln(1 + \tfrac{1}{2}x^2) \approx \tfrac{1}{2}x^2 \ ,$$

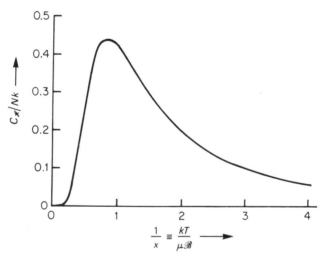

Fig. 3.5. Magnetic contribution to heat capacity versus $(kT/\mu\mathscr{B})$.

so that Eq. (3.15) becomes

$$S = Nk \ln 2 = k \ln(2^N) \ . \tag{3.16}$$

N dipoles, each possessing two possible states, possess 2^N different states altogether. Eq. (3.16) implies that in the weak field/high temperature limit all these states are equally probable, i.e. the orientation of each dipole — up or down — is random.

In the strong field/low temperature limit, when $x \gg 1$,

$$\cosh x \approx \tfrac{1}{2} e^x \ , \qquad \tanh x \approx 1 \ ,$$

so that Eq. (3.15) becomes

$$S = 0 \ , \tag{3.17}$$

i.e. the statistical weight $\Omega = \exp(S/k)$ is unity: there is only one microstate available for the system, that in which all magnetic dipoles are aligned parallel to the field.

★ 3.3 AN ISOLATED PARAMAGNETIC SOLID

The treatment of a system at constant temperature, as in sections 3.1 and 3.2, is usually much more convenient than at constant energy. All the same, to illustrate the approach of section 2.3 we shall now consider our simple model of a paramagnetic solid consisting of N dipoles in a magnetic field $\mathscr{B}$, and we shall assume that the total energy of interaction of the dipoles with the field has a fixed value. It follows from Eq. (2.1) that a given energy $E(n)$ corresponds to a given number n of 'spin up' dipoles, hence by Eq. (2.2) to a given statistical weight

$$\Omega(n) = \frac{N!}{n!(N-n)!}$$

and to an entropy

$$S(n) = k \ln \frac{N!}{n!(N-n)!} \ . \tag{3.18}$$

In order to use this formula, we employ the simple form of Stirling's approximation for the factorial function, Appendix A1, Eq. (A.2),

$$\ln(n!) = n \ln n - n \ ,$$

which is valid since n is very large (in practice of the order of 10^{23}). For the same reason we may treat n like a continuous variable, e.g. differentiate with respect to n. Substituting Stirling's approximation in Eq. (3.18) for all three factorials leads to

$$S(n) = k\,[N\ln N - n\ln n - (N-n)\ln(N-n)] \ . \tag{3.19}$$

The temperature is given by Eq. (2.9), i.e.

$$\frac{1}{T} = \left(\frac{\partial S}{\partial E}\right)_{\mathscr{B},N} \ , \tag{3.20}$$

the field $\mathscr{B}$ and not the volume V being the appropriate variable in this problem.* Using Eq. (3.19) and Eq. (2.1) for $E(n)$, we obtain

$$\frac{1}{T} = \frac{\partial S(n)}{\partial n}\,\frac{\partial n}{\partial E} = \left[k\ln\left(\frac{N-n}{n}\right)\right] \cdot \left[\frac{-1}{2\mu\mathscr{B}}\right] \ . \tag{3.21}$$

Solving Eq. (3.21) for n leads to

$$\frac{n}{N} = \frac{1}{Z_1}\,e^x \tag{3.22}$$

where Z_1 and x are defined by Eqs. (3.1) and (3.2). Our result (3.22) is identical with Eq. (3.3), since n/N is just the probability p_+ that a dipole should be in the 'spin up' state. All other results then follow as in sections 3.1 and 3.2.

★ 3.4   NEGATIVE TEMPERATURE

In section 2.3 we defined the temperature of a system in terms of its entropy or statistical weight by

$$\frac{1}{T} = \frac{\partial S}{\partial E} = k\,\frac{\partial \ln\Omega}{\partial E} \ , \tag{3.23}$$

and we stated that the absolute temperature T is positive; a consequence of the fact that S and Ω are monotonic increasing functions of E. Earlier in this chapter we had equations—for example, Eqs. (3.3) or (3.21)—which obviously can lead to negative temperatures. In this section we shall illustrate this in detail and we shall then discuss what it means.

To see how Eq. (3.3) can lead to negative temperatures, we substitute Eq. (3.1) for Z_1 in Eq. (3.3) and solve for $x \equiv \mu\mathscr{B}$ $\beta \equiv \mu\mathscr{B}/kT$:

$$\beta \equiv \frac{1}{kT} = \frac{1}{2\mu\mathscr{B}}\ln\frac{n}{N-n} \ . \tag{3.24}$$

*This intuitively reasonable result will be derived in Appendix C.

We see from this equation that we have $T \gtrless 0$ according as $n \gtrless \frac{1}{2}N$. In a state of negative temperature, *more* than half the dipoles are aligned *antiparallel* to the magnetic field. We know that even as T becomes infinite, the 'up-spin' and 'down-spin' populations only become equal. Hence a negative temperature is hotter than $T = \infty$: it is a more energetic state of the system. Clearly a negative temperature state cannot be obtained in any ordinary way by placing a system in a heat bath. We shall return to the question of how to obtain negative temperatures at the end of this section.

Qualitatively, a negative temperature means a negative derivative in Eq. (3.23), i.e. S and Ω must be decreasing functions of E. This will occur for a system which possesses a state of finite maximum energy E_{max}. This is qualitatively illustrated in Fig. 3.6. For energies in the range E' to E_{max} the temperature of the system is negative, while for energies between the ground state energy E_0 and E' the temperature is positive.

Strictly speaking, that is if one considers *all* aspects of a system, there exists no system with an upper bound to its energy, for the kinetic energy of a group of particles is always unbounded above. Correspondingly the statistical weight $\Omega(E)$ is a monotonic increasing function of E for all energies. (For a perfect gas this is shown explicitly in section 7.7, Eq. (7.71).)

The situation is very different if for a system one particular aspect interacts only very weakly with all other aspects of the system. This one aspect may then reach *internal* equilibrium without necessarily being in equilibrium with the other aspects. One can then define a temperature for the one aspect which is not necessarily equal to the temperatures related to other aspects. If the one aspect possesses an upper bound for its energy, one can associate a negative temperature with it.

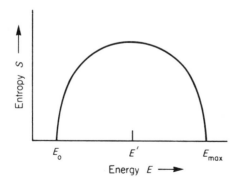

Fig. 3.6. The entropy S as a function of the energy E, for a system possessing negative temperatures. For $E_0 < E < E'$ the temperature $T = (\partial S / \partial E)^{-1}$ is positive; for $E' < E < E_{max}$, it is negative.

This is precisely the situation for the magnetic spin systems considered earlier in this chapter. The interaction between spins and other aspects, such as the lattice vibrations, in a crystalline solid is very weak. If the spin–lattice relaxation time (i.e. the time for the spins and the lattice vibrations to reach equilibrium) is long compared with the spin–spin relaxation time (i.e. the time for the spins to reach equilibrium amongst themselves), then one can talk of the spin temperature of the system.

Our system of N dipoles in a magnetic field possesses a unique microstate with the minimum energy $E_0 = -N\mu\mathcal{B}$, and a unique microstate with the maximum energy $E_{max} = +N\mu\mathcal{B}$. Both these states have the entropy $S = 0$. For intermediate energies the entropy–energy curve has the general shape shown in Fig. 3.6. Quantitatively it is given by

$$S(n) = k \ln \frac{N!}{n!(N-n)!} \tag{3.18}$$

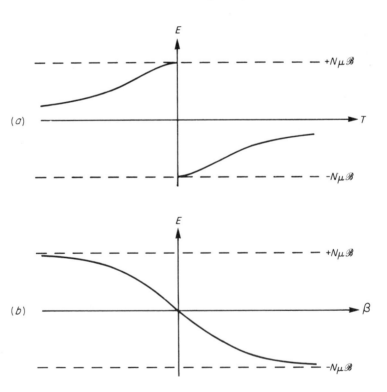

Fig. 3.7. The energy of a system of N spin $\frac{1}{2}$ dipoles, possessing magnetic moment μ, placed in a magnetic field $\mathcal{B}$. (a) as a function of the temperature T. (b) as a function of the temperature parameter $\beta \equiv 1/kT$.

[see also Eq. (3.19)] and

$$E(n) = (N - 2n)\mu\mathscr{B} \ . \tag{2.1}$$

The corresponding relation between energy and temperature is given by Eq. (3.5b) [which of course also follows from Eq. (3.24)] and is shown in Fig. 3.7(a). As the energy increases from $-N\mu\mathscr{B}$ to $+N\mu\mathscr{B}$, the spin temperature increases from $T = 0$ to $T = \infty$, jumps to $T = -\infty$ ($T = \pm\infty$ both correspond to $E = 0$), and increases through negative values to $T = 0$. The same physics looks less spectacular if E is plotted against the temperature parameter β, Fig. 3.7(b). As E increases monotonically from $-N\mu\mathscr{B}$ to $+N\mu\mathscr{B}$, β decreases monotonically from $+\infty$ to $-\infty$.

Negative temperatures were first realized for nuclear spin systems by Purcell and Pound.* Like atoms, nuclei possess magnetic moments. Purcell and Pound aligned the nuclear spins in a lithium fluoride crystal by means of a strong magnetic field. They then reversed the direction of the magnetic field so rapidly that the spins were unable to adjust themselves, resulting in the majority of spins being aligned antiparallel to the applied field. In this way an inverted population is obtained which, as we have seen, can be described by a negative spin temperature. The spin system can be considered in internal equilibrium in this case since the spin–spin relaxation time is very short compared with the spin–lattice relaxation time.

PROBLEMS 3

3.1 A crystal contains N atoms which possess spin 1 and magnetic moment μ. Placed in a uniform magnetic field $\mathscr{B}$ the atoms can orient themselves in three directions: parallel, perpendicular and antiparallel to the field. If the crystal is in thermal equilibrium at temperature T find an expression for its mean magnetic moment $\mathscr{M}$, assuming that only the interactions of the dipoles with the field $\mathscr{B}$ need be considered.

What is the magnetic moment of the crystal (i) when placed in a weak field at high temperature, (ii) when placed in a strong field at low temperature?

3.2 A magnetic specimen consists of N atoms, each having a magnetic moment μ. In an applied homogeneous magnetic field $\mathscr{H}$ each atom may orient itself parallel or antiparallel to the field, the energy of an atom in the two orientations being

$$\varepsilon_{\pm} = \mp(\mu\mu_0\mathscr{H} + k\theta\mathscr{M}/\mu N)$$

respectively. Here θ is a temperature characteristic of the specimen, $\mathscr{M}$ is its magnetic moment, and μ_0 is the permeability of the vacuum. Show that the magnetic moment of the specimen at temperature T is given by

$$\frac{\mathscr{M}}{\mu N} = \tanh\left[\frac{1}{kT}(\mu\mu_0\mathscr{H} + k\theta\mathscr{M}/\mu N)\right] \ .$$

*E. M. Purcell and R. V. Pound, *Phys. Rev.*, **81**, 279 (1951).

If $\mathscr{M} \ll \mu N$, obtain the temperature dependence of the magnetic susceptibility per atom, $\chi = \mathscr{M}/\mathscr{H}N$.

Sketch the temperature dependence of the magnetic moment in zero field ($\mathscr{H} = 0$).

What sort of material is represented by these equations?

3.3 A paramagnetic crystal contains N magnetic ions which possess spin $\frac{1}{2}$ and a magnetic dipole moment μ. The crystal is placed in a heat bath at temperature T and a magnetic field $\mathscr{B}$ is applied to the crystal. Use the fact that under these conditions the Helmholtz free energy is a minimum to find the net magnetic moment of the crystal.

3.4 The magnetic moment of nuclei is of the order of $10^{-26}\,\mathrm{A\,m^2}$. Estimate the magnetic field required at 0.01 K to produce appreciable alignment of such nuclei.

The second law of thermodynamics II

In Chapter 2 we introduced the second law. Section 4.1 contains two important developments: we derive an equation for the second law valid for infinitesimal changes, and we discuss the significance of reversible and irreversible processes. With section 4.1 we conclude the development of the main ideas which are needed later in this book (see the flow diagram on the inside-front cover).

In the remainder of this chapter thermodynamics is developed further. In section 4.2 the principle of the increase of entropy is generalized to systems which are not completely isolated. In section 4.3 some simple applications illustrate the entropy concept. Sections 4.4 to 4.6 deal with the equilibrium conditions and related topics for systems subject to different kinds of constraints. Sections 4.7 and 4.8 are devoted to the third law of thermodynamics, i.e. the properties of systems as their absolute temperature tends to zero kelvin.

4.1 THE SECOND LAW FOR INFINITESIMAL CHANGES

In Chapter 2 we saw that the second law gives a criterion whether in an isolated system a change from a state 1 to a state 2 may or may not occur: *the change may only occur if it corresponds to an increase of entropy.*

We now want to calculate the difference in entropy between two equilibrium states of a system which lie infinitesimally close together. We shall derive a formulation of the second law for such infinitesimal changes. We shall consider a fluid. The two states will be specified by (β, V) and $(\beta + d\beta, V + dV)$ respectively, where β is the temperature parameter (2.21). The particle number N will be constant throughout so we shall omit this label. The two states of the system thus correspond to the system being contained in enclosures of slightly different volumes and in thermal equilibrium with heat baths of slightly different temperatures. The arrangement for the state (β, V) is illustrated in Fig. 2.10.

The change in energy of the system in going from state (β, V) to $(\beta + d\beta, V + dV)$ is given from Eq. (2.26), which gives the energy of a system in equilibrium in a heat bath, as

$$dE = \sum_r p_r dE_r + \sum_r E_r dp_r \ . \tag{4.1}$$

(As we are considering a macroscopic system, we omit the 'bar' over the energy, as discussed in section 2.5.) With our choice of independent variables, corresponding to a system of given volume in a heat bath, E and dE are functions of β and V. It is important to be clear how these dependences arise. As discussed in section 2.5, following Eq. (2.26), the energy levels E_r do not depend on β but solely on the external parameters, in our case the volume: $E_r = E_r(V)$. The probabilities p_r, Eq. (2.22), on the other hand, depend on β and—through the E_r—on V.

It is possible to eliminate the p_r from Eq. (4.1) and rewrite it in terms of purely macroscopic, i.e. thermodynamic, quantities.

We first consider the term $\sum_r E_r dp_r$. With

$$E_r = - \frac{1}{\beta} (\ln Z + \ln p_r) \ , \tag{4.2}$$

which follows from Eq. (2.22), and $\sum dp_r = 0$, we obtain

$$\sum_r E_r dp_r = - \frac{1}{\beta} \sum_r dp_r \ln p_r \ . \tag{4.3}$$

From the general definition of entropy, Eq. (2.33), it follows that

$$dS = - k \sum_r dp_r \ln p_r \ . \tag{4.4}$$

where we again used $\sum dp_r = 0$; comparison of Eqs. (4.3) and (4.4) leads to

$$\sum_r E_r dp_r = T dS \ . \tag{4.5}$$

This is our first result. It attributes part of the energy change dE in Eq. (4.1) to a change in probability distribution over the various levels $E_1, E_2, \ldots,$ and relates this to an entropy change of the system.

The second term in Eq. (4.1) can be written

$$\sum_r p_r dE_r = \sum_r p_r \frac{dE_r}{dV} dV \ . \tag{4.6}$$

Now if we know that the system is in the state r, *and that it will stay in this state for all times*, then a change of the volume from V to $V + dV$ produces an energy change

$$dE_r = \frac{dE_r}{dV} dV \equiv -P_r dV \ ; \tag{4.7}$$

i.e. this is the work done on the system in the state r to produce this volume change and it defines the pressure P_r of the system in the state r. Suppose it is not certain that the system is in the state r but, as is the case for the system in the heat bath, we only have a probability distribution p_r. The pressure P is then given by the corresponding average

$$P = \sum_r p_r \left(-\frac{dE_r}{dV} \right) \tag{4.8}$$

so that Eq. (4.6) becomes

$$\sum_r p_r dE_r = -P dV \ . \tag{4.9}$$

This is our second result. But to establish that P as here defined is in fact the pressure still needs careful discussion of one point. In our derivation we tacitly assumed that the sole effect of changing the volume is to change the energy levels. But this is not so. According to quantum mechanics any perturbation of the system, such as changing its volume, induces transitions between the different energy levels of the system. Thus even if we started with the system for certain in the state r, it might not stay in that state as the volume is changed; a transition might have occurred. In this case Eq. (4.7) does not represent the work in changing the volume of the system which was initially in the state r. In the same way, Eq. (4.9) would, if transitions are induced, not represent the work done in changing the volume of the system in the heat bath. Now it can be shown that the transition rate to other states becomes negligible provided the time variation of the perturbation — in our case the rate of change of volume — is sufficiently slow. (The transition rate becomes zero for an infinitely slowly varying perturbation.) This result is known as Ehrenfest's principle.* In other words, the change must be a

*For a derivation see R. Becker,[2] pp. 170–173. There is also a good discussion in Wannier's *Statistical Physics*,[20] section 5.3.

quasistatic one. This is precisely how pressure and the work associated with it were defined in section 1.3: Eq. (1.15a) was shown to be valid for reversible changes only. Thus we have established that Eq. (4.8) defines the pressure of the system and that Eq. (4.9) is the work associated with a reversible change of volume. We note, for later use, that analogously to Eq. (2.26) (where the partial derivative is now to be taken at constant volume) we can write the pressure (4.8) as

$$P = \frac{1}{\beta} \left(\frac{\partial \ln Z}{\partial V} \right)_\beta . \tag{4.10}$$

This result follows directly by differentiation of Eq. (2.23). In terms of the Helmholtz free energy $F = -kT \ln Z$, Eq. (2.36), the pressure (4.10) becomes

$$P = - \left(\frac{\partial F}{\partial V} \right)_T . \tag{4.11}$$

Finally, we combine Eqs. (4.1), (4.5) and (4.9) to obtain the *fundamental thermodynamic relation*

$$dE = T dS - P dV . \tag{4.12}$$

It follows from our derivation that *Eq. (4.12) is generally valid for neighbouring equilibrium states, infinitesimally close together.*

If we compare Eq. (4.12) with the first law, Eq. (1.14), i.e.

$$dE = đQ + đW , \tag{1.14}$$

it is tempting to identify $đW$ with $(-P dV)$ and, consequently, $đQ$ with $T dS$. This is correct for *reversible changes* and *only for these* as we see from Eq. (1.15c). Thus for *infinitesimal reversible changes*

$$đW = -P dV \quad \text{(reversible)} \tag{1.15a}$$

and, from Eqs. (4.12) and (1.14),

$$dS = \frac{đQ}{T} \quad \text{(reversible)} . \tag{4.13a}$$

Put into words, this equation states that if we supply an infinitesimal amount of heat $đQ$ to a system *reversibly* at the temperature T (for the heat transfer to be reversible the heat bath and system must both be at the *same* uniform temperature T, i.e. there must be no temperature gradients which would result in heat conduction and other irreversible processes), the entropy of the system increases by $\mathrm{d}Q/T$.

If the infinitesimal change is irreversible, Eq. (1.15a) is replaced by the inequality

$$\mathrm{d}W > -P\mathrm{d}V \quad \text{(irreversible)} \tag{1.15b}$$

and correspondingly Eq. (4.13a) by the inequality

$$\mathrm{d}S > \frac{\mathrm{d}Q}{T} \quad \text{(irreversible)} . \tag{4.13b}$$

A simple example of an irreversible process is the free adiabatic expansion of a gas. In this case we know (see section 1.3) that $\mathrm{d}Q = \mathrm{d}W = \mathrm{d}E = 0$ for an infinitesimal change of volume. However $P\mathrm{d}V > 0$, and Eq. (4.12) gives $T\mathrm{d}S > 0$, i.e. the entropy increases.

We want to stress that unlike Eqs. (1.15a) and (4.13a) which hold for reversible changes only, *the fundamental thermodynamic relation* (4.12) *is always valid for infinitesimal changes, whether reversible or not*. This follows from our derivation of Eq. (4.12). It can also be seen from the fact that Eq. (4.12) only involves functions of state or differentials of such functions. On the other hand we know that $\mathrm{d}Q$ is not the differential of a function of state (hence our notation). The factor $1/T$ in Eq. (4.13a) turns $\mathrm{d}Q$ into the differential of a function of state, the entropy.*

As an example we shall calculate $\mathrm{d}Q/T$ for a reversible change in a perfect monatomic gas of N atoms. We have the equation of state

$$PV = NkT. \tag{1.9}$$

For simplicity we shall anticipate the result, to be proved in section 7.4 (or see Flowers and Mendoza,[26] Eq. (5.12)), that the internal energy E for such a gas is given by

$$E = \tfrac{3}{2}NkT.$$

(We saw in the discussion of Joule's experiment, in section 1.3, that E does not depend on the volume of the gas.)

---

*$1/T$ is called an integrating factor of the inexact differential $\mathrm{d}Q$, turning it into the exact (or total) differential $\mathrm{d}S$, i.e. into the differential of a function of state.

In classical thermodynamics, one of the formulations of the second law is just the statement that there exists an integrating factor for $\mathrm{d}Q$; this integrating factor then essentially defines the absolute temperature. For a discussion of this approach, due to Carathéodory, see, for example, Adkins,[1] Pippard,[9] Wannier,[20] or Zemansky and Dittman.[22]

Consider an infinitesimal reversible change between two definite neighbouring states of the gas. For such a change we have from Eqs. (1.14) and (1.15a) that

$$\mathrm{d}Q = \mathrm{d}E + P\,\mathrm{d}V .$$

We know from the discussion in section 1.3 that $\mathrm{d}Q$ (like $P\,\mathrm{d}V$) is not the differential of a function of state. (For example, for any finite change the heat supplied to the gas — like the work done on it — depends on how the change is performed.)

Instead of $\mathrm{d}Q$ we consider $\mathrm{d}Q/T$. From the last two equations and the equation of state (1.9) we obtain at once

$$\frac{\mathrm{d}Q}{T} = \frac{3}{2}Nk\frac{\mathrm{d}E}{E} + Nk\frac{\mathrm{d}V}{V} = \mathrm{d}\{Nk\ln(VE^{3/2})\} ,$$

confirming that $\mathrm{d}Q/T$ is the differential of a function of state, namely of

$$S(E,V,N) = \text{const.} + \{Nk\ln(VE^{3/2})\} .$$

Thus we have verified Eq. (4.13a). (This is not a derivation of the entropy of a perfect monatomic gas since we assumed the expression for the internal energy and the equation of state.)

From the fundamental relation (4.12) a host of relations between various partial derivatives follow. It is not our intention to derive an exhaustive list of such relations but only to illustrate the methods which enable one easily to obtain any relation one may need. Setting one of the three differentials in Eq. (4.12) equal to zero, leads to a differential coefficient for the other two ('either way up'). Thus one obtains

$$T = \left(\frac{\partial E}{\partial S}\right)_V ; \qquad P = -\left(\frac{\partial E}{\partial V}\right)_S . \tag{4.14}$$

The first of these equations is equivalent to Eq. (2.9), as follows from the rules of partial differentiation.* Since

$$\frac{\partial^2 E}{\partial V \partial S} = \frac{\partial^2 E}{\partial S \partial V} , \tag{4.17}$$

---

*We derive the most useful of these rules. If $z = z(x,y)$, then

$$\mathrm{d}z = \left(\frac{\partial z}{\partial x}\right)_y \mathrm{d}x + \left(\frac{\partial z}{\partial y}\right)_x \mathrm{d}y . \tag{4.15}$$

it follows from Eqs. (4.14) by 'cross-differentiation', i.e. forming the two derivatives (4.17), that

$$\left(\frac{\partial T}{\partial V}\right)_S = -\left(\frac{\partial P}{\partial S}\right)_V . \qquad (4.18)$$

This is one of Maxwell's four thermodynamic relations. (The others will be given in Eqs. (4.45), (4.51) and (4.56).) These relations are very useful in relating different thermodynamic quantities to each other. We want to stress that thermodynamic relations, such as Eqs. (4.14) or (4.18), hold for *any* system in equilibrium. It is this generality which makes the thermodynamic methods so extremely powerful.

We briefly note an alternative derivation of the fundamental relation (4.12). Considering the partition function Z as a function of β and V, we have

$$d\ln Z = \left(\frac{\partial \ln Z}{\partial \beta}\right)_V d\beta + \left(\frac{\partial \ln Z}{\partial V}\right)_\beta dV$$

$$= -E d\beta + \beta P dV ,$$

where the last line follows from Eq. (2.26) and Eq. (4.10). (Note that Eq. (4.10) follows directly from Eq. (2.23), i.e. we are not arguing in circles. Of course we still require the above justification for calling P the pressure!) From Eqs. (2.36) and (2.37), we have

$$\ln Z = -\beta E + S/k.$$

Differentiating this equation and eliminating $d\ln Z$ between the resulting equation and the previous equation for $d\ln Z$ leads to Eq. (4.12).

4.2   THE CLAUSIUS INEQUALITY

In section 2.3 we discussed Clausius's entropy principle for an *isolated* system. It provides a criterion for the direction of natural processes in such

---

Hence for $dy = 0$ one obtains

$$\left(\frac{\partial z}{\partial x}\right)_y = \left[\left(\frac{\partial x}{\partial z}\right)_y\right]^{-1} . \qquad (4.16a)$$

For $dz = 0$ one obtains from Eq. (4.15), using Eq. (4.16a),

$$\left(\frac{\partial x}{\partial y}\right)_z \left(\frac{\partial y}{\partial z}\right)_x \left(\frac{\partial z}{\partial x}\right)_y = -1 . \qquad (4.16b)$$

Note that each derivative in this relation is obtained from the preceding one by cyclically permuting x, y and z.

a system and, in particular, for equilibrium. We now want to generalize the results of section 2.3 to obtain corresponding criteria for systems which are not completely isolated but interact with the surrounding in some specified manner.

For an isolated system, Clausius's principle states that for any natural change from a state 1 to a state 2 the entropy increases

$$\Delta S \equiv S_2 - S_1 > 0 \ . \tag{4.19}$$

A reversible change is a theoretical abstraction, a limiting case of a real process. If in an isolated system a process from state 1 to state 2 is reversible, it can occur equally well in the reverse direction from state 2 to state 1. It follows that for a reversible process the inequality (4.19) is replaced by the equality

$$\Delta S \equiv S_2 - S_1 = 0 \ . \tag{4.20}$$

We combine Eqs. (4.19) and (4.20) into

$$\Delta S \equiv S_2 - S_1 \geqslant 0 \tag{4.21}$$

where the > and = signs hold for irreversible and reversible processes respectively.

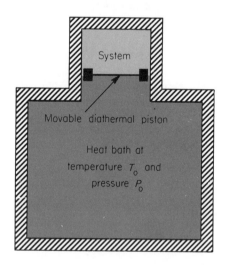

Fig. 4.1. System in thermal and mechanical contact with a heat bath but not necessarily in temperature and pressure equilibrium with it. The heat bath is so large that its temperature T_0 and pressure P_0 remain constant as heat transfer occurs across the diathermal piston and as the piston moves.

Consider now the isolated composite system shown in Fig. 4.1. It consists of a system coupled both thermally and mechanically to a heat bath. The latter is so large that its temperature T_0 and pressure P_0 remain constant in spite of heat transfer across, and movement of, the diathermal piston. We are now considering a situation where the system and heat bath are not necessarily in thermal or pressure equilibrium. Indeed the system may well not have a uniform temperature or a uniform pressure. (The conditions here envisaged are those under which many experiments are done: a system in thermal contact with the atmosphere and subject to atmospheric pressure.) If we put

$$S_{\text{total}} = S + S_0 \qquad (4.22)$$

where S, S_0 and S_{total} are the entropies of the system, heat bath and composite system respectively, then for any finite change of the composite system we have from Eq. (4.21) that

$$\Delta S_{\text{total}} = \Delta S + \Delta S_0 \geqslant 0 \ . \qquad (4.23)$$

Suppose that in this change an amount of heat Q is transferred to the system from the heat bath whose volume may also have changed. Because of the very large size of the heat bath, its temperature T_0 and pressure P_0 will have remained constant during this change. Hence the change is a reversible one as far as the heat bath is concerned and from Eq. (4.13a) we have

$$\Delta S_0 = - \frac{Q}{T_0} \ . \qquad (4.24)$$

Hence Eq. (4.23) becomes

$$\Delta S - \frac{Q}{T_0} \geqslant 0 \ . \qquad (4.25)$$

From the first law applied to the system we have

$$\Delta E = Q + W \qquad (4.26)$$

where ΔE is the change in energy of the system and W the work done on it. If the only work done on the system is that done by the heat bath then

$$W = - P_0 \Delta V \qquad (4.27)$$

where ΔV is the change in volume of the system. Combining Eqs. (4.25) to (4.27), we obtain

$$\Delta S - \frac{\Delta E + P_0 \Delta V}{T_0} \geqslant 0 \ . \qquad (4.28)$$

The $>$ and $=$ signs apply to irreversible and reversible processes respectively.

Eq. (4.28) is our final result. It is the extension of Clausius's entropy principle (4.21) to a system in a surrounding at constant temperature T_0 and constant pressure P_0. We shall refer to Eq. (4.28) (or particular cases of it, such as Eq. (4.21)) as *Clausius's inequality*. Eq. (4.28) states a criterion for the direction of natural processes and for equilibrium. In general Eq. (4.28) depends on the system *and its surrounding*. Only in special circumstances is it a function of the system only; for example if we consider isothermal processes at a temperature T equal to that of the environment ($T = T_0$), the volume of the system being kept constant ($\Delta V = 0$). In this case Eq. (4.28) becomes

$$\Delta S - \frac{\Delta E}{T} \geqslant 0 \qquad (4.29)$$

which depends only on properties of the system.

For a totally isolated system where $\Delta E = \Delta V = 0$ (i.e. constant energy and volume) Eq. (4.28) reduces to our original statement (4.21): $\Delta S \geqslant 0$: the entropy of an isolated system tends to increase; in equilibrium it is a maximum. Equilibrium conditions for other constraints are easily derived from Eq. (4.28) (see sections 4.4 and 4.5).

These equilibrium conditions are expressed concisely in terms of the availability A defined by

$$A \equiv E + P_0 V - T_0 S \qquad (4.30)$$

which is a property of the system *and its environment*. In terms of A Clausius's inequality (4.28) becomes

$$\Delta A \leqslant 0 : \qquad (4.31)$$

the availability of a system in a given surrounding tends to decrease (i.e. this is the direction of natural changes); it has a minimum value in equilibrium, i.e. no further natural changes are possible. The equality sign in Eq. (4.31) corresponds to reversible changes.

The considerations which led to Clausius's inequality (4.28) can be generalized to allow for additional 'irreversibilities' in the process. For example, not all the work done by the heat bath need be transferred to the system; some may be expended against frictional forces. Eq. (4.27) is then replaced by an inequality. The effect of all such additional 'irreversibilities' is to make the *in*equality in (4.28) even stronger.

4.3 SIMPLE APPLICATIONS

In this section we calculate entropy changes in some very simple situations, to familiarize the reader with these ideas. More elaborate applications are considered in Chapters 5 and 8.

The entropy is a function of state. Hence the difference in the entropies of two states 1 and 2 of a system depends only on these states. If the system is changed from state 1 to state 2 by means of different processes, then the entropy of the system will always change by the *same* amount, but the entropy of the environment of the system will in general change by *different* amounts, depending on the particular process. A reversible process is distinguished by the fact that the changes in the system are accompanied by compensating changes in the system's surrounding such that the entropy of system plus surrounding remains constant. For a reversible process we know how to calculate the entropy change for the system. According to Eq. (4.13a) the entropy difference is given by

$$\Delta S \equiv S_2 - S_1 = \int_1^2 \frac{dQ}{T} .$$

(4.32)

Here the integral must be taken along a *reversible* path connecting the two states. If the initial and final volumes of the system are the same, one would probably choose a process at constant volume, and dQ in Eq. (4.32) is then replaced by $C_V dT$, where C_V is the heat capacity of the system at constant volume. If initial and final pressures, instead of volumes, are the same, C_V would be replaced by C_P, the heat capacity at constant pressure. If initial and final temperatures are the same, one would choose a reversible isothermal path. To make any of these processes reversible, one must put the system in contact with a succession of heat baths whose temperatures differ only infinitesimally from the instantaneous temperature of the system. To produce a finite temperature change reversibly requires, in principle, an infinite number of such heat baths. More realistically one would use a temperature regulated bath and cause the temperature of the bath to change in the way required.

4.3.1 Heating water

Suppose 1000 g of water are heated from 20 °C to 80 °C. What is the entropy change of the water, given that its specific heat may be assumed to have the constant value $4.2 \, J \, g^{-1} (°C)^{-1}$?

If the heating is carried out reversibly by placing the water in thermal contact with a succession of heat baths of steadily increasing temperatures, then the entropy increase is, from Eq. (4.32), given by

$$\Delta S = Mc \int_{T_1}^{T_2} \frac{dT}{T} = Mc \ln \frac{T_2}{T_1} = 782 \, J \, K^{-1} ,$$

where $M = 1{,}000 \, g$, $c = 4.2 \, J \, g^{-1} K^{-1}$, $T_1 = 293 \, K$, $T_2 = 353 \, K$.

4.3.2   Melting ice

A beaker contains a mixture of 50 g of ice and 50 g of water. By heating the beaker 10 g of ice are melted. The ice-water mixture is in equilibrium at a pressure of one atmosphere both initially and finally. What change of entropy of the ice-water mixture has occurred?

The required entropy difference depends only on the initial and final states. The heating of the beaker will almost certainly not be a reversible procedure: the ice-water mixture will not be at a constant uniform temperature, equal to that of the ice point, throughout the process, and so on. This is of no concern to us. To find the entropy change we specifically choose a reversible process. At one atmosphere the melting point of ice is 273.15 K and the latent heat of fusion of ice is 334 J/g. We melt 10 g of ice in a reversible process by supplying 3340 J from a heat bath at 273.15 K, the ice-water mixture being in equilibrium at a pressure of one atmosphere throughout the process. The entropy of the ice-water mixture has thus increased by $3340/273.15 = 12.2$ J/K, as follows directly from Eq. (4.32) with $T = 273.15$ K throughout the process of melting.

4.3.3   Temperature equalization

As our next example, we shall show that temperature equalization is accompanied by an increase of entropy.

Consider a system put in thermal contact with a heat bath at temperature T_0. If the system is initially at a temperature $T_1 < T_0$, it will warm up till it reaches the temperature T_0. For simplicity let us assume that the heat capacity C of the system is constant over the temperature interval (T_1, T_0). (This would be a good approximation for many solids and temperatures near room temperature.) With $dQ = C dT$, Eq. (4.32) gives for this case

$$\Delta S = C \int_{T_1}^{T_0} \frac{dT}{T} = C \ln \left(\frac{T_0}{T_1} \right) . \tag{4.33}$$

(This is essentially the result of section 4.3.1.)

The corresponding entropy change of the heat bath is

$$\Delta S_0 = - \frac{Q}{T_0} = \frac{-C(T_0 - T_1)}{T_0} \tag{4.34}$$

since the heat transferred to the system from the heat bath is $Q = C(T_0 - T_1)$. Thus the net entropy change for system plus heat bath is

$$\Delta S_{\text{total}} = \Delta S + \Delta S_0 = C \frac{T_1}{T_0} f \left(\frac{T_0}{T_1} \right) \tag{4.35}$$

where we introduced the abbreviation

$$f(x) \equiv x\ln x - x + 1 \ . \tag{4.36}$$

Now $f(1) = 0$ and $f(x) > 0$ for $x \neq 0$ (as follows at once from $f'(x) = \ln x$ and $f(1) = 0$). Hence for all initial temperatures $T_1 < T_0$, the resulting temperature equalization produces an entropy increase.

The same procedure, involving a slightly more elaborate calculation, shows that if two systems of finite heat capacities and different temperatures are put into thermal contact, the total entropy increases as the temperatures equalize.

4.3.4 Isothermal compression of a perfect gas

We consider the reversible isothermal compression of a perfect gas at temperature T_1 from state 1 to state 2, as illustrated in Fig. 4.2. We use the fact, discussed in section 1.3, that the internal energy E of a perfect gas is a function of its temperature T only but is independent of the volume and the pressure of the gas:

$$E = E(T) \ . \tag{4.37}$$

It follows that along an isotherm E is constant. Substituting

$$dQ = dE + P dV \ , \tag{4.38}$$

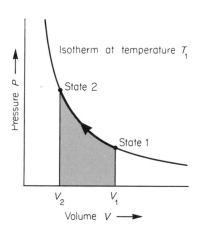

Fig. 4.2. P–V diagram for isothermal compression of a perfect gas.

valid for reversible processes, into Eq. (4.32) and writing $P = NkT_1/V$ (we are considering a gas consisting of N molecules) we obtain

$$\Delta S = S_2 - S_1 = \frac{1}{T_1} \int_1^2 P\,dV = -Nk\ln\left(\frac{V_1}{V_2}\right)\ ,\qquad(4.39)$$

i.e. the entropy of the gas has decreased during the reversible compression.
During the compression an amount of work

$$W = -\int_1^2 P\,dV = NkT_1\ln\left(\frac{V_1}{V_2}\right)\qquad(4.40)$$

is done on the gas. This is given by the shaded area in Fig. 4.2. But since the energy of the gas is not altered, a positive amount of heat $(-Q) = W$ is transferred to the heat bath at temperature T_1 with which the system is in contact during the compression. Hence the entropy of the heat bath increases by $\Delta S_0 = (-Q)/T_1$ which exactly compensates the entropy decrease (4.39) of the system, as it must for a reversible change.

The entropy difference (4.39) is independent of the path joining states 1 and 2 in Fig. 4.2. The work W done on the system, Eq. (4.40), given by the shaded area of Fig. 4.2, depends on the path. (On account of Eq. (4.37), this work must all be transferred to the heat bath with which the system is in contact during the process.) One can easily verify these statements for some different path, for example the path $1 \rightarrow 3 \rightarrow 2$ shown in Fig. 4.3. The work done on the system is given by the shaded rectangle in Fig. 4.3 and is less

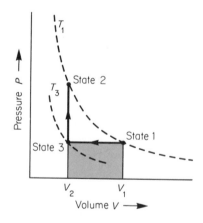

Fig. 4.3. Compression of a perfect gas from state 1 to state 2 along path $1 \rightarrow 3 \rightarrow 2$. The dashed curves are the two isotherms at temperatures $T_1 = T_2$ and T_3.

than for the isothermal process. For the entropy change we have from Eqs. (4.32) and (4.38)

$$S_2 - S_1 = \left(\int_1^3 + \int_3^2 \right) \frac{dE + P\,dV}{T}$$

$$= \int_1^2 \frac{dE}{T} + \int_1^3 \frac{P\,dV}{T} \ .$$

The first integral vanishes since E is a function of T only and $T_1 = T_2$. If in the second integral we substitute $P = NkT/V$ it reduces to the entropy change (4.39) since the volume of the system in state 3 is V_2.

We saw in section 1.3 that in an adiabatic free expansion the temperature of a perfect gas remains constant. Hence we can change the state of a perfect gas from the state 2 to the state 1 of Fig. 4.2 by means of such an adiabatic free expansion. The states 2 and 1 are both equilibrium states although during the expansion the conditions of the gas deviate greatly from equilibrium. This process is irreversible and cannot be described by any path on the $P - V$ diagram. But the entropy change $(S_1 - S_2)$ depends only on the initial and final equilibrium states 2 and 1, and is hence given by Eq. (4.39), except for a minus sign since Eq. (4.39) refers to compression of the gas.

4.4 THE HELMHOLTZ FREE ENERGY

We shall now apply the general results of sections 4.1 and 4.2 to systems under various conditions starting, in this section, with a system at constant volume in contact with a heat bath, as shown in Fig. 4.4. Consider a process in which the temperature T of the system in the initial and final states is the same as that of the heat bath: $T = T_0$. For this change we have from Eqs. (4.30) and (4.31), where we must now put $T_0 = T$ and $\Delta V = 0$, that

$$\Delta A = \Delta E - T\Delta S = \Delta(E - TS) = \Delta F \leqslant 0 \qquad (4.41)$$

where F is the Helmholtz free energy which was defined in section 2.5:

$$F = E - TS \ . \qquad (2.37)$$

We see that for the change envisaged, the availability which in general depends on *the system and its environment* becomes a property of *the system alone*, i.e. of the Helmholtz free energy of the system.

We had the result (4.41), in slightly different form, already earlier in Eq. (4.29). For simplicity, we there considered an isothermal process, i.e. the system is at the same temperature $T = T_0$ *throughout the process*. This is

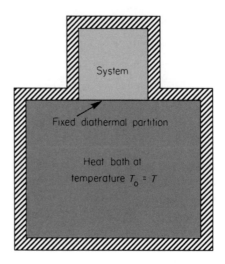

Fig. 4.4 A system in thermal contact with a heat bath
at temperature $T_0 = T$.

unnecessarily restrictive. We only require that the *initial* and *final* states should
be at the same temperature T; *during* the process the temperature of the
system may well change. This is frequently the case in practice. Consider,
for example, a chemical reaction

$$A + B \to AB.$$

The two substances A and B are in a container of fixed volume immersed
in a heat bath at temperature T. Initially they are separated by a partition
and are at the temperature T. If the partition is removed the chemical reaction
will occur, frequently with production of heat so that the system will warm
up. Eventually the reaction is over and the system has cooled down to
temperature T again, having given up the excess energy to the heat bath.

Just as the process does not have to be isothermal for Eq. (4.41) to apply
(we only require the initial and final temperatures to be T) so we could
generalize the process to which we apply Eq. (4.41) further: we only require
the *initial* and *final* volumes the *same*: $\Delta V = V_2 - V_1 = 0$, but *during* the
process the volume could vary.

The type of argument which we have here applied to temperature and
volume is generally valid for Clausius's inequality (4.30)–(4.31): it deals with
functions of state of the system in the *initial* and *final* states only. It does
not refer to the intermediate states while the process occurs.

We conclude from Eq. (4.41) that for a system of constant volume, initially
at temperature T, in contact with a heat bath at the same temperature T,

spontaneous (i.e. natural) changes occur in the direction of *decreasing Helmholtz free energy*; hence equilibrium corresponds to *minimum Helmholtz free energy*. As previously the $<$ and $=$ signs in Eq. (4.41) correspond to irreversible and reversible changes respectively.

The fundamental thermodynamic relation (4.12) can be rewritten in a form appropriate to T and V as independent variables rather than E and V. If we differentiate Eq. (2.37)

$$dF = dE - TdS - SdT \qquad (4.42)$$

and substitute from Eq. (4.12) for dE, we obtain the fundamental relation in the form

$$dF = -SdT - PdV . \qquad (4.43)$$

From this equation one obtains at once

$$S = -\left(\frac{\partial F}{\partial T}\right)_V , \qquad P = -\left(\frac{\partial F}{\partial V}\right)_T , \qquad (4.44)$$

and by cross-differentiation of the last two relations

$$\left(\frac{\partial S}{\partial V}\right)_T = \left(\frac{\partial P}{\partial T}\right)_V . \qquad (4.45)$$

This is the second Maxwell relation.

Substituting for the entropy from Eqs. (4.44) into Eq. (2.37), leads to the Gibbs-Helmholtz equation

$$E = F - T\left(\frac{\partial F}{\partial T}\right)_V = -T^2\left(\frac{\partial}{\partial T}\frac{F}{T}\right)_V \qquad (4.46)$$

which is frequently used in applications.

4.5   OTHER THERMODYNAMIC POTENTIALS

In sections 4.1 and 4.2 we considered the entropy S as characterizing an isolated system. In the last section we saw that F plays an analogous role for a constant-volume system in a heat bath. We singled out this case because we shall base our development of statistical mechanics on it. But there exist other thermodynamic potentials corresponding to different constraints. These we shall now consider. Since the methods are those used in section 4.4 we shall be quite brief.

Firstly we consider a system in thermal and mechanical contact with a constant-temperature constant-pressure heat bath. This is the situation shown in Fig. 4.1, but we shall now assume that both initially and finally the temperature T and pressure P of the system equal those of the heat bath: $T = T_0$, $P = P_0$.

From Eqs. (4.30) and (4.31) we obtain

$$\Delta A = \Delta(E + PV - TS) = \Delta G \leqslant 0 \qquad (4.47)$$

where the Gibbs free energy G is defined by

$$G \equiv E + PV - TS \ . \qquad (4.48)$$

Thus during natural processes in a constant-temperature and constant-pressure surrounding G *decreases and in equilibrium is a minimum*. The $<$ and $=$ signs in Eq. (4.47) correspond to irreversible and reversible processes respectively.

Differentiating Eq. (4.48) and substituting the fundamental relation (4.12) gives

$$dG = -SdT + VdP \ , \qquad (4.49)$$

whence

$$S = -\left(\frac{\partial G}{\partial T}\right)_P \ , \qquad V = \left(\frac{\partial G}{\partial P}\right)_T \ , \qquad (4.50)$$

and by cross-differentiation

$$\left(\frac{\partial S}{\partial P}\right)_T = -\left(\frac{\partial V}{\partial T}\right)_P \ , \qquad (4.51)$$

which is the third Maxwell relation. Comparing Eqs. (4.43) with (4.49), and (4.44) with (4.50), we see that F and G, and V and P, play analogous roles for systems specified by (T, V) and by (T, P) respectively.

Lastly we consider a thermally isolated system held at a constant pressure P, i.e. in mechanical equilibrium with an environment at constant pressure P. Fig. 4.5 shows this situation. Eq. (4.38) with $dQ = 0$ now leads to

$$dH = 0 \ , \qquad H = \text{const.} \qquad (4.52)$$

where the enthalpy H is defined by

$$H \equiv E + PV \ . \qquad (4.53)$$

The condition of constant enthalpy, Eq. (4.52), replaces that of constant energy for a totally isolated system.

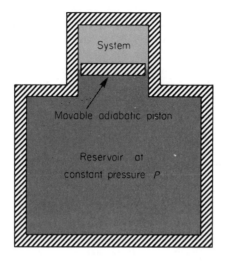

Fig. 4.5. Thermally isolated system in mechanical equilibrium with an environment at constant pressure P.

From Eqs. (4.30), (4.31), (4.52) and (4.53) we see that spontaneous processes in a system subject to the present constraints (constant pressure, constant enthalpy) increase the entropy, and equilibrium again corresponds to *maximum entropy*, as for a totally isolated system. (But note that now we have the subsidiary condition $H = \mathrm{const.}$ instead of $E = \mathrm{const.}$)

Differentiating Eq. (4.53) and combining it with Eq. (4.12) gives

$$dH = T\,dS + V\,dP \ . \tag{4.54}$$

Hence we obtain

$$T = \left(\frac{\partial H}{\partial S}\right)_P \ , \qquad V = \left(\frac{\partial H}{\partial P}\right)_S \ , \tag{4.55}$$

and Maxwell's fourth relation follows by cross-differentiation:

$$\left(\frac{\partial T}{\partial P}\right)_S = \left(\frac{\partial V}{\partial S}\right)_P \ . \tag{4.56}$$

Note the similarities between Eqs. (4.12), (4.14) and (4.54), (4.55): analogous roles are played by E and H and by V and P, with S playing the same part, in the two situations of thermally isolated constant-volume and constant-pressure systems.

In the cases considered, equilibrium has sometimes corresponded to the maximum of a thermodynamic potential and sometimes to a minimum. This is a matter of definition, whether we operate with the inequality (4.28) or with (4.31). Thus instead of using F for a system with given V and T, leading to $\Delta F \leqslant 0$, some authors prefer to use the entropy-like potential* $J = S - E/T$ for these constraints leading to $\Delta J \geqslant 0$. In addition, one has a choice of *which* variables are held *constant* and *which* variable is *maximized or minimized*, so that the four cases we have considered can be characterized in many different ways. For example it follows from Eq. (4.28) that instead of specifying a totally isolated system by

$$E = \text{const.}, \qquad V = \text{const.}, \qquad \Delta S \geqslant 0 , \qquad (4.57a)$$

we could specify it by

$$S = \text{const.}, \qquad V = \text{const.}, \qquad \Delta E \leqslant 0 , \qquad (4.57b)$$

i.e. the energy at constant volume and constant entropy is a minimum in equilibrium.

The above development again shows that to each thermodynamic potential belong certain *natural variables*:

$$
\begin{aligned}
S &= S(E, V, N) \\
S &= S(H, P, N) \\
F &= F(T, V, N) \\
G &= G(T, P, N) .
\end{aligned}
\qquad (4.58)
$$

The significance of this connection of potentials and variables shows up in Eqs. (4.12), (4.54), (4.43) and (4.49). *From a potential, expressed in terms of its natural variables, all other thermodynamic properties, in particular the equation of state, follow.* Examples of this are Eq. (4.44) for the pressure and Eq. (4.50) for the volume. In section 2.5 we saw that a similar relationship exists in the statistical description. We had $\Omega = \Omega(E, V, N)$ and $Z = Z(T, V, N)$ which lead to S and F respectively. Similar connections exist in all cases.

★ 4.6   MAXIMUM WORK

As in section 4.2, we again consider a system in a surrounding at given temperature T_0 and given pressure P_0 but not necessarily in thermal or

---

*We repeat an earlier warning to the reader that names and notations of thermodynamic quantities are not completely standardized. Thus the Gibbs free energy G is also called the Gibbs function or the thermodynamic potential and the symbol Φ is used for it; the Helmholtz free energy F is quite often denoted by A; *and so on.*

pressure equilibrium with the surrounding. It follows from Eq. (4.25) and the first law (1.13) that for any change of the system in its environment we have

$$T_0 \Delta S \geqslant \Delta E - W \ . \tag{4.59}$$

Previously we considered the cases $W = 0$ (constant-volume system: no work can be done on the system) and $W = -P_0 \Delta V$ (all the work done on the system is done by the surrounding 'atmosphere' at the constant pressure P_0). A more interesting situation arises if in addition to the coupling to the surrounding the system is also mechanically coupled to a second system — we shall call it the body — which is thermally isolated. The system can then do work on the body, so that $(-W)$, the total work done *by* the system, consists of two parts: firstly the work done against the surrounding given by $P_0 \Delta V$ where ΔV is the volume change of the system; and secondly the work done by the system on the body which we shall denote by $(-W_u)$. Thus we have

$$(-W) = P_0 \Delta V + (-W_u) \ . \tag{4.60}$$

We shall call $(-W_u)$ the *useful work* done *by* the system since it is that part of the work which can be usefully extracted in the given change. This is in contrast to the work $P_0 \Delta V$ which is *necessarily and uselessly* performed by the system in pushing back the surrounding 'atmosphere'.

Substituting Eq. (4.60) into Eq. (4.59) gives

$$(-W_u) \leqslant -(\Delta E + P_0 \Delta V - T_0 \Delta S) = -\Delta A \tag{4.61}$$

where we introduced the availability from Eq. (4.30). Thus *for a given change of the system* (i.e. from a definite state 1 to a definite state 2: hence ΔE, ΔV and ΔS are given) *in a given environment* (i.e. P_0 and T_0 are given) *the maximum useful work* $(-W_u)_{max}$ *which can be done by the system is obtained in a reversible change and is given by the corresponding decrease in the availability*:

$$(-W_u)_{max} = -\Delta A \ . \tag{4.62}$$

The derivation of the inequality (4.61) can be generalized. In addition to the work $P_0 \Delta V$ performed in pushing back the surrounding atmosphere, more 'useless' work may be done, for example against frictional forces. As a result the *in*equality in (4.61) becomes even stronger. (Compare the corresponding comment made at the end of section 4.2 in connection with Clausius's inequality (4.28).)

As in sections 4.4 and 4.5 we can consider special cases of Eq. (4.62).

If in the initial and final states (called 1 and 2 above) the system has the same volume ($\Delta V = 0$) and the same temperature, equal to that of the surrounding ($T = T_0$), then Eq. (4.62) becomes

$$(-\Delta W_u)_{max} = -\Delta F .$$ (4.63)

Thus under these conditions the maximum useful work is obtained in a reversible change and equals the decrease in the Helmholtz free energy of the system. Note that we did *not* have to assume that the *volume of the system remains constant throughout the process* or that the process is *isothermal*, but only that *in the initial and final states volume and temperature assume the prescribed values.*

Correspondingly, if the pressure and temperature of the system in both the initial and final states are those of the surrounding ($T = T_0$, $P = P_0$) then Eq. (4.62) gives

$$(-\Delta W_u)_{max} = -\Delta G .$$ (4.64)

Thus the maximum useful work is obtained in a reversible change and equals the decrease in the Gibbs free energy of the system. The comments following Eq. (4.63) are, *mutatis mutandis*, also appropriate here.

In connection with Eqs. (4.63) and (4.64) it should be mentioned that if the system is sufficiently simple (e.g. a homogeneous body) so that for a given mass (V, T) or (P, T) determine its state completely, then no process is possible at all and so no useful work can be obtained. Hence for Eqs. (4.63) and (4.64) to be non-trivial, the system must have some additional degrees of freedom such as the concentrations of several components, etc.

To illustrate the concept of maximum useful work consider a thermally isolated system, kept at constant volume and consisting of two identical bodies which initially are at the temperatures T_1 and T_2. Let us find the maximum useful work which can be obtained from this system.

Any work gained from this system is compensated for by a corresponding decrease in the energy of the system. In other words, we want to decrease the energy of the system as much as possible, i.e. we desire the lowest possible final temperature for the system. But we cannot make the final temperature as small as we please; that would violate the second law as formulated by Kelvin. (We shall show in section 5.1 how Kelvin's statement follows from the entropy principle.) From Clausius's inequality (4.61) we know that the best we can do is to perform the process reversibly, that is at constant entropy since the system is thermally isolated and has constant volume. From Eq. (4.62) we have for the maximum useful work

$$(-W_u)_{max} = -(\Delta E)_{S,V} ,$$

where the suffixes indicate that the entropy S and the volume V of the system are constant during the process. For a process at constant entropy, the entropy changes of the two bodies compensate, i.e.

$$dS = C\left[\frac{dT'}{T'} + \frac{dT''}{T''}\right] = 0 \ ,$$

where T' and T'' are the temperatures of the two bodies, and C is the heat capacity at constant volume of one of the bodies. (We are taking C as constant.) By integrating the last equation, one sees that during the process $(T'T'')$ is constant and hence equal to the initial value $(T_1 T_2)$, so that the final temperature T_f of the system is $T_f = (T_1 T_2)^{1/2}$. Hence the maximum useful work is given by

$$(-W_u)_{max} = -\{2C(T_1 T_2)^{1/2} - C(T_1 + T_2)\} \ ,$$

which is always positive. If the two bodies are simply put into thermal contact without extracting any work, then $(-\Delta W_u) = \Delta E = 0$, and the final temperature of the system is $\frac{1}{2}(T_1 + T_2)$ which is always higher than the temperature $(T_1 T_2)^{1/2}$ reached in a reversible process.

4.7   THE THIRD LAW OF THERMODYNAMICS

The third law of thermodynamics deals with the entropy of a system as its absolute temperature tends to zero. It fixes the absolute value of the entropy which is left undetermined by such purely thermodynamic relations as Eq. (4.32) which only give entropy differences. The third law is of more limited use than the first and second laws and for this reason we shall deal with it only briefly.* Its main applications are in chemistry where it originated in the study of chemical reactions by Nernst (1906) — it is also known as Nernst's theorem — and in low temperature physics.

If one starts from statistical mechanics rather than phenomenological thermodynamics the value of the entropy of any system is completely determined. If we enumerate the energy levels of a system *without* repetition

$$E_1 < E_2 < \ \ldots \ < E_r < \ \ldots \ , \tag{4.65}$$

the general level E_r being g_r-fold degenerate, then the probability $p(E_r)$ that the system is in a state with energy E_r, when at temperature T, is given by Eq. (2.25). If the temperature is sufficiently low so that

$$E_2 - E_1 \gg kT \tag{4.66}$$

*For detailed treatments see, for example, Wilks[38] or ter Haar.[17]

then only the first term in Eq. (2.25) is appreciably different from zero, i.e.

$$p(E_1) \approx 1 , \qquad p(E_r) \approx 0 \quad \text{for} \quad r > 1 , \tag{4.67}$$

and the system is necessarily in a state of lowest energy E_1. This level is g_1-fold degenerate so that the entropy at sufficiently low temperatures, i.e. so that (4.66) holds, is, from Boltzmann's definition, given by

$$\underset{T \to 0}{\text{Lim}} \ S = k \ln g_1 . \tag{4.68}$$

(The last result also follows from Eq. (2.33) with $p_r = 1/g_1$ for $r = 1, 2, \ldots \ldots, g_1$, and $p_r = 0$ for $r > g_1$.) Our argument depended essentially on the discreteness of the energy levels of our system, so that it is a consequence of quantum theory. If the energy levels would form a continuum, as was assumed in classical statistical physics, we could not write down an inequality like (4.66) and our whole discussion would break down.

It is generally believed by theoretical physicists that the ground state of any system is non-degenerate, $g_1 = 1$; there is no proof of this conjecture but it holds for all cases where it has been checked. If this conjecture is correct it follows that the entropy of any system vanishes at zero temperature*

$$\underset{T \to 0}{\text{Lim}} \ S = 0 . \tag{4.69}$$

The importance of this result is the fact that the zero-temperature entropy is independent of any other properties of the system, such as its volume, pressure, etc. On the other hand from the point of view of thermodynamics the actual value of the zero-temperature entropy is not significant; we could on the right-hand side of Eq. (4.69) replace zero by any absolute constant value S_0. It is then obviously simplest to take $S_0 \equiv 0$.

We shall next derive some of the consequences of Eq. (4.69).

It follows from Eq. (4.32) that the heat capacity of a system vanishes at $T = 0 \, \text{K}$. For changes at constant volume Eq. (4.32) may be written

$$S(T_2, V) = S(T_1, V) + \int_{T_1}^{T_2} C_V \frac{dT}{T} . \tag{4.70}$$

If we take the limit $T_1 \to 0$ in this equation, it follows from Eq. (4.69) that the integral in Eq. (4.70) must tend to a finite limit as $T_1 \to 0$, and

---

*We saw in section 4.3.4 that the entropy of a system of N particles is of the order of kN. With $N \sim 10^{23}$, $k \ln g_1$ is utterly negligible compared with kN, even for $g_1 \sim N$; hence Eq. (4.69) would always hold for practical purposes. It would break down for $\ln g_1 \propto N$.

for this we require $C_V \to 0$ as $T \to 0$. (Note that this derivation only requires that $S(0, V)$ is finite and not the much stronger statement of the third law that $S(0, V)$ is independent of V.) In the same way one concludes that $C_P \to 0$ as $T \to 0$. This property of heat capacities, that they tend to zero as $T \to 0$, was experimentally well established by measurements of Nernst and his school in the period 1910 to 1912. Some typical results of modern measurements are shown in Fig. 4.6. We shall consider the specific heats of solids further in Chapter 6.

Other consequences of the third law follow from Maxwell's relations. Consider Eq. (4.51):

$$\left(\frac{\partial S}{\partial P} \right)_T = - \left(\frac{\partial V}{\partial T} \right)_P . \tag{4.51}$$

From $S(T, P) \to 0$ as $T \to 0$ for all values of P, it follows that as $T \to 0$ the left-hand side of Eq. (4.51) tends to zero, and hence

$$\mathop{\text{Lim}}_{T \to 0} \left(\frac{\partial V}{\partial T} \right)_P = 0 , \tag{4.71}$$

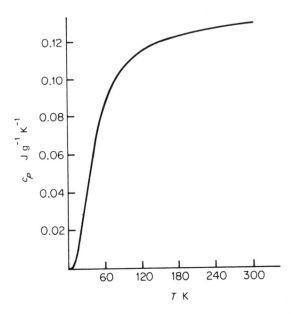

Fig. 4.6. Specific heat at constant pressure of gold. (Data from: *A Compendium of the Properties of Materials at Low Temperature, Part II: Properties of Solids.* Wadd Technical Report 60–56, Part II, 1960.)

i.e. the thermal expansion coefficient at constant pressure vanishes at $T = 0$. Similarly

$$\lim_{T \to 0} \left(\frac{\partial P}{\partial T} \right)_V = 0 \tag{4.72}$$

follows from Eq. (4.45).

We next consider the magnetization of a body in an applied uniform magnetic field $\mathscr{H}$.* The work in an infinitesimal reversible change is now given by Eq. (1.45), i.e. by

$$\mathrm{d}W = -\mu_0 \mathscr{M} \mathrm{d}\mathscr{H} , \tag{1.45}$$

where $\mathscr{M}$ is the component of $\mathscr{M}$ in the direction of the field $\mathscr{H}$, and the fundamental thermodynamic relation (4.12) is replaced by

$$\mathrm{d}E = T \mathrm{d}S - \mu_0 \mathscr{M} \mathrm{d}\mathscr{H} . \tag{4.73}$$

In this equation we have omitted the work term $(-P \mathrm{d}V)$ which is usually negligible for solids. If we are interested in the temperature dependence of the magnetization at constant magnetic field, we require T and $\mathscr{H}$ as independent variables, not S and $\mathscr{H}$. To change variables, we use the same trick as in sections 4.4 and 4.5 and rewrite Eq. (4.73) in the form

$$\mathrm{d}(E - TS) = -S \mathrm{d}T - \mu_0 \mathscr{M} \mathrm{d}\mathscr{H} . \tag{4.74}$$

We see from this equation that the Helmholtz free energy

$$F = E - TS \tag{4.75}$$

is the appropriate thermodynamic potential when T and $\mathscr{H}$ are the independent variables. One then obtains from Eq. (4.74) by cross-differentiation

$$\left(\frac{\partial S}{\partial \mathscr{H}} \right)_T = \mu_0 \left(\frac{\partial \mathscr{M}}{\partial T} \right)_{\mathscr{H}} . \tag{4.76}$$

Again, from the third law, the entropy of the specimen goes to zero as $T \to 0$ at any constant field $\mathscr{H}$; hence the left-hand side of (4.76) vanishes as $T \to 0$ and therefore

---

*This example uses results from section 1.4.

$$\text{Lim}_{T \to 0} \left(\frac{\partial \mathscr{M}}{\partial T} \right)_{\mathscr{H}} = 0 \; . \tag{4.77}$$

Eq. (4.77) is an interesting result derived by very general means. The magnetization $\mathscr{M}$, Eq. (3.7), of our simple model of a paramagnetic solid satisfies Eq. (4.77). On the other hand no substance could satisfy Curie's law ($\mathscr{M} \propto 1/T$) down to the lowest temperatures since this implies

$$\left(\frac{\partial \mathscr{M}}{\partial T} \right)_{\mathscr{H}} \propto \frac{1}{T^2} \; , \tag{4.78}$$

which violates Eq. (4.77).

★ 4.8 THE THIRD LAW (CONTINUED)

In the last section we discussed the third law and some simple consequences of it. We now want to consider in more detail the assumptions which underlie the applications of the third law.

Our derivation of Eq. (4.69) assumes that the system is in thermodynamic equilibrium. But at low temperatures some processes occur so slowly that the relevant relaxation times are practically infinite, as discussed in section 1.2, so that equilibrium is never reached. Nevertheless the third law can frequently be applied. The reason is that the properties of a system can usually be attributed to different aspects which hardly affect each other. Each aspect makes its own contribution to the energy and to the entropy of the system. Some of these aspects will be in internal thermodynamic equilibrium, others won't. The third law applies to those that are in equilibrium and states that the corresponding contributions to the entropy vanish at absolute zero. This is essentially the formulation due to Simon (1927).

Consider, as an example, the problem of a binary alloy, treated in section 1.2. The ordering process in the solid state is so slow that it can be ignored as the temperature is lowered. Unless the alloy was specially prepared, the atoms will be randomly arranged on the sites of the crystal lattice, and they will so remain as the temperature is lowered. There will be a corresponding entropy of disorder. On the other hand the thermal lattice vibrations decrease as the temperature is lowered until all the atoms are in their vibrational ground states. The vibrational contributions to the entropy and to the specific heat then vanish as $T \to 0$. (This problem will be treated in Chapter 6. See also, Fig. 4.6 above.)

A second example is given by a paramagnetic solid, considered in Chapters 2 and 3. We again have the lattice vibrations. At a temperature of a few kelvin these make negligible contributions to the total entropy and to the specific heat. On the other hand the magnetic dipoles will be oriented essentially at random unless a strong magnetic field is applied.

A similar but more extreme situation exists with respect to the magnetic moments of the atomic nuclei. These are about a factor 2,000 smaller than the Bohr magneton, Eq. (3.11), since the electronic mass in this equation must be replaced by the proton mass. Although it is possible to align nuclear magnetic moments solely by cooling this usually requires temperatures of about 10^{-5} K or less; at higher temperatures they are for most purposes oriented at random and give a constant contribution S_0 to the entropy. If a nucleus has spin J (where, according to quantum theory, J can only assume one of the values $J = 0, \frac{1}{2}, 1, \frac{3}{2}, 2, \ldots$) it can orient itself in $(2J+1)$ ways in a magnetic field, and this leads to a spin entropy of $S_0 = Nk\ln(2J+1)$ for N such nuclei. This agrees with $S_0 = Nk\ln 2$ for spin $\frac{1}{2}$ (i.e. $J = \frac{1}{2}$), Eq. (3.16). We may then ignore the nuclear spin entropy (of course in some cases it is zero, i.e. if $J = 0$) for most purposes since it remains constant above about 10^{-5} K.

The third law enables us to relate the entropies of systems in very different states; for example, of the same solid existing in different allotropic forms: diamond and graphite, grey and white tin. Let us look at the last example. White tin is a metal; grey tin an insulator. The two forms have quite different crystal structures. At $T_0 = 292$ K the two forms coexist in equilibrium in any proportions, and a heat of transformation L is required to transform one mole of grey tin into white tin. Grey tin and white tin are the stable forms below and above the transition temperaure T_0 respectively. Near room temperature, grey and white tin transform into each other in a few hours, but the transition rate is very slow if the white tin is at lower temperatures and if no grey tin is present. Hence if one cools white tin rapidly from above to below the transition temperature, then it will continue to exist indefinitely as a metastable state. We can write the entropies of white and grey tin at temperature T_0 as

$$S_w(T_0) = S_w(0) + \int_0^{T_0} C_w(T)\,\frac{\mathrm{d}T}{T} \tag{4.79a}$$

$$S_g(T_0) = S_g(0) + \int_0^{T_0} C_g(T)\,\frac{\mathrm{d}T}{T} \tag{4.79b}$$

where the suffixes w and g refer to white and grey tin respectively; C is the molar heat capacity at constant pressure. To convert one mole of grey tin into white tin at the transition temperature T_0 one must supply a quantity of heat L at this temperature, i.e. the entropy of the tin is increased by L/T_0 so that

$$S_w(T_0) = S_g(T_0) + \frac{L}{T_0}. \tag{4.80}$$

Now the third law tells us that

$$S_g(0) = S_w(0) , \qquad (4.81)$$

so that we can combine Eqs. (4.79a) to (4.81) to give

$$\int_0^{T_0} \{C_w(T) - C_g(T)\} \frac{\mathrm{d}T}{T} = \frac{L}{T_0} . \qquad (4.82)$$

Lange (1924) measured the specific heats of both forms of tin as a function of temperature. He also measured the heat of transformation L. In this way he was able to calculate both sides of Eq. (4.82) independently. He found very good agreement, the values of the left- and right-hand sides being 7.3 and 7.5 (both in joule per kelvin per mole) respectively. Similar agreement has also been obtained in other cases, and this affords a direct verification of the third law.

Above we derived some low temperature properties of matter ($C \to 0$, etc., as $T \to 0$) from the third law. These followed from the fact that at $T = 0$ the system is in its ground state. It has only in the last few decades been realized that the observation of these low-temperature behaviours *at experimentally readily accessible temperatures* cannot be explained in this simple way. Certainly for specific heat measurements such as are shown in Fig. 4.6 very many energy levels are occupied: not just one or two. The same is true even at the lowest experimentally accessible temperatures. An explanation has been proposed by Kramers and Casimir (1963).* It depends on the *rate* at which the density of states $f(E)$ (cf. Eq. (2.32)), and hence the entropy $S(E)$, go to zero as E approaches the ground state energy E_1.

SUMMARY

In this chapter we have completed the derivation of all basic results required later on. We summarize these, including some obtained in Chapters 1 and 2.

The first law

$$dE = \mathrm{d}Q + \mathrm{d}W . \qquad (1.14)$$

Boltzmann's definition of entropy for equilibrium

$$S(E, V, N) = k \ln \Omega(E, V, N) . \qquad (2.34)$$

Partition function

$$Z(T, V, N) = \sum_r \exp(-\beta E_r) , \qquad (2.23)$$

where the summation is over all microstates of the system.

*For a full discussion and references, see ter Haar,[17] section 9.2.

Helmholtz free energy

$$F(T, V, N) = -kT \ln Z(T, V, N) \ . \tag{2.36}$$

Thermodynamic potentials

(i) enthalpy	$H \equiv E + PV$	(4.53)
(ii) Helmholtz free energy	$F \equiv E - TS$	(2.37)
(iii) Gibbs free energy	$G \equiv E + PV - TS \ .$	(4.48)

Fundamental thermodynamic relations

Valid for *infinitesimal* changes between equilibrium states. The change need NOT be reversible:

$$dE = T dS - P dV \tag{4.12}$$

$$dH = T dS + V dP \tag{4.54}$$

$$dF = -S dT - P dV \tag{4.43}$$

$$dG = -S dT + V dP \ . \tag{4.49}$$

The second law

For an isolated system:

$$\Delta S \geqslant 0 \ . \tag{4.21}$$

Clausius's inequality

(i) For a system in constant-temperature surrounding:

$$\Delta S - \frac{Q}{T_0} \geqslant 0 \ . \tag{4.25}$$

(ii) For system in constant-temperature constant-pressure surrounding:

$$\Delta S - \frac{\Delta E + P_0 \Delta V}{T_0} \geqslant 0 \tag{4.28}$$

or

$$\Delta A \leqslant 0 \ , \tag{4.31}$$

where

Availability

$$A \equiv E + P_0 V - T_0 S \ . \tag{4.30}$$

Special equilibrium conditions (sections 4.4 and 4.5)

$\Delta E = \Delta V = 0 \ ,$	$S = \text{maximum}$
$\Delta H = \Delta P = 0 \ ,$	$S = \text{maximum}$
$\Delta T = \Delta V = 0 \ ,$	$F = \text{minimum}$
$\Delta T = \Delta P = 0 \ ,$	$G = \text{minimum} \ .$

The four Maxwell relations

$$\left(\frac{\partial T}{\partial V}\right)_S = -\left(\frac{\partial P}{\partial S}\right)_V \qquad (4.18)$$

$$\left(\frac{\partial S}{\partial V}\right)_T = \left(\frac{\partial P}{\partial T}\right)_V \qquad (4.45)$$

$$\left(\frac{\partial S}{\partial P}\right)_T = -\left(\frac{\partial V}{\partial T}\right)_P \qquad (4.51)$$

$$\left(\frac{\partial T}{\partial P}\right)_S = \left(\frac{\partial V}{\partial S}\right)_P . \qquad (4.56)$$

PROBLEMS 4

4.1 1,000 g of water at 20 °C are placed in thermal contact with a heat bath at 80 °C. What is the change in entropy of the total system (water plus heat bath) when equilibrium has been re-established?

If the 1,000 g of water at 20 °C are heated by successively placing the water into heat baths at 50 °C and 80 °C (waiting each time until equilibrium is reached) what is the entropy change of the total system?

(Specific heat of water $= 4.2 \text{ J g}^{-1} (°C)^{-1}$.)

4.2 Two vessels A and B each contain N molecules of the same perfect monatomic gas. Initially the two vessels are thermally isolated from each other, the two gases being at the same pressure P, and at temperatures T_A and T_B respectively. The two vessels are now brought into thermal contact, the pressure of the gases being kept constant at the value P. Find the change in entropy of the system after equilibrium is established, and show that this change is non-negative.

4.3 Two vessels contain the same number N of molecules of the same perfect gas. Initially the two vessels are isolated from each other, the gases being at the same temperature T but at different pressures P_1 and P_2. The partition separating the two gases is removed. Find the change in entropy of the system when equilibrium has been re-established, in terms of the initial pressures P_1 and P_2. Show that this entropy change is non-negative.

4.4 When 1 mole of nitrous oxide decomposes into nitrogen and oxygen, the system being at 25 °C and 1 atm both initially and finally, the entropy S of the system increases by 76 J/K and its enthalpy H decreases by 8.2×10^4 J. Calculate the change in the Gibbs free energy in this process.

Comment on the fact that nitrous oxide appears to be stable under these conditions of temperature and pressure.

4.5 One mole of superheated water at 110 °C and at 1 atm is evaporated to steam at the same temperature and pressure. Calculate the change in entropy given that:

(i) In the pressure range 1 atm to 1.4 atm the entropy of liquid water at 110 °C can be considered constant.

(ii) At 110 °C liquid water and steam are in equilibrium at a pressure of 1.4 atm, and the latent heat of evaporation under these conditions is 4×10^4 J/mol.

(iii) In the pressure range 1 atm to 1.4 atm water vapour at 110 °C behaves as a perfect gas.

Calculate the error introduced by the last assumption if in the range 1 atm to 1.4 atm and at 110 °C one has, for one mole of steam,

$$\left(\frac{\partial V}{\partial T}\right)_P = \frac{R}{P} + 0.46 \text{ cm}^3/\text{K} \ .$$

4.6 Phosphine exists in three allotropic forms, known as the α, β and γ forms. The α and β forms are in equilibrium at 49.43 K, the α and γ forms at 30.29 K. Obtain the molar heat of transformation for the γ form changing to the α form at 30.29 K from the following data:

(i) The entropy of the α form at 49.43 K is 34.03 J mol^{-1} K^{-1}.

(ii) The entropy change in heating the γ form from 0 K to 30.29 K is 11.22 J mol^{-1} K^{-1}.

(iii) The entropy change in heating the α form from 30.29 K to 49.43 K is 20.10 J mol^{-1} K^{-1}.

4.7 What is the maximum useful work which can be obtained by cooling 1 mole of a perfect gas at constant volume from a temperature T to the temperature T_0 of the surrounding?

4.8 What is the maximum useful work which can be obtained by expanding 1 mole of a perfect gas from pressure P_1 to pressure P_2 at a constant temperature T_0 equal to that of the surrounding? The surrounding is at pressure P_0.

5

Simple thermodynamic systems

In this chapter the thermodynamic principles and methods which have been developed will be applied to simple systems. The material is arranged so that *the various sections may be read independently of each other in any order. Any of the sections may be omitted.*

★ 5.1 OTHER FORMS OF THE SECOND LAW

In section 2.3 we gave Clausius's principle of the increase of entropy and we have taken this as the basic formulation of the second law. We shall now briefly show that it implies the earlier formulations of Clausius and Kelvin stated in section 2.1.

Clausius's original form—that heat cannot by itself pass from a colder to a hotter body—follows easily. Consider two systems A and B at temperatures T_1 and T_2 ($< T_1$) respectively. Suppose it were possible to devise a mechanism M which transfers an infinitesimal amount of heat đQ from B to A (see Fig. 5.1), there being no other changes of any sort except the infinitesimal resulting temperature changes of A and B. The net entropy change is

$$\Delta S = \text{đ}Q \left(\frac{1}{T_1} - \frac{1}{T_2} \right) < 0 \qquad (5.1)$$

which violates Clausius's principle of increase of entropy.

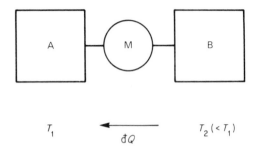

Fig. 5.1. A heat transfer process which violates the
second law.

The example of temperature equalization in section 4.3.3 is essentially equivalent to this proof.

We next consider Kelvin's form of the second law: a process whose only effect is the complete conversion of heat into work cannot occur. Fig. 5.2 represents such a process schematically. The complete system consists of a heat bath at temperature T coupled to a mechanism M which extracts an amount of heat Q from the heat bath and completely converts it into work $W = Q$ which is performed on a third body B, for example by lifting a weight. Since the complete conversion of heat into work is to be the only effect, the device M must be operating in a cycle at the end of which it must be in its original state, i.e. its entropy is not changed. Similarly the work performed on the body B will change some external parameter of B, e.g. in the lifting of a body, but will not alter its internal properties such as its entropy. Hence the net entropy change results from the heat extracted from the heat bath and is given by

$$\Delta S = - Q/T < 0 \; , \tag{5.2}$$

again violating Clausius's entropy principle.

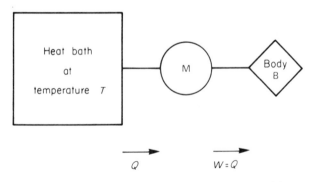

Fig. 5.2. A heat engine which violates the second law.

★ 5.2 HEAT ENGINES AND REFRIGERATORS

A perfect heat engine which converts thermal energy completely into work would involve a decrease in entropy. A heat engine which does not violate the second law must allow for a compensating change so that the entropy does not decrease during a complete cycle. The simplest such engine was first considered by Carnot (1824) and is called a Carnot engine. It operates beween two temperatures T_1 and T_2 ($< T_1$). The energy changes that occur in one cycle are illustrated in Fig. 5.3. During one cycle, a quantity of heat Q_1 is extracted from the heat bath at the higher temperature T_1, the mechanism M does an amount of work W on the body B, and an amount of heat

$$Q_2 = Q_1 - W \qquad (5.3)$$

is rejected to the heat bath at the lower temperature T_2. The net entropy change during a cycle is therefore

$$\Delta S = -\frac{Q_1}{T_1} + \frac{Q_2}{T_2} \qquad (5.4)$$

and from the second law we must have

$$\Delta S \geqslant 0 \ . \qquad (5.5)$$

The efficiency of a heat engine is defined by

$$\eta \equiv \frac{W}{Q_1} \ ; \qquad (5.6)$$

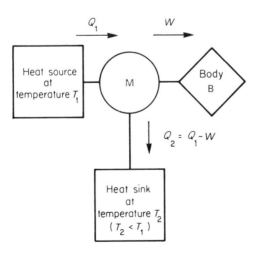

Fig. 5.3. A Carnot heat engine.

it is the ratio of the work gained to the heat taken from the heat bath at the higher temperature in operating the engine. One would have liked to be able to have $\eta = 1$: all the heat is converted into work. The second law tells us that this is not possible. By combining Eqs. (5.3) to (5.6) we see that the efficiency is limited to

$$\eta \leqslant \frac{T_1 - T_2}{T_1} \, . \tag{5.7}$$

The *maximum theoretical efficiency* is given by the *equality* sign in Eq. (5.7) and is attained in a reversible process. We see that the efficiency depends only on the temperatures of the two heat baths. For an engine working between 0 and 100 °C, the maximum efficiency is $\eta_{max} = 100/373 = 0.27$. To improve the efficiency requires larger temperature differences. In practice the temperature T_2 of the heat sink is limited to that of the surrounding of the engine, $T_2 \approx 300$ K say. The temperature T_1 of the heat source of steam engines is raised by working with saturated steam at high pressures. In this way temperatures of $T_1 \approx 500$ °C have been obtained for which $\eta_{max} = 473/773 = 0.6$. The efficiencies which are obtained in practice for real engines are only about 50 per cent of these maximum efficiencies for ideal reversible engines.

We next consider the Carnot cycle in more detail. It consists of four distinct stages: (1) An isothermal process at temperature T_1. (2) A reversible adiabatic process which cools the working substance of the device M (Fig. 5.3) from temperature T_1 to T_2. (Being reversible and adiabatic this process occurs at constant entropy—it is known as an isentropic process.) (3) An isothermal process at temperature T_2, and (4) an isentropic process which heats the working substance from T_2 to T_1 and takes it back to its initial state at the beginning of the cycle. A Carnot cycle is best illustrated by means of an entropy–temperature diagram, as shown in Fig. 5.4. The four stages described above are represented by the paths AB, BC, CD and DA respectively. It follows from Eq. (4.13a) that the heat absorbed by M from the heat source during stage 1 (path AB) is

$$Q_1 = T_1(S_1 - S_2) \, , \tag{5.8a}$$

and that rejected from M to the heat sink during stage 3 (path CD) is

$$Q_2 = T_2(S_1 - S_2) \, , \tag{5.8b}$$

whence $Q_2/Q_1 = T_2/T_1$ in agreement with Eq. (5.7).

The above results are quite independent of the details of the heat engine, for example of what the working substance is, as long as the cycle is reversible.

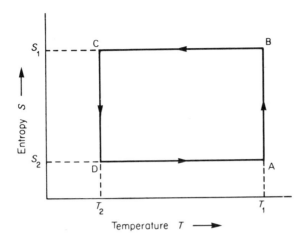

Fig. 5.4. A Carnot cycle represented on a (S, T) diagram.

Of course, the details are particularly simple if the working substance is a perfect gas where one possesses a simple analytic equation of state. The four stages of the Carnot cycle in this case are illustrated in Fig. 5.5. They are (1) an isothermal expansion of the gas at the higher temperature, (2) an isentropic expansion which cools the gas, (3) an isothermal compression at the lower temperature, and (4) an isentropic compression returning the gas to its original state. Fig. 5.6 shows the corresponding (P, V) diagram. In these two figures, temperatures are measured on the perfect gas scale which, for the moment, we shall not assume identical with the absolute Kelvin scale. From the properties of a perfect gas and the equation defining an adiabatic process (see problem 1.1) one easily shows that $Q_2/Q_1 = \Theta_2/\Theta_1$ (see problem 5.2) and hence that *the perfect gas and absolute temperature scales are identical.* This fills in a gap which was left in section 2.3 where we stated this identity but did not prove it. A different and from our approach more satisfying proof will occur in Chapter 7 where we shall derive the equation of state of a perfect gas using the statistical temperature definition of Chapter 2 and we shall obtain the usual equation $PV = NkT$.

It is also possible to operate a heat engine backwards, as illustrated in Fig. 5.7, i.e. to do work W on the system in order to extract heat Q_2 from the heat bath at the lower temperature T_2 and transfer an amount of heat $Q_1 = Q_2 + W$ to the heat bath at the higher temperature T_1. This is the principle on which a refrigerator works. The work W supplied is frequently electrical work, the refrigerator cabinet is cooled (heat Q_2 extracted) and the heat Q_1 is given out into the room. The efficiency of a reversible refrigerator is $\eta = Q_2/W = T_2/(T_1 - T_2)$; thus it is more efficient for smaller temperature

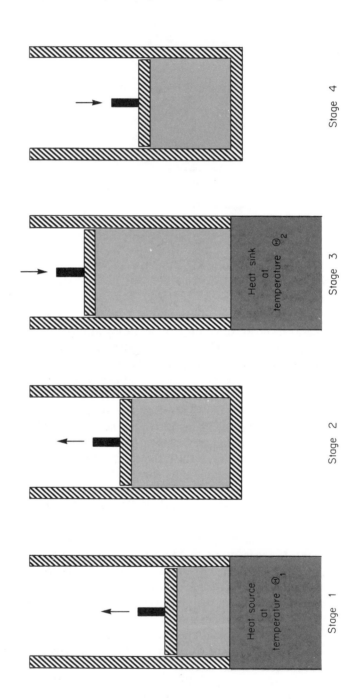

Fig. 5.5. The four stages of a Carnot cycle using a perfect gas as working substance.

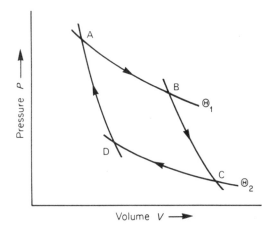

Fig. 5.6. A Carnot cycle represented on a (P, V) diagram. The temperatures Θ_1 and Θ_2 of the isotherms AB and CD are measured on the perfect gas scale.

differences. Thus for a domestic refrigerator one might have $T_2 = 270$ K, $T_1 = 300$ K, giving $\eta = 9$, showing why refrigerators are cheap to run.

A heat pump works on the same principle as a refrigerator but the object is to warm up the heat bath at the higher temperature; for example, to heat the Royal Festival Hall in London by cooling the water of the river Thames a little. The ideal efficiency is now defined by $\eta = Q_1/W = T_1/(T_1 - T_2)$ since this measures the heat supplied for a given input of work. Although their theoretical efficiency is so high, in practice heat pumps are only marginally economically viable and have only rarely been used because of their fairly high capital cost.

We conclude this section by discussing the petrol engine's mode of operation. We shall consider an idealized cycle which will thus provide an upper bound for the efficiency of a real petrol engine.

The cycle of a petrol engine consists of six stages. (1) A mixture of air and petrol is drawn into the cylinder. (2) The mixture is compressed rapidly, causing a temperature rise. (3) The mixture is exploded. During the explosion the piston hardly moves leading to a sharp increase in pressure and temperature. (4) The hot gases push the piston back, doing mechanical work. (5) The outlet valve is opened and the hot exhaust gases rush out into the atmosphere. (6) The piston forces out the remaining exhaust gases. The outlet valve is now shut, the inlet valve opened, and the cycle can be repeated.

An idealized version of this cycle, known as the air standard Otto cycle, assumes that the working substance is air throughout, instead of a mixture of air and other gases with a composition which changes during the cycle.

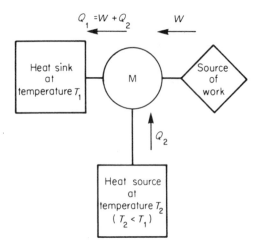

Fig. 5.7. A Carnot heat engine being driven backwards, i.e. operating as a refrigerator or heat pump.

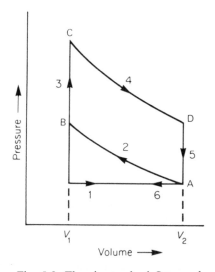

Fig. 5.8. The air standard Otto cycle.

It is assumed that the air obeys the perfect gas equation and has a constant specific heat, that the amount of air is constant throughout the cycle, and that the cycle can be represented by the path in the (P, V) diagram shown in Fig. 5.8. Stages 2 and 4 represent reversible adiabatic compression and expansion of the gas. During stages 3 and 5 the gas is heated and cooled at constant volume. The energy transfer to and from the gas during stages

3 and 5 depends only on the initial and final temperatures of the gas in each of these stages but is independent of how the heat transfer is accomplished. In particular, it is not necessary to carry stages 3 and 5 out reversibly using an infinite number of heat baths at appropriately matched temperatures. It is left as an exercise for the reader (see problem 5.3) to show that the theoretical maximum efficiency of this cycle is given by

$$\eta = 1 - r^{(1-\gamma)} \tag{5.9}$$

where $r \equiv (V_2/V_1)$ is the compression ratio of the engine and $\gamma \equiv C_P/C_V$ is the ratio of the specific heats of air. The compression ratio in a modern motor car engine is about $r = 10$ and for air γ has the value 1.4, giving the ideal efficiency $\eta = 0.6$. The efficiency of a real engine is more nearly half of this value.

★ 5.3   THE DIFFERENCE OF HEAT CAPACITIES

The second law enables us to relate the heat capacities of a substance at constant volume and at constant pressure. If we take T and V as independent variables, the first law, Eq. (1.17), becomes

$$dQ = C_V \, dT + \left[P + \left(\frac{\partial E}{\partial V} \right)_T \right] dV \tag{5.10}$$

where $C_V = (\partial E/\partial T)_V$ is the heat capacity at constant volume. The second term $(P \, dV)$ is the work that must be done against the external forces. The last term, $(\partial E/\partial V)_T dV$, is the change in the internal potential energy of the gas which results from changing the mean spacing between molecules. In other words, this last term derives from the intermolecular cohesive forces. For an ideal gas this term vanishes since E depends on the temperature only, as discussed in connection with Eq. (1.19).

To evaluate the square bracket in Eq. (5.10) we write the fundamental thermodynamic relation (4.12), with T and V as independent variables, in the form

$$dS = \frac{1}{T} \left(\frac{\partial E}{\partial T} \right)_V dT + \frac{1}{T} \left[P + \left(\frac{\partial E}{\partial V} \right)_T \right] dV \ . \tag{5.11}$$

Cross-differentiation at once gives

$$\frac{\partial}{\partial V}\left(\frac{1}{T}\frac{\partial E}{\partial T}\right) = \frac{\partial}{\partial T}\left\{\frac{1}{T}\left[P+\left(\frac{\partial E}{\partial V}\right)_T\right]\right\} \cdot ^* \qquad (5.12)$$

Since the two 'mixed' second-order derivatives in this equation are the same, it follows that

$$T\left(\frac{\partial P}{\partial T}\right)_V = P+\left(\frac{\partial E}{\partial V}\right)_T \cdot \qquad (5.15)$$

Hence we obtain for the heat capacity at constant pressure from Eqs. (5.10) and (5.15)

$$C_P = \left(\frac{dQ}{\partial T}\right)_P = C_V + \left[P+\left(\frac{\partial E}{\partial V}\right)_T\right]\left(\frac{\partial V}{\partial T}\right)_P \qquad (5.16a)$$

$$= C_V + T\left(\frac{\partial P}{\partial T}\right)_V\left(\frac{\partial V}{\partial T}\right)_P \cdot \qquad (5.16b)$$

This last equation allows us to express the difference of heat capacities, $C_P - C_V$, entirely in terms of quantities which are easily measured (unlike the derivative $\partial E/\partial V$ of the internal energy). By means of Eqs. (4.16) we write

$$\left(\frac{\partial P}{\partial T}\right)_V = -\left(\frac{\partial P}{\partial V}\right)_T\left(\frac{\partial V}{\partial T}\right)_P \qquad (5.17)$$

and we introduce the expansion coefficient β_0 and the isothermal compressibility K_T through

$$\beta_0 \equiv \frac{1}{V}\left(\frac{\partial V}{\partial T}\right)_P \qquad (5.18)$$

and

$$K_T \equiv -\frac{1}{V}\left(\frac{\partial V}{\partial P}\right)_T \cdot \qquad (5.19)$$

---

*It is most useful to be able to go 'at once' from Eq. (5.11) to Eq. (5.12), and the reader should aim to acquire this facility. In detail, the omitted steps are as follows.
Comparison of Eq. (5.11) with

$$dS = \left(\frac{\partial S}{\partial T}\right)_V dT + \left(\frac{\partial S}{\partial V}\right)_T dV \qquad (5.13)$$

gives

$$\left(\frac{\partial S}{\partial T}\right)_V = \frac{1}{T}\left(\frac{\partial E}{\partial T}\right)_V , \qquad \left(\frac{\partial S}{\partial V}\right)_T = \frac{1}{T}\left[P+\left(\frac{\partial E}{\partial V}\right)_T\right] ; \qquad (5.14)$$

and since

$$\frac{\partial^2 S}{\partial V \partial T} = \frac{\partial^2 S}{\partial T \partial V} ,$$

Eqs. (5.14) at once lead to Eq. (5.12).

If we substitute Eqs. (5.17) to (5.19) in (5.16b) we obtain

$$C_P - C_V = TV\beta_0^2/K_T \ . \tag{5.20}$$

Since all the quantities on the right of this equation can be measured, this equation can be used to obtain C_V from C_P. The latter is the more easily measured quantity whereas it is C_V which is obtained more easily theoretically. Since K_T is always positive, it follows that C_P is never less than C_V, the two heat capacities being equal if $\beta_0 = 0$, for example for water at 4 °C.

If we apply Eq. (5.16b) to a perfect gas we obtain

$$C_P - C_V = R \tag{5.21}$$

for the molar heat capacities, in agreement with Eq. (1.20).

★ 5.4 SOME PROPERTIES OF PERFECT GASES

5.4.1 The entropy

If we take V and T as independent variables, we have for an infinitesimal change of state, from Eq. (4.12),

$$dS = \frac{1}{T}\left(\frac{\partial E}{\partial T}\right)_V dT + \frac{1}{T}\left[P + \left(\frac{\partial E}{\partial V}\right)_T\right] dV \ . \tag{5.22}$$

For a perfect gas the last term vanishes since the internal energy E is a function of T only (see Eq. (1.19)). For a perfect gas of N molecules at temperature T and occupying a volume V, we then obtain, using $PV = NkT$,

$$S(T,V,N) = N\int^T c_V(T')\frac{dT'}{T'} + Nk\ln V + C' \tag{5.23a}$$

where $c_V(T')$ is the heat capacity at constant volume per molecule and C' is a constant of integration. (It is essentially the entropy in some reference state.) The specific heat of a perfect gas is not constant, but it is approximately so over quite large temperature ranges. If we restrict ourselves to temperature variations within such a range, we can rewrite Eq. (5.23a) as

$$S(T,V,N) = Nc_V\ln T + Nk\ln V + C' \ , \tag{5.23b}$$

where c_V is a constant heat capacity per molecule. It follows from our discussion of the third law in section 4.7 that Eq. (5.23b) cannot be correct as

$T \to 0$: we cannot have a constant heat capacity and must have $c_V \to 0$ as $T \to 0$. Eqs. (5.23a) or (5.23b) show directly that entropy is a function of state. The entropy changes which we calculated in section 4.3.4 are special cases of these formulas.

We saw in section 2.3 that the entropy is an extensive quantity. It is proportional to the size of the system. Of the terms on the right-hand side of Eq. (5.23b), the first term is proportional to the size (it is directly proportional to N), but the second term is not proportional to the size, since on doubling the size of the system both N and V are doubled. The integration constant C' which resulted in going from Eq. (5.22) to Eq. (5.23a) is independent of T and V but may depend on N. To make the entropy, as defined in Eq. (5.23b) extensive, C' must be of the form

$$C' = -Nk \ln N + NC \tag{5.24}$$

where C is a constant which does *not* depend on N. If we substitute Eq. (5.24) into Eq. (5.23b) it becomes

$$S(T, V, N) = N \left\{ c_V \ln T + k \ln \frac{V}{N} + C \right\}, \tag{5.23c}$$

which is a manifestly extensive quantity.

5.4.2 The entropy of mixing

Consider the mixing by diffusion of two different perfect gases, which do not interact with each other. Initially the two gases are at the same temperature T and the same pressure P. They occupy volumes V_1 and V_2 and are separated from each other by a partition AB, as shown in Fig. 5.9(a). The partition is removed, the gases mix and eventually equilibrium is established in which the molecules of each gas are uniformly distributed throughout the whole volume $V_1 + V_2$, as indicated in Fig. 5.9(b). This process is clearly irreversible: the two gases are not going to separate again spontaneously. The process is, of course, isothermal. This follows from the fact that the total internal energy of the two gases remains constant in this process and is independent of volume and pressure since we are dealing with two perfect non-interacting gases.

Because the two gases are of different species and non-interacting, the entropy of the mixture, occupying the volume $V_1 + V_2$, is equal to the sum of the entropies of each gas occupying the whole volume: each microstate of the one gas occupying the whole volume by itself can be combined with each microstate of the other gas occupying the whole volume by itself. Hence we can at once apply Eq. (5.23c) to the states before and after diffusion and

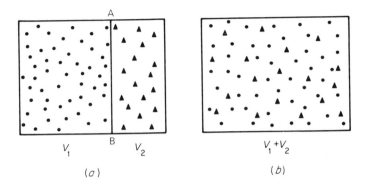

Fig. 5.9. The mixing of two different gases. (a) The two gases are in separate enclosures separated by the partition AB. (b) The two gases have reached equilibrium after removal of the partition AB.

thus find the entropy of mixing. Initially, before mixing, each gas is at pressure P, temperature T, and occupies a volume V_1 or V_2 respectively. Hence the number of molecules of each kind is given by

$$N_i = PV_i/kT, \qquad (i=1,2) \;, \tag{5.25}$$

where, as throughout, the index $i(=1,2)$ labels quantities referring to the two gases. It follows that

$$\frac{V_1}{N_1} = \frac{V_2}{N_2} = \frac{V}{N} \tag{5.26}$$

where

$$V = V_1 + V_2, \qquad N = N_1 + N_2 \;. \tag{5.27}$$

(The fact that the gases do not interact also leads to Dalton's law of partial pressures: after mixing, each gas exerts a partial pressure P_i as though it occupies the whole space alone: $P_i = N_i kT/V = N_i P$.) From Eq. (5.23c) we obtain for the entropy before mixing

$$S = \sum_{i=1}^{2} N_i \left\{ c_V^{(i)} \ln T + k \ln \frac{V_i}{N_i} + C_i \right\} \;. \tag{5.28}$$

After mixing, each gas occupies the whole volume, so the entropy of the mixture is

$$S' = \sum_{i=1}^{2} N_i \left\{ c_V^{(i)} \ln T + k \ln \frac{V}{N_i} + C_i \right\} \;. \tag{5.29}$$

Hence the entropy of mixing is given by

$$\Delta S = S' - S = \sum_{i=1}^{2} N_i k \ln \frac{V}{V_i} = \sum_{i=1}^{2} N_i k \ln \frac{N}{N_i} . \qquad (5.30)$$

As expected ΔS is positive: the state when the two gases are separated is a more highly ordered state than when they are mixed.

We can verify Eq. (5.29) by means of the idealized process shown in Fig. 5.10. A cylinder contains four pistons A, B, C and D. Pistons A and B are fixed, while C and D can move being fastened together by a rod R so that the distance CD remains constant and equal to the distance AB. The whole system is thermally isolated and the space above the piston D is evacuated. Pistons B and C are made of semipermeable materials, permeable only to gases 1 and 2 respectively. Materials which approximate such semipermeable membranes exist. For example, hot palladium foil is permeable to hot hydrogen but not to air. The pistons A and D are opaque to both gases. Initially the piston D is in contact with B, and C with A. The space AB is filled with a mixture of gases 1 and 2. The coupled pistons CD are now slowly withdrawn. As a result gas 1 will flow into the space between B and D, and gas 2 into the space between A and C. When the pistons have been withdrawn till the pistons C and B touch, the two gases have been separated, each occupying the same volume as it did originally. During the process no heat flows into the system and no work is done since there is a net downwards pressure on piston C equal to the upwards pressure on piston D. (Remember that as far as gas 2 is concerned, piston C might as well not be there.) Hence the energy and temperature of the gases is not changed by the separation. Since the process is reversible it is also isentropic, i.e. the entropy of the mixture in volume V equals the sum of the entropies of each gas occupying the same volume V by itself.

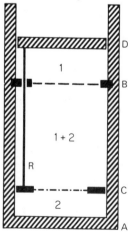

Fig. 5.10. The reversible separation of a mixture of gases by means of semipermeable membranes. The movable pistons C and D are held a fixed distance apart by means of the rod R. With the device shown, no work is done in this process.

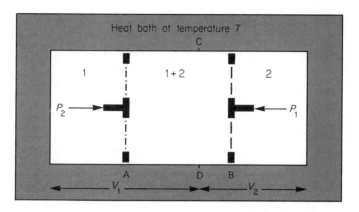

Fig. 5.11. The isothermal reversible mixing of gases by means of semipermeable membranes. With the device shown, work is done in this process.

We can similarly obtain the entropy of mixing, Eq. (5.30), directly by means of the idealized reversible process shown in Fig. 5.11. An enclosure of volume V is immersed in a heat bath at temperature T. The two movable pistons A and B are semipermeable to gases 1 and 2 respectively. Initially they are touching at CD; correspondingly, gas 1 occupies a volume V_1, gas 2 a volume V_2. The two pistons are now slowly withdrawn until they are at the extremities of the enclosure and the gases are mixed in the common volume $V(= V_1 + V_2)$. Throughout the process, appropriately adjusted pressures P_1 and P_2, equal to the partial pressures of gases 1 and 2 in the common region AB, are applied to the pistons B and A from the outside to keep them in equilibrium. The work done by the gases during this reversible isothermal mixing is given by

$$W = \int_{V_1}^{V} P_1 \, dV + \int_{V_2}^{V} P_2 \, dV = \sum_{i=1}^{2} N_i k T \ln \frac{V}{V_i} \ .$$

Since the internal energy of the gases remains constant in this isothermal process, an amount of heat $Q = W$ must flow into the enclosure from the heat bath at temperature T during the process, i.e. the entropy of the gases has increased by Q/T, in agreement with Eq. (5.30).

Suppose now that the two gases 1 and 2 are identical (so that $C_1 = C_2$, $c_V^{(1)} = c_V^{(2)}$). We again consider the situation shown in Figs 5.9(a) and (b), i.e. the states before and after removal of the partition AB, except that now there is only one kind of molecule. The entropy after removal of the partition is now

$$S' = N \left\{ c_V^{(1)} \ln T + k \ln \frac{V}{N} + C_1 \right\} , \tag{5.31}$$

while the entropy with the partition in place is still given by Eq. (5.28). This last expression and Eq. (5.31) have the same value on account of Eq. (5.26): the entropy of mixing $\Delta S (\equiv S' - S)$ is zero now, as expected. Removing and then replacing the partition does not alter the macroscopic behaviour of the gas; it is a reversible process. The total number of accessible microstates is the same whether the partition is in place or not. Note that there is no sense in which we can *gradually* let the properties of two kinds of molecules tend to each other and hence the entropy difference (5.30) tend to zero. Eq. (5.30) depends only on the numbers N_1 and N_2 and not on any such property. *The molecules either are identical or they are different.* In the latter case *their difference in properties can be used to actually effect a separation* by means of semipermeable membranes, as described above, and similar devices.

Historically, failure to appreciate the significance of the identity of the particles in a system led to certain inconsistencies, known as the Gibbs paradox. We shall not discuss this paradox since a careful reader of this book should not be troubled by it.

★ 5.5 SOME PROPERTIES OF REAL GASES

5.5.1 The Joule effect

In section 1.3 we considered the Joule effect, i.e. the free expansion of a gas into a vacuum, the whole system being thermally isolated. We shall now calculate the temperature change produced in a free expansion using thermodynamics. Although this is an irreversible process (as one can confirm by calculating the entropy change for it; cf. the end of section 4.3) it can be treated by thermodynamics as the initial and final states are equilibrium states. During the Joule expansion the energy of the gas is conserved, so that $dE(T, V) = 0$. This condition can be written

$$\left(\frac{\partial E}{\partial T} \right)_V dT + \left(\frac{\partial E}{\partial V} \right)_T dV = 0 \tag{5.32}$$

whence

$$\alpha_J \equiv \left(\frac{\partial T}{\partial V} \right)_E = - \frac{(\partial E / \partial V)_T}{(\partial E / \partial T)_V} \tag{5.33a}$$

$$= - \frac{1}{C_V} \left[T \left(\frac{\partial P}{\partial T} \right)_V - P \right] \tag{5.33b}$$

where the last line follows from Eq. (1.18a), which defines the heat capacity at constant volume C_V, and from Eq. (5.15).

We see at once, from either of Eqs. (5.33), that for a perfect gas the Joule coefficient α_J vanishes. We can try and represent the equation of state of a real gas as a power series expansion in the density of the gas. It is usually written, for one mole of gas, in the form

$$PV = RT\left[1 + \frac{B_2}{V} + \frac{B_3}{V^2} + \cdots\right].\qquad(5.34)$$

B_i is called the ith virial coefficient and is a function of T: $B_i = B_i(T)$. At low densities one need only retain the second virial coefficient. At higher densities more terms must be included.

Fig. 5.12 shows the second virial coefficient for argon, but all gases show the same general behaviour. We see that at S.T.P. $B_2 \approx -22\ \mathrm{cm^3/mol}$, so that $B_2/V \approx -10^{-3}$.

A knowledge of the appropriate number of virial coefficients and of C_V allows one to find α_J from Eqs. (5.33b) and (5.34). If we deal with a gas at low density and retain B_2 only, we can easily find $(\partial P/\partial T)_V$ from Eq. (5.34), and substituting this value into Eq. (5.33b) we obtain for the Joule coefficient

$$\alpha_J \equiv \left(\frac{\partial T}{\partial V}\right)_E = -\frac{1}{C_V}\frac{RT^2}{V^2}\frac{dB_2}{dT}.\qquad(5.35)$$

(C_V is the molar heat capacity since we are referring to one mole of gas

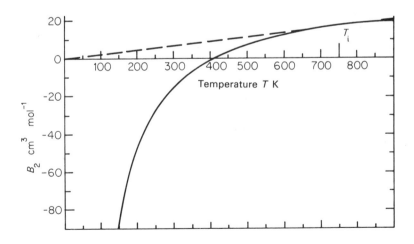

Fig. 5.12. The second virial coefficient $B_2(T)$ for argon. T_i = inversion temperature. (*American Institute of Physics Handbook*, 2nd edn., McGraw-Hill, New York, 1963, pp. 4–162.)

throughout.) At S.T.P. for argon we obtain, with $C_V = \frac{3}{2}R$ and $dB_2/dT \approx 0.25 \, \text{cm}^3 \, \text{K}^{-1} \, \text{mol}^{-1}$ from Fig. 5.12,

$$\alpha_J \approx -2.5 \times 10^{-5} \, \text{K mol cm}^{-3} \ . \tag{5.36}$$

To find the temperature change ΔT for a finite increase in volume from V_1 to V_2, we should integrate Eq. (5.35):

$$\Delta T \equiv T_2 - T_1 = \int_{V_1}^{V_2} \left(\frac{\partial T}{\partial V} \right)_E dV \ . \tag{5.37}$$

To estimate the order of magnitude, we shall treat α_J as constant. Doubling the volume ($V_2 - V_1 = 22.4 \times 10^3 \, \text{cm}^3/\text{mol}$) then gives $\Delta T \approx -0.6 \, °\text{C}$; a very small effect. Without relying on these approximate calculations, it follows from Eq. (5.33a) that the free expansion always produces a cooling. For $(\partial E/\partial V)_T$ depends only on the potential energy of the molecules. (The kinetic energy is constant at a given temperature.) The intermolecular potential is repulsive for very small intermolecular distances; it is attractive for large separations. The intermediate separation at which the potential has its minimum is the equilibrium separation in the solid. Hence in the gaseous phase the intermolecular force is necessarily attractive, so that $(\partial E/\partial V)_T > 0$ follows.

5.5.2   The Joule–Thomson effect

The Joule–Thomson (or Joule–Kelvin) effect is the change of temperature which occurs in the expansion of a gas through a throttle from high pressure P_1 to low pressure P_2. It is a continuous steady-state flow process, as indicated in Fig. 5.13, with gas entering at a constant rate from the left-hand side. The change in pressure leads to a change in temperature, from T_1 to T_2, which we shall calculate. Like the Joule effect, this is an irreversible process but again the initial and final states are equilibrium states, so we can apply thermodynamics. For this purpose, we consider the arrangement shown in Fig. 5.14 instead of Fig. 5.13. The two pistons, moving at the appropriate rates, ensure that the pressures on the two sides of the valve stay

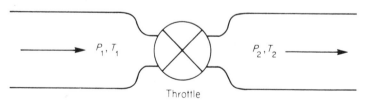

Fig. 5.13. The Joule–Thomson expansion through a throttle.

at the constant values P_1 and P_2. Let one mole of gas occupy a volume V_1 at pressure P_1 on the left-hand side of the valve, and a volume V_2 at pressure P_2 after expansion through the valve. The work done by the gas on the pistons in this process is

$$W = P_2 V_2 - P_1 V_1 \ . \tag{5.38}$$

Since the whole device shown in Fig. 5.14 is thermally isolated, no heat enters the system and the work W must have been performed at the expense of the internal energy of the gas, i.e. if E_1 and E_2 are the internal energy per mole on the left- and right-hand sides of the valve:

$$E_1 + P_1 V_1 = E_2 + P_2 V_2 \tag{5.39}$$

or in terms of the enthalpy $H \equiv E + PV$, Eq. (4.53),

$$H_1 = H_2 \ . \tag{5.40}$$

Thus the Joule–Thomson expansion is a process that conserves the enthalpy.* This is typical of steady-state continuous flow processes.

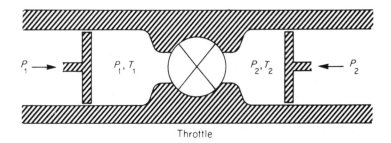

Throttle

Fig. 5.14. The Joule–Thomson expansion through a throttle. An idealized arrangement to allow easier analytical treatment. The two pistons ensure constant pressures on the two sides of the throttle.

The calculation of the temperature change in the Joule–Thomson effect is quite similar to that in the Joule effect, with H replacing E and P replacing V. Since the Joule–Thomson expansion is a process at constant enthalpy, we have, analogously to Eqs. (5.32) to (5.33b),

$$\left(\frac{\partial H}{\partial T}\right)_P dT + \left(\frac{\partial H}{\partial P}\right)_T dP = 0 \tag{5.41}$$

*We are assuming that the gas flow is slow enough for the kinetic energy of the flow to be neglected.

whence

$$\alpha_{JT} \equiv \left(\frac{\partial T}{\partial P}\right)_H = -\frac{(\partial H/\partial P)_T}{(\partial H/\partial T)_P} \;.\qquad(5.42a)$$

From the definition of the enthalpy H and of the heat capacity C_P, Eq. (1.18b), it follows that

$$\alpha_{JT} = \frac{-\left(\dfrac{\partial E}{\partial P}\right)_T - \left(\dfrac{\partial(PV)}{\partial P}\right)_T}{C_P} \;.\qquad(5.42b)$$

The quantity $(\partial E/\partial P)_T$ in this relation cannot be obtained directly, from measurements or from the equation of state, and a more convenient form for α_{JT} follows from the relation

$$\left(\frac{\partial H}{\partial P}\right)_T = V - T\left(\frac{\partial V}{\partial T}\right)_P \;.\qquad(5.43)$$

The derivation of this relation is analogous to that of Eq. (5.15) and is left as an exercise to the reader (see problem 5.5). Using Eq. (5.43), we obtain

$$\alpha_{JT} = \frac{1}{C_P}\left[T\left(\frac{\partial V}{\partial T}\right)_P - V\right] \;.\qquad(5.42c)$$

For an ideal gas α_{JT} vanishes. Eq. (5.42b) shows particularly clearly how the Joule–Thomson effect depends on deviations from ideal gas behaviour. The first term $[-(\partial E/\partial P)_T]$ on the right-hand side of Eq. (5.42b) depends on departures from Joule's law (1.19); the second term $[-(\partial(PV)/\partial P)_T]$ depends on deviations from Boyle's law: $PV = \text{const.}$ at constant temperature. Deviations from Joule's law always lead to a cooling of a real gas on expansion, as discussed in connection with the Joule effect (see the end of section 5.5.1). Deviations from Boyle's law can result in heating or cooling, depending on the temperature and pressure of the gas. Hence unlike the Joule coefficient α_J, the Joule–Thomson coefficient α_{JT} can have either sign. It will vanish when cooling due to deviations from Joule's law exactly cancels heating due to deviations from Boyle's law, i.e. when

$$\left(\frac{\partial H}{\partial P}\right)_T = \left(\frac{\partial E}{\partial P}\right)_T + \left(\frac{\partial(PV)}{\partial P}\right)_T$$

$$= V - T\left(\frac{\partial V}{\partial T}\right)_P = 0 \;.\qquad(5.44)$$

This equation defines a curve in the (T, P) plane. (Think of H as a function of T and P: $H = H(T, P)$, to see this most easily.) This curve is known as the *inversion curve*. Fig. 5.15 which is typical of all gases shows the inversion curve, drawn boldly. Also shown are a series of isenthalps, that is, curves of constant enthalpy: $H(T, P) = $ const. We see that inside the inversion curve T increases with P along an isenthalp, that is the Joule–Thomson coefficient $(\partial T/\partial P)_H$ is positive and the gas is cooled on expansion. Outside the inversion curve, on the other hand, $(\partial T/\partial P)_H$ is negative and the gas warms up on expansion. We see from Fig. 5.15 that for temperatures $T > T_i$ a Joule-Thomson expansion necessarily warms the gas. T_i is called the *inversion temperature*. Table 5.1 gives the inversion temperature of several

Table 5.1. Inversion temperature of some gases

gas	He	H_2	N_2	A	O_2
$T_i(K)$	23.6	195	621	723	893

gases. For most gases T_i is well above room temperature. Helium and hydrogen are exceptions and they must be precooled below their inversion temperature if they are to be liquefied by Joule–Thomson expansion. The precooling is frequently done with other liquids. Hydrogen, for example, is precooled with liquid nitrogen whose normal boiling point is 77.3 K.

To find the temperature change $\Delta T = T_2 - T_1$ produced in a finite pressure drop $\Delta P = P_2 - P_1 (< 0)$ we must integrate Eq. (5.42a):

$$\Delta T = T_2 - T_1 = \int_{P_1}^{P_2} \left(\frac{\partial T}{\partial P}\right)_H dP \ . \tag{5.45}$$

The temperature T_1 can be read off directly from an enthalpy diagram such as Fig. 5.15. Given the initial condition (T_1, P_1) of the gas (point A on Fig. 5.15) and the final pressure P_2 (this is usually atmospheric pressure), the final temperature T_2 lies on the same isenthalp as the point (T_1, P_1), i.e. the final state corresponds to the point B on the figure.

An approximate approach which gives the correct general features but is quantitatively not accurate is to use an approximate analytic equation of state for a real gas, such as van der Waals' or Dieterici's equations. (See Flowers and Mendoza,[26] section 7.4.4, where van der Waals' equation is used in this way. See also problem 5.9.)

For a gas at low density we may use the virial expansion (5.34), keeping terms up to B_2 only. To this order, Eq. (5.34) can be written

$$PV = RT + B_2(T)P \tag{5.46}$$

whence

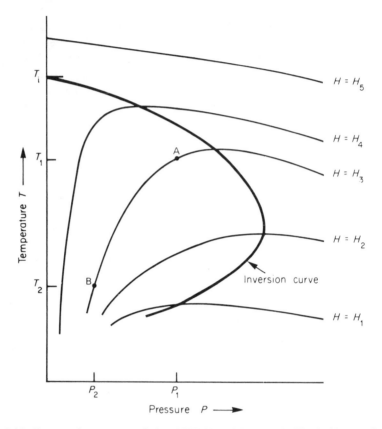

Fig. 5.15. Curves of constant enthalpy $H(T,P) = H_i(= \text{const.})$. The bold curve is the inversion curve. On it $(\partial T / \partial P)_H = 0$. Inside (outside) it, the gas is cooled (warmed) on expansion. T_i is the inversion temperature.

$$\alpha_{JT} = \frac{T \dfrac{dB_2(T)}{dT} - B_2(T)}{C_P} .$$
(5.47)

The vanishing of the numerator of the last expression defines the inversion temperature T_i in this approximation.* We see from Fig. 5.12 that it has a very direct geometrical interpretation, and that for $T < T_i$ and $T > T_i$ cooling and warming of the expanded gas results. For higher pressures further terms in the virial expansion would have to be retained.

---

*In this approximation there exists only one unique temperature, independent of pressure, for which α_{JT} vanishes. This is in contrast to the behaviour of a real gas, shown in Fig. 5.15. Since we are considering the limit of low pressure, this temperature is indeed the inversion temperature T_i as follows from Fig. 5.15.

The Joule–Thomson effect is also quite small. Typical values for α_{JT} are of the order of a few tenths of a degree Celsius per atmosphere. But since it is a steady-state process one can make the effect cumulative by using the cooled expanded gas in a heat exchanger to precool the incoming gas. (This problem will be considered further in section 5.5.3.) For this reason and because of its simplicity, the Joule–Thomson effect is of great practical importance in the liquefaction of gases and in studying the properties of matter.

5.5.3 The counter-current heat exchanger

The simplest counter-current heat exchanger, shown in Fig. 5.16, consists of two concentric circular metal tubes. Two streams of gas flow in opposite directions in the inner tube and in the annular region. If the two streams of gas enter the heat exchanger at different temperatures, heat transfer through the wall of the inner tube will occur between the two streams as they traverse the heat exchanger. As a result the warmer gas is cooled and the colder gas is warmed. In practice, heat exchangers are usually of more complicated design so as to achieve maximum heat transfer, small pressure drops, and so on.

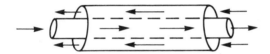

Fig. 5.16. Simple counter-current heat exchanger. The arrows mark the directions of flow of the two streams.

Fig 5.17 shows the flow diagram of a simple liquefier of the type first used by Linde (1895) for air liquefaction. It consists of a compressor, a heat exchanger and an expansion valve V. Gas from the compressor enters the heat exchanger at A and leaves it at B having been cooled by the colder gas flowing from C to D. The gas from B is expanded through the throttle V. Provided the inlet temperature T_i' at A is below the inversion temperature T_i of the gas some cooling will occur. When starting the liquefier up this cooling will not suffice to cause liquefaction. But the expanded cooled gas precools the 'next lot' of high-pressure gas in the heat exchanger. In this way the temperature at B is continually lowered until liquefaction occurs as the gas expands through the valve V. A steady state is reached in which a fraction ε of the incident gas is liquefied, collecting in the container shown, and the remaining fraction $(1 - \varepsilon)$ of gas returns up the heat exchanger CD.

In this system, no work is done by the system on the surrounding and no heat is exchanged with it. (We assume the system thermally isolated.) Hence

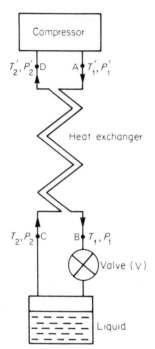

Fig. 5.17 Flow diagram of a simple Linde liquefier.

in the steady state enthalpy must be conserved: the enthalpy of the inflowing high pressure gas must equal the sum of the enthalpies of the outflowing low pressure gas and the liquid produced. (See Eqs. (5.39) and (5.40), derived for a simpler but similar flow process.) We therefore require

$$H_A = (1 - \varepsilon)H_D + \varepsilon H_L \ ,$$

whence the liquefier efficiency ε becomes

$$\varepsilon = \frac{H_D - H_A}{H_D - H_L} \tag{5.48}$$

where $H_A(T_1', P_1')$, $H_D(T_2', P_2')$ and $H_L(T_2, P_2)$ are the molar enthalpies of the gas at A, of the gas at D, and of the liquid respectively. We see from Fig. 5.15 that, for a given T_1', P_1' should be chosen so that $H_A(T_1', P_1')$ lies on the inversion curve. This minimizes H_A and hence maximizes the liquefier efficiency ε. P_2' is usually atmospheric pressure and, with small pressure drops in the exchanger, T_2 is approximately the normal boiling point of the liquid. Using enthalpy tables, the efficiency ε is easily found for any given operating conditions.

★ 5.6   ADIABATIC COOLING

The production of low temperatures represents an interesting example of thermodynamic reasoning. We know from the third law (section 4.7) that the entropy of a system approaches zero as the temperature goes to zero. In other words, a cooling process 'squeezes' the entropy out of a system.

Adiabatic cooling processes depend on the fact that the entropy S of a system depends on its temperature and on other parameters, such as volume or applied magnetic field, etc. We shall label such parameters α. During a reversible adiabatic process the entropy is constant

$$S(T, \alpha) = \text{const.}$$

Hence if in such a process we alter α in the right sense a compensating temperature drop will occur in order to keep S constant. The statistical interpretation of this is as follows. The order of a system, measured by its entropy, represents a compromise between two opposing tendencies: the disrupting effects of thermal motion and the ordering effects of such things as applied magnetic fields. If such ordering influences are reduced the

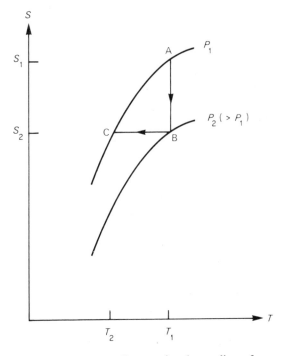

Fig. 5.18. The entropy–temperature diagram for the cooling of a gas by adiabatic expansion.

tendency of the system is to become more disordered, i.e. for the entropy to increase. If experimentally one can keep the order, i.e. the entropy, constant then the temperature must drop to compensate for this reduction in the ordering influences.

The adiabatic expansion of a gas is a simple example of adiabatic cooling. The two stages of the process are shown in the entropy–temperature diagram, Fig. 5.18. For the parameter which was called α in the general discussion we now take the pressure of the gas. During the first stage the gas is in thermal contact with a heat bath at temperature T_1 and is isothermally compressed from pressure P_1 to P_2 (line AB in Fig. 5.18). The heat bath is now removed and the gas thermally isolated. On reversible adiabatic expansion to the lower pressure P_1 the gas cools to the temperature T_2 (line BC in Fig. 5.18). This cooling by adiabatic expansion is the basis of the Claude process for liquefying gases (see, for example, White[36] or Hoare et al.,[31] Chapter 3).

Adiabatic demagnetization is a very powerful method of achieving low temperatures. The disorder (i.e. the magnetic entropy) associated with the alignment of the magnetic dipoles of a paramagnetic salt depends on the temperature and on the magnetic field. If the magnetic field $\mathscr{B}$ is reduced, while keeping the magnetic entropy $S(T,\mathscr{B})$ constant, then the temperature must fall in order to retain the same degree of magnetic order. To obtain substantial temperature changes one must of course work in a region where the entropy depends strongly on the magnetic field. We know from section 3.1 that this requires low temperatures, of the order of 1 K, and high fields, of the order of one tesla (10^4 gauss).

Adiabatic demagnetization is a two-stage process, in principle very similar to adiabatic expansion. During the first stage the paramagnetic sample, in contact with a heat bath at about 1 K, is isothermally magnetized by switching on an applied magnetic field $\mathscr{B}_1$. During this stage there is heat transfer from the paramagnetic specimen to the heat bath. The specimen is now thermally isolated and the applied magnetic field is switched off adiabatically and reversibly. In order to be reversible, this switching off must be done slowly. In practice it is sufficient to take a few seconds over this stage, unless one is cooling another body in contact with the magnetic salt when one must allow for the time required for the heat flow between the body and the salt.

Fig. 5.19 indicates the experimental arrangement. The paramagnetic specimen P is suspended inside a cryostat cooled by liquid helium boiling under reduced pressure. During the first, isothermal stage the space G contains helium gas to allow heat exchange between specimen and liquid helium. During the second stage the space G is evacuated to provide adiabatic conditions.

Fig. 5.20 shows qualitatively how the magnetic entropy of N 'spin $\frac{1}{2}$' magnetic dipoles depends on the temperature and on the magnetic field. (In section 3.2 we derived the entropy for this case, Eq. (3.15), but we do not

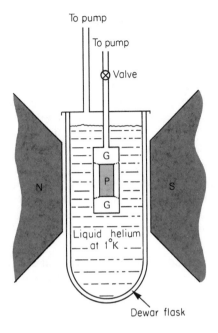

Fig. 5.19. Schematic diagram of apparatus for cooling
by adiabatic demagnetization. P is the paramagnetic
salt. The space G contains He gas or is evacuated. N
and S are the poles of the magnet.

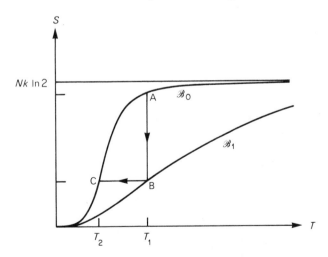

Fig. 5.20. The entropy–temperature curves for a paramagnetic salt, for different
applied magnetic fields.

require the detailed expression now.) At sufficiently high temperatures the entropy tends to $Nk \ln 2$, corresponding to random orientations of the dipoles (see section 3.2). As $T \to 0$, the entropy tends to zero.

The field which determines the alignment of the dipoles, and hence the magnetic entropy, is the *local* magnetic field, as we discussed in section 3.1. Even when the *external applied* field is zero, there is an effective *residual* field $\mathscr{B}_0$ due to various causes such as the interactions between the dipoles. This residual field is typically of the order of 100 gauss (10^{-2} tesla) and is very small compared with the applied field $\mathscr{B}_1$ used to align the dipoles. Fig. 5.20 shows the entropy as a function of the temperature for zero applied field (i.e. only the residual field $\mathscr{B}_0$ acts) and for the applied field $\mathscr{B}_1$. The path ABC represents the whole process: AB represents the isothermal magnetization at temperature T_1; BC represents the adiabatic reversible demagnetization (i.e. the entropy remains constant) leading to the lower temperature T_2.

To estimate the magnitude of this effect, we shall describe the paramagnetic salt by the model of section 3.1. From Eqs. (3.1) and (3.5b) we obtain for the magnetic entropy

$$
\begin{aligned}
S &= (E - F)/T \\
&= (E + NkT \ln Z_1)/T \\
&= Nk\{\ln (2 \cosh x) - x \tanh x\}, \quad \left(x \equiv \frac{\mu \mathscr{B}}{kT} \right) .
\end{aligned}
\tag{5.49}
$$

We see that the entropy is a function of $\mathscr{B}/T$ only. Hence this ratio remains constant during the adiabatic reversible demagnetization, and we obtain for the final temperature

$$
T_2 = T_1 \frac{\mathscr{B}_0}{\mathscr{B}_1} .
\tag{5.50}
$$

Thus with $T_1 \approx 1$ K and $\mathscr{B}_0/\mathscr{B}_1 \approx 1/100$, one reaches a final temperature $T_2 \approx 10^{-2}$ K.

So far we have only discussed the magnetic aspect of this problem. In addition there are the lattice vibrations which have their own entropy associated with them. During the adiabatic demagnetization it is only the *total* entropy which is conserved. So the magnetic entropy could increase at the expense of the vibrational entropy. However the vibrational entropy of course also goes to zero as the temperature tends to zero, and at $T_1 \approx 1$ K is already completely negligible. Hence during the adiabatic demagnetization the magnetic entropy by itself is essentially constant.

The statistical analysis of this process is also very interesting. Fig. 5.21 shows the energy levels and their populations (given by Eq. (3.3)) for the three states of the system labelled A, B and C in Fig. 5.20.

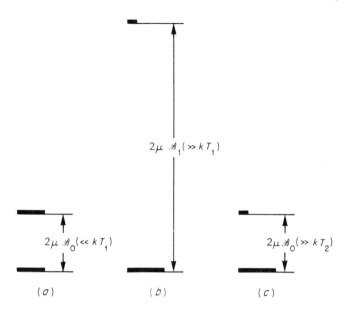

Fig. 5.21. The energy levels and their populations (shown solid black) in the adiabatic demagnetization of a 'spin $\frac{1}{2}$' system. The situations (a), (b) and (c) correspond to points A, B and C on the S–T diagram, Fig. 5.20.

Fig. 5.21(a) shows state A. For the weak residual field $\mathscr{B}_0$ one has $\mu\mathscr{B}_0 \ll kT_1$. One sees from the Boltzmann factors $\exp(\pm\mu\mathscr{B}_0/kT_1)$ of the two levels that their populations are equal.

Fig. 5.21(b) shows state B after the isothermal magnetization. The level separation $2\mu\mathscr{B}_1$ is very much greater than in state A. The Boltzmann factors $\exp(\pm\mu\mathscr{B}_1/kT_1)$ and hence the populations of the two levels are now very different since $\mu\mathscr{B}_1 \gg kT_1$.

Fig. 5.21(c) shows the state C after adiabatic demagnetization. The level spacing is as in Fig. 5.21(a). But the populations of the levels are the same in states B and C, since these states have the same degree of order. (Mathematically expressed: the populations are, from Eq. (3.3), given by $e^{\pm x}/(2\cosh x)$, with $x \equiv \mu\mathscr{B}/kT$, and we have seen that $\mathscr{B}/T$ remains constant during the change BC.) But the same degree of order *without* applied magnetic field must mean a lower temperature, i.e. $\mu\mathscr{B}_0 \gg kT_2$.

We mention in passing that the process of adiabatic demagnetization can also be applied to the magnetic moments of nuclei. In this way temperatures of the order of 10^{-6} K have been obtained (see Kittel,[29] pp. 412–413, and McClintock *et al.*,[34] section 7.4).

The general question arises whether it is possible to *reach* the absolute zero of temperature by means of a succession of adiabatic cooling processes. The

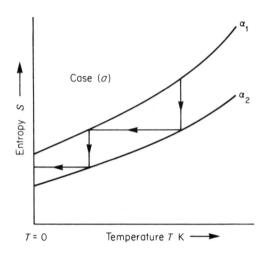

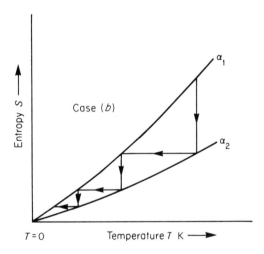

Fig. 5.22. The entropy $S(T, \alpha)$ versus temperature for different values of α. Case (a): 0 K can be reached in a finite number of steps. Case (b): 0 K cannot be reached in a finite number of steps. Case (a) *violates* the third law and so cannot occur.

answer is provided by the third law: it is *not* possible to reach the absolute zero of temperature by means of a *finite* number of adiabatic cooling processes. To see this we consider the two situations shown in Figs. 5.22(*a*) and (*b*). These show the entropy $S(T, \alpha)$ plotted against T for different values of the variable α. If the entropy curves are as shown in (*a*) it is possible to reach 0 K by means of a finite number of steps operating between the values α_1 and α_2 of the variable α; if the entropy curves are as shown in (*b*), $T = 0$ K cannot be reached in a finite number of such steps. Now according to the third law (section 4.7).

$$\lim_{T \to 0} S(T, \alpha) = 0,$$

for *all* values of α. In other words, the entropy curves must be as shown in Fig. 5.22(*b*), and not as in Fig. 5.22(*a*), and it is therefore not possible to reach the absolute zero in a finite number of steps. This statement of the unattainability of the absolute zero is therefore seen to be an alternative statement of the third law of thermodynamics.

PROBLEMS 5

5.1 The efficiency of a heat engine operating between temperatures T_1 and $T_2 (< T_1)$ is η'. Show that the inequality

$$\eta' > \frac{T_1 - T_2}{T_1}$$

would imply a violation of the second law of thermodynamics.

5.2 A Carnot engine using a perfect gas as working substance operates between the temperatures Θ_1 and $\Theta_2 (< \Theta_1)$, measured on the ideal gas temperature scale. Show that the efficiency of the engine is $(\Theta_1 - \Theta_2)/\Theta_1$.

5.3 Derive the maximum efficiency of the standard Otto cycle, Eq. (5.9).

5.4 C_P and C_V are the atomic heat capacities of lead at constant pressure and constant volume. Find $(C_P - C_V)/3R$ at 300 K given the following data: atomic volume = 18.27 cm³/mol, linear coefficient of thermal expansion = 2.89×10^{-5} K⁻¹, isothermal compressibility = 2.237×10^{-11} m²/N.

5.5 Derive Eq. (5.43).

5.6 Consider the adiabatic Joule–Thomson expansion of one mole of a perfect gas from pressure P_1 to pressure P_2. Obtain the change of entropy of the gas for this expansion.

5.7 Near STP argon may be described by the virial expansion

$$PV = RT + B(T)P,$$

with $B = -22$ cm³/mol and $dB/dT = 0.25$ cm³ K⁻¹ mol⁻¹. Obtain an approximate value of the Joule–Thomson coefficient of argon at STP.

5.8 In the Joule–Thomson expansion of a gas from temperature T_1 and pressure P_1 to temperature T_2 and pressure P_2, T_1 and P_2 are in practice essentially fixed:

T_1 by the liquid used to precool the gas before expansion, while P_2 is atmospheric pressure. What initial pressure P_1 leads to the lowest temperature T_2 of the expanded gas?

5.9 For a gas obeying van der Waals' equation

$$P = \frac{RT}{V-b} - \frac{a}{V^2} :$$

(i) obtain the isobaric coefficient of thermal expansion α;

(ii) show that the energy E satisfies

$$\left(\frac{\partial E}{\partial V}\right)_T = \frac{a}{V^2} ;$$

(iii) show that the molar heat capacity at constant volume is a function of the temperature only;

(iv) obtain the difference of the molar heat capacities at constant pressure and at constant volume: $(C_P - C_V)$;

(v) obtain the change in entropy dS of the gas for infinitesimal changes in volume dV and in temperature dT;

(vi) obtain the Joule–Thomson coefficient $(\partial T/\partial P)_H$;

(vii) obtain the equation of the inversion curve as a function of volume and temperature.

5.10 A paramagnetic salt whose magnetic ions have spin $\frac{1}{2}$ is at a temperature of 0.6 K in a magnetic field of 4 tesla. The field is slowly reduced to 6×10^{-3} tesla, the salt being thermally isolated during the process. What is the final temperature of the salt?

*5.11 The Gibbs free energy for a body in an applied uniform magnetic intensity $\mathscr{H}$ is defined by $G = E - TS + \mu_0 \mathscr{H} \mathscr{M}$, where $\mathscr{M}$ is the component of the magnetic moment $\mathscr{M}$ of the body in the direction of $\mathscr{H}$. Show that $dG = -S dT + \mu_0 \mathscr{H} d\mathscr{M}$, and hence derive the Maxwell relation

$$(\partial S/\partial \mathscr{M})_T = -\mu_0 (\partial \mathscr{H}/\partial T)_{\mathscr{M}} .$$

For a sample of an ideal paramagnetic material the magnetic moment $\mathscr{M}$ depends on the applied uniform magnetic intensity $\mathscr{H}$ and the temperature T through the ratio $\mathscr{H}/T$ only, i.e. $\mathscr{M} = f(\mathscr{H}/T)$. Show that for this sample the magnetic contribution to the heat capacity at constant magnetization vanishes.

---

*This problem presupposes knowledge of section 1.4 on magnetic work.

CHAPTER

6

The heat capacity of solids

6.1 INTRODUCTORY REMARKS

In this chapter we shall apply statistical methods to calculate the heat capacity of solids or, more precisely, the heat capacity associated with the lattice vibrations of crystalline solids. In general there are other contributions to the heat capacities of solids. In paramagnetic salts there is a contribution from the orientational ordering (Chapter 3). In metals there is a contribution from the conduction electrons (Chapter 11). These different aspects may be treated separately since they only interact quite weakly with each other. We shall only consider the vibrational heat capacity in this chapter.

The two basic experimental facts about the heat capacity of solids which any theory must explain are:

(1) Near room temperature the heat capacity of most solids is close to $3k$ per atom, so that for molecules consisting of n atoms the molar heat capacity is $3nR$ where R is the gas constant, Eq. (1.6). This is essentially the law of Dulong and Petit (1819). A few values for molar heat capacities at constant pressure (1 atmosphere) and 298 K are shown in Table 6.1. These values should be compared with $3R = 24.9 \, \text{J mol}^{-1} \, \text{K}^{-1}$. Dulong and Petit's law only holds approximately and can be quite wrong.

(2) At low temperatures the heat capacities decrease and vanish at $T = 0$. Typical experimental results are shown in Figs. 4.6, 6.1 and 6.2. The latter two figures show the low-temperature behaviour, up to about 4 K, in detail.

Table 6.1. Molar heat capacity C_P (J mol^{-1} K^{-1}) at 298 K

Element	C_P	Element	C_P
Al	24.4	S	22.7
Au	25.4	Si	19.9
Cu	24.5	C (diamond)	6.1

They are plots of C_V/T versus T^2, and show that the heat capacities are of the form

$$C_V = \alpha T^3 + \gamma T$$

where α and γ are constants. For potassium chloride, (Fig. 6.1), we have $\gamma = 0$ and a pure T^3-behaviour of the heat capacity which is typical of insulators. For copper the heat capacity also contains a term linear in T (corresponding to the intercept of the straight line at $T=0$ in Fig. 6.2: $\gamma \neq 0$). This linear term is a contribution to the heat capacity from the conduction electrons. The T^3 term is in both cases due to the lattice vibrations, as we shall see in section 6.3.

In classical statistical mechanics, the equipartition theorem which we shall discuss in section 7.9.1 (see also Flowers and Mendoza,[26] section 5.3) leads to the constant atomic heat capacity $3R$ *at all temperatures*, in violation of experiments and of the third law.

The fundamental step to resolve this discrepancy between classical theory and experiment was taken by Einstein, in 1907, who treated the lattice vibrations according to quantum theory. Einstein's theory, which will be considered in section 6.2, qualitatively reproduces the observed features. Einstein treated a simplified model. He did not expect detailed agreement with experiment and pointed out the kind of modifications which the model requires. The improvements of Einstein's theory will be discussed in section 6.3 where Debye's theory of heat capacities will be developed.

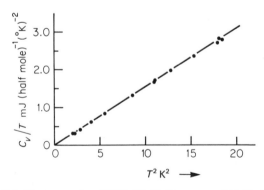

Fig. 6.1. Heat capacity of KCl. (P. H. Keesom and N. Perlman, *Phys. Rev.*, **91**, 1354 (1953).)

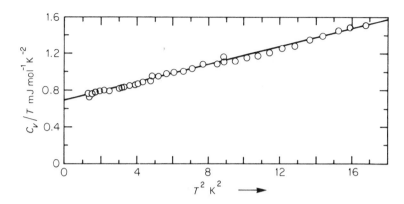

Fig. 6.2. Atomic heat capacity of copper. (W. S. Corak, M. P. Garfunkel, C. B. Satterthwaite and Aaron Wexler, *Phys. Rev.*, **98**, 1699 (1955).)

6.2   EINSTEIN'S THEORY

6.2.1   Derivation of Einstein's result

In this subsection we shall derive Einstein's expression for the heat capacity of a solid on the basis of his model. In this model the atoms in a crystal are treated as vibrating independently of each other about fixed lattice sites. The vibrations are assumed simple harmonic, all with the same frequency $v_E = \omega_E/2\pi$, where ω_E is the circular frequency. The equation of motion of one of the atoms,

$$\ddot{\mathbf{r}} = -\omega_E^2 \mathbf{r}$$

(where $\mathbf{r}$ is the displacement of the atom from its lattice site), decomposes into three mutually independent equations, of the same form, for the x-, y- and z-components of $\mathbf{r}$. Hence a solid consisting of N atoms is equivalent to $3N$ harmonic oscillators, vibrating independently of each other, all with the same circular frequency ω_E. The value of this frequency ω_E depends on the strength of the restoring force acting on the atom. We shall consider values for ω_E below.

This is Einstein's model. It is, of course, a great oversimplification. A crystal does not consist of atoms vibrating totally independently of each other about fixed lattice sites. Rather, there is strong coupling between the atoms. The restoring force on each atom, i.e. the potential in which it vibrates, is provided by the surrounding atoms.* A macroscopic mechanical analogue

*This is so for molecular crystals such as solid argon, or covalent ones such as diamond. In some crystals, such as sodium chloride (NaCl), we must consider the ions, rather than atoms. In metals, the picture is more complicated since the conduction electrons play an essential role in producing the crystal. For a discussion of different types of crystals see, for example, Blakemore,[24] Flowers and Mendoza,[26] Hall,[27] or Kittel.[29]

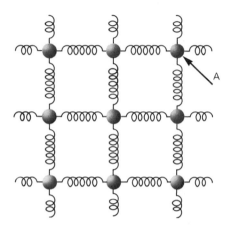

Fig. 6.3. Two-dimensional analogue of a crystal lattice,
consisting of billiard balls connected by springs.

of a crystal lattice would consist of billiard balls tied together with identical springs. Fig. 6.3 shows a two-dimensional picture of this. If you set one ball vibrating, say the one labelled A in Fig. 6.3, a disturbance will be propagated through the whole system till all the balls vibrate. So the lattice vibrations of a crystal are really very complicated coupled oscillations of all the atoms. These aspects are neglected in Einstein's model.

Einstein treats the $3N$ independent harmonic oscillators of circular frequency ω_E using quantum theory. According to quantum mechanics a single harmonic oscillator of circular frequency ω_E can only exist in a discrete set of states which we shall label $0, 1, \ldots r, \ldots$. In the state r the energy of the oscillator is

$$\varepsilon_r = \hbar\omega_E(r + \tfrac{1}{2}), \qquad r = 0, 1, \ldots . \qquad (6.1)$$

(Note that we are labelling the ground state by $r = 0$, not by $r = 1$ as so far. This is in order to obtain the usual form of Eq. (6.1).) In Eq. (6.1) $\hbar$, Planck's constant h divided by 2π, has the value

$$\hbar \equiv \frac{h}{2\pi} = 1.055 \times 10^{-34} \text{ Js} . \qquad (6.2)$$

Planck introduced the constant h in 1900 to explain the observed spectral distribution of black-body radiation. We shall usually employ $\hbar$. The levels (6.1) are non-degenerate. In the nomenclature of Eq. (2.24), we have $g(E_r) = 1$ for all r. The first four levels of the level scheme (6.1) are shown in Fig. 6.4. The ground state ($r = 0$), contrary to one's expectation from classical physics, possesses a 'zero-point' energy $\varepsilon_0 = \tfrac{1}{2}\hbar\omega_E$. This is a

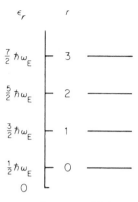

Fig. 6.4. The first four energy levels $\varepsilon_r = \hbar\omega_E(r+\frac{1}{2})$, $r=0,1,\ldots 3$, of harmonic oscillator of circular frequency ω_E.

consequence of Heisenberg's uncertainty principle, according to which we cannot know simultaneously the position and velocity of a particle completely. In the present context this means that we cannot ensure that an atom is at rest at its lattice site where its potential energy is a minimum. Since we are assuming that the oscillators do not interact with each other (except, as always, for a very weak interaction which allows thermal equilibrium to establish itself) we can consider any one atom in a solid at temperature T as in a heat bath at temperature T.

This completes the discussion of the underlying physics. Using the mathematics of section 2.5, the results follow easily. From Eqs. (2.23) and (6.1) we obtain for the partition function of one oscillator

$$Z_1 = \sum_{r=0}^{\infty} \exp(-\beta\varepsilon_r)$$

$$= \sum_{r=0}^{\infty} \exp\left[-\beta\hbar\omega_E(r+\tfrac{1}{2})\right]$$

$$= e^{-x/2} \sum_{r=0}^{\infty} e^{-xr} \qquad (6.3)$$

where

$$x \equiv \beta\hbar\omega_E \equiv \frac{\hbar\omega_E}{kT} . \qquad (6.4)$$

Since

$$(1-t)^{-1} = 1 + t + t^2 + \ldots \qquad \text{for } |t| < 1 ,$$

Eq. (6.3) gives immediately

$$Z_1 = \frac{e^{-x/2}}{1 - e^{-x}} , \qquad (6.5)$$

so that the Helmholtz free energy, Eq. (2.36), becomes

$$F_1 = -kT\ln Z_1 = \frac{1}{2}\hbar\omega_E + \frac{1}{\beta}\ln[1 - \exp(-\beta\hbar\omega_E)] . \qquad (6.6)$$

From Eqs. (2.26) and (6.6) we obtain for the mean energy $\bar{\varepsilon}$ per oscillator

$$\bar{\varepsilon} = -\frac{\partial \ln Z_1}{\partial \beta}$$

$$= -\frac{\partial}{\partial \beta}(-\beta F_1)$$

$$= \frac{\partial}{\partial \beta}\{\tfrac{1}{2}\beta\hbar\omega_E + \ln[1 - \exp(-\beta\hbar\omega_E)]\}$$

$$= \frac{1}{2}\hbar\omega_E + \frac{\hbar\omega_E}{\exp(\beta\hbar\omega_E) - 1}. \tag{6.7}$$

The first term, $\tfrac{1}{2}\hbar\omega_E$, is the zero-point energy of the oscillator which it possesses even at the absolute zero of temperature. Since the energy is an extensive quantity (see section 2.3), we obtain at once the energy of the $3N$ oscillators, i.e. of the solid consisting of N atoms,

$$E = 3N\bar{\varepsilon} \tag{6.8}$$

and for the heat capacity at constant volume, using Eq. (6.7),

$$C_V \equiv C_V(T) = \left(\frac{\partial E}{\partial T}\right)_V = 3N\left(\frac{\partial \bar{\varepsilon}}{\partial \beta}\right)_V \frac{d\beta}{dT}$$

$$= 3Nk\frac{x^2 e^x}{(e^x - 1)^2}, \tag{6.9}$$

where the parameter x, Eq. (6.4), can also be written in the form

$$x \equiv \frac{\hbar\omega_E}{kT} \equiv \frac{\Theta_E}{T}. \tag{6.10}$$

Θ_E is known as the *Einstein temperature* and ω_E as the *Einstein* (circular) *frequency* of the solid. The reader may wonder why Eq. (6.9) is the heat capacity at constant volume. Where has V been kept constant in the differentiation in Eq. (6.9)? The answer is that ω_E depends on V, since it depends on the restoring force which acts on the vibrating atom, and this force depends on the mean spacing between atoms, i.e. on the lattice spacing.

Eq. (6.9) is the basic result of Einstein's analysis. It depends on a single parameter, the Einstein temperature Θ_E, which can be chosen differently for each solid.

Since C_V, Eq. (6.9), depends on the temperature through $x = \Theta_E/T$ only, we obtain two limiting regimes according as $T \gg \Theta_E$ or $T \ll \Theta_E$.

(i) *High-temperature limit.* For $T \gg \Theta_E$, i.e. $x \ll 1$, Eq. (6.9) reduces directly to*

$$C_V = 3Nk, \qquad (T \gg \Theta_E) \; . \qquad\qquad (6.11)$$

At sufficiently high temperatures Einstein's result reduces to Dulong and Petit's law.

(ii) *Low-temperature limit.* For $T \ll \Theta_E$, i.e. $x \gg 1$, we have $e^x \gg 1$ so that Eq. (6.9) goes over into

$$C_V = 3Nk \left(\frac{\Theta_E}{T}\right)^2 \exp\left(-\frac{\Theta_E}{T}\right), \qquad (T \ll \Theta_E) \; . \qquad (6.12)$$

The specific heat goes to zero as T decreases, as required by the third law of thermodynamics (see section 4.7).

The high-temperature limit agrees with the classical equipartition theorem. This theorem will be derived in section 7.9.1 (see also Flowers and Mendoza,[26] section 5.3) but we shall quote it here in order to be able to discuss the present problem. A three-dimensional oscillator has three degrees of freedom (i.e. we need three independent coordinates to specify the position of a particle in space, for example its Cartesian coordinates). For N oscillators we can write the total energy as

$$E = \sum_{i=1}^{N} \left[\frac{1}{2m} (p_{xi}^2 + p_{yi}^2 + p_{zi}^2) + \tfrac{1}{2} m\omega_E^2 (x_i^2 + y_i^2 + z_i^2) \right]$$

where $x_i, y_i, z_i, p_{xi}, p_{yi}, p_{zi}$ are the position and momentum coordinates of the ith oscillator. According to the equipartition theorem each squared term in the energy E contributes an energy $\tfrac{1}{2}kT$ to the mean energy of a system in a heat bath at temperature T. A 'vibrational' degree of freedom contributes two squared terms to the energy E, one each from the kinetic and potential energy, i.e. a mean energy kT in all, giving $\bar{E} = 3NkT$ and $C_V = 3Nk$ for the system of N atoms. We see now that according to quantum theory this is only a high-temperature result and one refers to the vibrational degrees of freedom in this case as *fully excited*. At lower temperatures where these oscillations are not fully excited they make a smaller contribution to the internal energy and to the heat capacity.

*For $x \to 0$

$$\frac{x^2 e^x}{(e^x - 1)^2} = \frac{x^2 (1 + x + \ldots)}{(x + \tfrac{1}{2}x^2 + \ldots)^2} \to 1 \; .$$

At very low temperatures, if kT is sufficiently much smaller than $\hbar\omega_E$, all oscillators will be in their ground state, in the quantum-mechanical treatment where the energy levels are discrete. The oscillators will then, for practical purposes, make no contribution to the heat capacity: $C_V/3N \ll k$. One calls such degrees of freedom *frozen*.

The situation becomes interesting if one deals with a system which possesses several vibrational frequencies $\omega_1, \omega_2, \ldots$. At a given temperature, some of these degrees of freedom will be fully excited, some partially excited and some frozen, and the proportions of each kind of degree of freedom change with temperature. It is this behaviour which determines, for example, the temperature variation of the specific heats of di- and polyatomic gases. At very low temperatures the rotational and vibrational degrees of freedom are frozen, and the gas behaves like a monatomic gas: its molar heat capacity at constant volume is $\frac{3}{2}R$. As the temperature is raised the rotational and vibrational degrees of freedom gradually become excited until they make their full contribution to the heat capacity. (See section 7.5 where further references are also given.)

Fig. 6.5 shows the atomic heat capacity at constant volume (in units of $3R$), as given by the Einstein expression (6.9), as a function of T/Θ_E up to $T/\Theta_E = 1$.

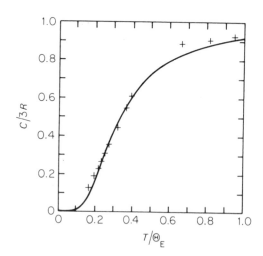

Fig. 6.5. Comparison of the observed molar heat capacity of diamond (+ experimental points with $\Theta_E = 1325$ K) with Einstein's model (full line). (After Einstein's original 1907 paper.)

6.2.2   Comparison of Einstein's result with experiment

Fig. 6.5 also shows the experimental heat capacities at constant pressure*
for diamond for several temperatures in the range 200 K to 1200 K, as given
in Einstein's historic 1907 paper. Einstein chose the value $\Theta_E = 1325$ K to
produce agreement for the experimental point at $T = 331.3$ K. The general
agreement of theory with experiment is seen to be good. We see that the large
deviation from the Dulong and Petit value, in Table 6.1, results from the
large Einstein temperature of diamond. For most substances Θ_E is about
200 to 300 K and since $C_V/3R = 0.92$ for $T = \Theta_E$ it follows that Dulong and
Petit's law usually holds at room temperature.

More detailed comparison of Einstein's theory with experiment shows the
agreement is only approximate. In particular the experimentally observed
decrease in heat capacity at low temperatures goes as T^3 and not
exponentially as predicted by Eq. (6.12). In view of the simplicity of Einstein's
model, the fact that the agreement is only of a qualitative nature is not
surprising. We shall discuss modifications of the Einstein theory in section 6.3.

The physics which distinguishes different substances from each other in
Einstein's theory is contained in the parameter Θ_E or ω_E, Eq. (6.10). The
natural frequency ω_E of the lattice vibrations depends on the restoring force
acting on the vibrating atoms and hence on the masses of the atoms, the elastic
properties of the solid and the interatomic spacing. One can estimate ω_E
from many different properties of solids and compare it with values of ω_E
obtained from heat capacity measurements. Properties which allow such
comparisons are the elasticity coefficients, the melting point or, in the case
of ionic crystals, the selective absorption or reflection of infrared radiation.[†]
Because of the oversimplifications of Einstein's model—the atoms do not
vibrate independently of each other about fixed sites, nor is there only one
natural frequency—one can only expect very approximate qualitative
agreement and this one finds.

The frequency ω_E is related to the elastic constants of the solid by the
following approximate expression:

$$\omega_E \approx A \left(\frac{aE}{m} \right)^{1/2} , \tag{6.13a}$$

where E is Young's modulus of the solid, m is the mass of each atom and
a is the interatomic spacing. The constant A depends on the crystal structure

---

*The difference $C_p - C_V$ is very small for solids at room temperature or below. For copper
at room temperature $(C_p - C_V)/C_V$ is less than 3 per cent and even for lead at 400 K it is less
than 8 per cent. An expression for $(C_p - C_V)$ was derived in section 5.3, Eq. (5.20).
[†]See Hall,[27] Chapter 2, or Kittel,[29] Chapter 10.

and on the approximations made in deriving the Einstein model from a real solid. A is of the order of magnitude unity.

We give a simple derivation of Eq. (6.13a). An atom displaced a distance x from its equilibrium position experiences a restoring force $m\omega_E^2 x$; this displacement leads to a stress $\sigma \approx m\omega_E^2 x/a^2$ and a strain $\varepsilon \approx x/a$. From the definition of Young's modulus $E = \sigma/\varepsilon$, eq. (6.13a) with $A = 1$ follows. With a more careful analysis, allowing for the motion of other atoms, Einstein obtained $A \approx 2$.

We can rewrite Eq. (6.13a) by eliminating a and m in terms of the density $\rho = m/a^3$ and the gram-atomic weight $M = mN_0$:

$$\omega_E \approx N_0^{1/3} \frac{E^{1/2}}{\rho^{1/6} M^{1/3}} . \tag{6.13b}$$

It follows from Eq. (6.13b) that diamond which is hard, has low density and small atomic weight should have a large Einstein frequency ω_E and large Einstein temperature Θ_E. Lead, for which all three factors work in the opposite direction, should have much smaller ω_E and Θ_E.

Table 6.2.

Element	$E/10^{10(a)}$ N/m^2	ρ g/cm^3	M g	$\omega_E/10^{12}$ sec^{-1}	Θ_E(el.) K	Θ_E(therm.) K
Diamond	83	3.52	12	172	1316	1220[b]
Al	6.9	2.7	27	39.6	303	240[c]
Pb	1.8	11.36	207.2	8.1	62	67[c]

Sources: (a) Cottrell[25], pp. 4–7.
   (b) Experimental specific heats quoted in A. Einstein, *Ann. Phys.*, **22**, 180 (1907).
   (c) *A Compendium of the Properties of Materials at Low Temperature, Part II: Properties of Solids* (Wadd Technical Report 60–56, Part II, 1960).

Table 6.2 shows values for diamond, lead and aluminium which is a more average element. The Einstein frequency ω_E is calculated from Eq. (6.13b) (with $A = 2$). Θ_E(el.) and Θ_E(therm.) are the Einstein temperatures derived from these elastic data and from thermal data (i.e. heat capacities) respectively.* We see that the agreement between Θ_E(el.) and Θ_E(therm.) is qualitatively quite good. Eqs. (6.13) afford an explanation of the large variations in Θ_E which are required to fit the observed heat capacities by Einstein's equation.

*We defined Θ_E(therm.) such that the Einstein and experimental heat capacities agree at $T = \Theta_E$(therm.). This illustrates the arbitrary element involved in fixing Θ_E which results from the fact that Einstein's expression (6.9) agrees only approximately with the experimental heat capacity.

Finally, in this section, we note that we can express ω_E in terms of the velocity v_0 of plane elastic waves in an isotropic medium (i.e. these are longitudinal waves in a thin rod), since*

$$v_0 = \sqrt{\frac{E}{\rho}}$$

so that Eq. (6.13a) becomes

$$\omega_E \approx A\frac{v_0}{a}. \qquad (6.14)$$

★ 6.3 DEBYE'S THEORY

6.3.1 Derivation of Debye's result

In this section we shall see how to improve on Einstein's theory of heat capacities. Consider a crystal consisting of N atoms. This system has $3N$ degrees of freedom corresponding to the $3N$ coordinates required to specify the positions of the atoms. But the atoms do not vibrate independently of each other about fixed sites. They execute very complicated coupled vibrations as discussed in section 6.2.1 (see Fig. 6.3). From classical mechanics one knows that the oscillations of such a system can be described in terms of $3N$ normal modes of vibration of the whole crystal, each with its own characteristic (circular) frequency $\omega_1, \omega_2, \ldots \omega_{3N}$. In terms of these normal modes, the lattice vibrations of the crystal are equivalent to $3N$ independent harmonic oscillators with these circular frequencies.

To illustrate this idea, consider a system of two degrees of freedom consisting of two coupled one-dimensional oscillators, described by the equations of motion

$$\left.\begin{array}{l} \ddot{q}_1 + \omega^2 q_1 + \lambda q_2 = 0 \\ \ddot{q}_2 + \omega^2 q_2 + \lambda q_1 = 0 \end{array}\right\}. \qquad (6.15)$$

q_1 and q_2 are the position coordinates of the two oscillators. The term in λ represents the coupling between them. For $\lambda = 0$ we would have two independent oscillators each with frequency ω. To find the normal modes of this coupled system we must find new coordinates Q_1 and Q_2 which satisfy the equations of simple harmonic motion. If we introduce

$$Q_1 = q_1 + q_2, \qquad Q_2 = q_1 - q_2, \qquad (6.16)$$

*See Cottrell,[25] Chapter 6.

then Eqs. (6.15) go over into

$$\left.\begin{aligned} \ddot{Q}_1 + (\omega^2 + \lambda)Q_1 = 0 \\ \ddot{Q}_2 + (\omega^2 - \lambda)Q_2 = 0 \end{aligned}\right\} \cdot \tag{6.17}$$

Q_1 and Q_2 are called the normal coordinates: in terms of them the system behaves like two independent oscillators with frequencies $(\omega^2 \pm \lambda)^{1/2}$, the characteristic frequencies of the system.

If we know these $3N$ characteristic frequencies $\omega_1, \omega_2, \ldots \omega_{3N}$, we can straightaway write down the energy of the crystal, since the system is equivalent to $3N$ independent oscillators of these frequencies. From Eq. (6.7), which gives the mean energy of one oscillator with frequency ω_E, we now obtain for the energy E of the crystal

$$E = \sum_{\alpha=1}^{3N} \bar{\varepsilon}_\alpha$$

$$= \sum_{\alpha=1}^{3N} \left\{ \frac{1}{2} \hbar\omega_\alpha + \frac{\hbar\omega_\alpha}{\exp(\beta\hbar\omega_\alpha) - 1} \right\} \cdot \tag{6.18}$$

This approach, of actually calculating the characteristic frequencies of a crystal, was initiated by Born and von Kármán (1912). It is a very complicated problem although it is capable of solution by means of large computers. Debye, also in 1912, suggested a very much simpler approximate treatment which is very successful.

Debye starts from the fact that we already know some of the normal modes of a crystal, namely the propagation of sound waves, which are elastic waves of low frequency, i.e. whose wavelength λ is very large compared with the interatomic spacing a of the crystal: $\lambda \gg a$. Under these conditions one can ignore the discrete atomic structure of the solid and describe it as a homogeneous elastic medium. For simplicity we shall assume it isotropic so that the velocity of propagation of elastic waves is independent of their direction. (It is possible to relax this assumption, as mentioned in Appendix B.) To find the normal modes for low frequencies, we then need only find the different modes of standing waves which are possible in this medium. For elastic waves in such a medium, of volume V, the number of modes with (circular) frequency between ω and $\omega + d\omega$ is given by the following expression:

$$f(\omega)\,d\omega = V \frac{\omega^2\,d\omega}{2\pi^2} \left[\frac{1}{v_L^3} + \frac{2}{v_T^3} \right] \tag{6.19a}$$

$$= V \frac{3\omega^2\,d\omega}{2\pi^2\,\bar{v}^3} \cdot \tag{6.19b}$$

This result is derived in Appendix B, Eq. (B.35). The problem of counting normal modes, of which Eq. (6.19) is a particular case, is of great importance in many branches of physics, and it occurs in many other contexts in this book. For this reason we give in Appendix B a general discussion of this problem. (The reader not familiar with these ideas is urged to study Appendix B at this stage.) Here v_L and v_T are the velocities of propagation, in this isotropic continuous solid, of longitudinal (compression) and transverse (shear) waves which, in general, are different. The factor 2 multiplying $1/v_T^3$ in Eq. (6.19a) is due to the fact that there are two independent transverse directions of polarization, just as for light. In Eq. (6.19b) we have defined a mean velocity $\bar{v}$:

$$\frac{3}{\bar{v}^3} \equiv \frac{1}{v_L^3} + \frac{2}{v_T^3} \ . \tag{6.20}$$

Eq. (6.19) applies to the low-frequency lattice vibrations in a crystal. Debye now makes the assumption that it applies to *all* frequencies. Since the crystal (of N atoms) possesses $3N$ modes in all, he introduced a *maximum* frequency ω_D called the *Debye frequency*, such that there are $3N$ modes altogether, i.e.

$$\int_0^{\omega_D} f(\omega)\,d\omega = 3N \ . \tag{6.21}$$

This 'cut-off' frequency ω_D occurs because at sufficiently high frequency, i.e. short wavelength, we can't ignore the atomic nature of the solid. A crystal with interatomic spacing a cannot propagate waves with wavelengths less than $\lambda_{min} = 2a$. In this case neighbouring atoms vibrate in antiphase. This is shown for transverse vibrations of a linear chain of atoms in Fig. 6.6.

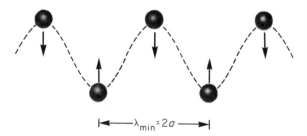

Fig. 6.6. Mode of oscillation of a linear chain of atoms corresponding to the minimum wavelength, i.e. maximum frequency. Neighbouring atoms vibrate in antiphase.

Substituting Eq. (6.19b) in (6.21) and performing the integration gives at once

$$\omega_D^3 = 6\pi^2 \frac{N}{V} \bar{v}^3 \tag{6.22}$$

so that the Debye frequency is given in terms of the velocity of sound waves in the solid. Using Eq. (6.22), we can rewrite Eq. (6.19b) as

$$f(\omega)\,d\omega = 9N \frac{\omega^2\,d\omega}{\omega_D^3}\,. \tag{6.23}$$

Using Eq. (6.23), we can now replace the sum over characteristic frequencies ω_α in Eq. (6.18) by an integral. In this way we obtain for the energy of the crystal

$$E = \frac{9}{8}\,N\hbar\omega_D + \frac{9N\hbar}{\omega_D^3} \int_0^{\omega_D} \frac{\omega^3\,d\omega}{\exp(\beta\hbar\omega) - 1} \tag{6.24a}$$

$$= \frac{9}{8}\,Nk\Theta_D + \frac{9NkT}{x_D^3} \int_0^{x_D} \frac{x^3\,dx}{e^x - 1}\,, \tag{6.24b}$$

where in the last line we changed the variable of integration:

$$x \equiv \beta\hbar\omega \equiv \frac{\hbar\omega}{kT}\,, \tag{6.25}$$

and put

$$x_D \equiv \beta\hbar\omega_D \equiv \frac{\hbar\omega_D}{kT} \equiv \frac{\Theta_D}{T} \tag{6.26}$$

which defines the Debye temperature Θ_D in analogy to the Einstein temperature Θ_E, Eq. (6.10).

From Eq. (6.24a) the heat capacity at constant volume follows by differentiation with respect to T. (T enters this expression only through β in the denominator.) In this way one obtains, again introducing x and x_D from Eqs. (6.25) and (6.26) after differentiation,

$$C_V = \left(\frac{\partial E}{\partial T}\right)_V = 3Nk\left\{\frac{3}{x_D^3} \int_0^{x_D} \frac{x^4 e^x dx}{(e^x - 1)^2}\right\}\,. \tag{6.27}$$

Eq. (6.27) is our final result. It is the heat capacity at constant volume as given by Debye's theory. It depends on only a single parameter, the Debye temperature Θ_D. It is not possible to evaluate the integral in Eq. (6.27) analytically but the function in curly parentheses is tabulated.

We can derive the high and low temperature limits from Eq. (6.27).

(i) *High-temperature limit.* This is defined by $T \gg \Theta_D$, i.e. $x_D \ll 1$. In this case we rewrite the integrand in Eq. (6.27) in the form

$$\frac{x^4 e^x}{(e^x - 1)^2} = \frac{x^4}{(e^x - 1)(1 - e^{-x})}$$

$$= \frac{x^4}{2(\cosh x - 1)}$$

$$= \frac{x^4}{2\left(\dfrac{x^2}{2!} + \dfrac{x^4}{4!} + \cdots\right)}. \qquad (6.28)$$

For the range of integration $0 \leqslant x \leqslant x_D$ in Eq. (6.27) and $x_D \ll 1$, we need retain only the leading term x^2 in the denominator of expression (6.28) so that Eq. (6.27) becomes

$$C_V = 3Nk \left\{ \frac{3}{x_D^3} \int_0^{x_D} x^2 \, dx \right\} = 3Nk, \qquad (T \gg \Theta_D), \qquad (6.29)$$

which is Dulong and Petit's law again. It is not surprising that Debye's theory gives this high-temperature limit, as it follows directly from our starting point,

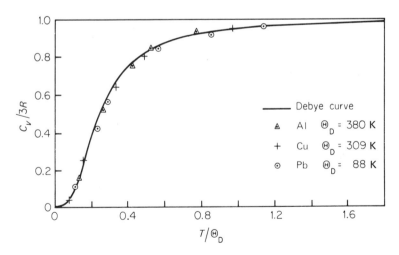

Fig. 6.7. Comparison of Debye's heat capacity curve, Eq. (6.27), with observations. Full line: Debye's curve. Experimental points for aluminium, copper and lead are plotted for the Debye temperatures shown. (Data from : *A Compendium of the Properties of Materials at Low Temperature, Part II: Properties of Solids.* Wadd Technical Report 60–56, Part II, 1960.)

Eq. (6.18), independently of the values of the frequencies $\omega_1, \omega_2, \ldots$ which enter this equation.

(ii) *Low-temperature limit.* This is defined by $T \ll \Theta_D$, i.e. $x_D \gg 1$. Since the integrand in Eq. (6.27) goes to zero rapidly for large values of x (it goes like $x^4 e^{-x}$), we can replace the upper limit by infinity in this case. Eq. (6.27) then becomes

$$C_V = 3Nk \left(\frac{T}{\Theta_D} \right)^3 \left[3 \int_0^\infty \frac{x^4 e^x dx}{(e^x - 1)^2} \right] \, , \qquad (T \ll \Theta_D) \; . \quad (6.30a)$$

The integral in square parentheses in this equation is, of course, just a number. Hence it follows from Eq. (6.30a) that the Debye heat capacity decreases like T^3 at low temperatures, in agreement with observation. It is therefore markedly superior to Einstein's theory. The term in square parentheses in Eq. (6.30a) has the value $4\pi^4/5$ (see Appendix A.2, Eq. (A.22)), so that we obtain

$$C_V = \frac{12}{5} \pi^4 Nk \left(\frac{T}{\Theta_D} \right)^3 \, , \qquad (T \ll \Theta_D) \; . \quad (6.30b)$$

Fig. 6.7 shows the Debye heat capacity (6.27) plotted over the whole temperature range as a function of T/Θ_D.

6.3.2 Comparison of Debye's result with experiment

Fig. 6.7 also shows the experimental molar heat capacities (with a small correction applied to convert them from C_P to C_V, see section 5.3) for aluminium, copper and lead for Debye temperatures Θ_D of 380 K, 309 K and 88 K respectively. These values give good fits to the experimental points over the whole temperature range. The excellent agreement for the T^3-dependence of the heat capacities at very low temperatures has already been noted in section 6.1, (see Figs. 6.1 and 6.2). Nevertheless Debye's theory is still only an approximation and this shows up in detailed comparisons with experiments. In particular, if one chooses Θ_D at *each* temperature T to fit the experimental heat capacity of a solid by the Debye formula (6.27), one finds that Θ_D is *not constant* but varies with T. However Θ_D rarely varies by as much as 20 per cent and in many cases by less than 10 per cent.

Debye's theory agrees so well with experiment, especially at low temperatures, because *at low temperatures and at high temperatures it is exact.* It is an interpolation formula between two correct limits. At high temperatures, this follows directly from Eq. (6.18) as already mentioned in connection with Eq. (6.29). At low temperatures, this follows since only those terms contribute to

$$\left\{ E - \sum_{\alpha=1}^{3N} \frac{1}{2} \hbar \omega_\alpha \right\} \quad (6.31)$$

in Eq. (6.18), and therefore to the specific heat $(\partial E/\partial T)_V$, for which $\hbar\omega_\alpha$ is less than or of the order of kT. (For $\hbar\omega_\alpha \gg kT$, the factor $[\exp(\beta\hbar\omega_\alpha)-1]^{-1}$ in Eq. (6.18) becomes $\exp(-\beta\hbar\omega_\alpha) \ll 1$.) In other words, in the low-temperature limit it is only the low frequency modes of oscillation which contribute to the heat capacity and for these, as we have seen, it is justified to replace the characteristic modes of the crystal by the elastic waves in a continuous medium.

Table 6.3 gives the Debye temperatures for a few elements obtained by fitting the Debye heat capacity to experiment at $T \approx \frac{1}{2}\Theta_D$. These temperatures and systematic variations are similar to those of the Einstein temperatures. This is not surprising since we can rewrite Eq. (6.22), with $V/N = a^3$, in the form

$$\omega_D = (6\pi^2)^{1/3} \frac{\bar{v}}{a} \approx 4\frac{\bar{v}}{a} \tag{6.32}$$

which is very similar to Eq. (6.14) for ω_E. The velocities of different types of elastic waves will, in general, show the same systematic variations with properties of the solid (i.e. its stiffness, density, etc.). Table 6.4 gives some values of Θ_D(el.), i.e. the Debye temperature calculated from Eq. (6.32), as well as Θ_D(therm.), the Debye temperature derived from thermal data. The agreement is remarkably good considering the approximate nature of Debye's theory. Alternatively, one can express $\bar{v}$ in Eq. (6.32) in terms of the elastic constants for an isotropic solid and in this way effect a comparison of thermal and elastic data.

Table 6.3. Debye temperatures $\Theta_D{}^{(a)}$

Element	Pb	Na	Ag	Cu	Al	Be	Diamond
Θ_D K	86	160	220	310	380	980	≈ 1950

[a] Source: M. Blackman, *Handbuch der Physik*, Springer, Berlin, 1955, Vol. VII (1), p. 325.

Table 6.4.

Element	$\bar{v}/10^{5(a)}$ cm/sec	a Å	$\omega_D/10^{12}$ sec^{-1}	Θ_D(el.) K	Θ_D(therm.) K
Al	3.4	2.5	52	399	380
Cu	2.6	2.3	44	335	310
Pb	0.8	3.1	9.8	75	86

[a] Source: *American Institute of Physics Handbook*, 2nd edn., McGraw-Hill, New York, 1963, pp. 3–88.

PROBLEMS 6

6.1 Qualitatively the melting point of a crystalline solid can be defined as the temperature T_m at which the amplitude of simple harmonic vibrations of the individual atoms about their lattice sites is about ten per cent of the interatomic spacing a. Show that on this picture the frequency v of the vibrations of a monatomic crystal (atomic weight M) satisfies the proportionality relation

$$v \propto \frac{1}{a} \left(\frac{T_m}{M} \right)^{1/2} \quad .$$

The Einstein temperatures of aluminium and of lead are 240 K and 67 K respectively. Guess orders of magnitude or look up data to decide whether these temperatures and the above formula are in reasonable agreement.

6.2 Calculate the entropy of the lattice vibrations of a monatomic crystal as described by the Einstein theory.

Obtain limiting expressions for the entropy valid at low and at high temperatures.

6.3 The same as the last problem but for the Debye theory.

6.4 The relation between the frequency v and the wavelength λ for surface tension waves on the surface of a liquid of density ρ and surface tension σ is

$$v^2 = 2\pi\sigma/(\rho\lambda^3) \quad .$$

Use a method analogous to the Debye theory of specific heats to find a formula, analogous to the Debye T^3 law, for the temperature dependence of the surface energy E of a liquid at low temperatures.

The surface tension of liquid helium at 0 K is 0.352×10^{-3} N/m and its density is 0.145 g cm^{-3}. From these data estimate the temperature range over which your formula for $E(T)$ is valid for liquid helium, assuming that each helium atom in the surface of the liquid possesses one degree of freedom.

You may assume

$$\int_0^\infty \frac{x^{4/3}}{e^x - 1} \, dx = 1.68 \quad .$$

CHAPTER

7

The perfect classical gas

In previous chapters we have often used a perfect gas to illustrate general results. In this chapter we shall study the perfect classical gas in more detail. In section 7.1 it will be exactly defined and in section 7.2 we shall obtain its partition function. In section 7.3 the conditions under which the classical approximation holds will be studied. In section 7.4 we derive the two equations which thermodynamically characterize the perfect classical gas: the equation of state $PV = NkT$, and Joule's law $E = E(T)$ which states that the internal energy E is independent of the volume and the pressure of the gas.

The next three sections, 7.5–7.7, deal with various properties of a perfect classical gas. In section 7.5 we give a brief discussion of its heat capacity (including vibrational and rotational properties of di- and polyatomic gases); in section 7.6 we obtain its entropy. Section 7.7 deals with the Maxwell velocity distribution.

The two remaining sections of this chapter are not concerned with the perfect classical gas but nevertheless are natural developments of the main ideas of this chapter. In section 7.8 we give a brief introductory discussion of real gases: how deviations from perfect gas behaviour first show up as the density is increased. This analysis is really necessary in order to see under what conditions a real gas behaves like a perfect gas. In section 7.9 we give a brief introduction to classical statistical mechanics, which appears as a natural generalization of the treatment of a perfect gas in section 7.2.

As indicated by the bold stars, section 7.5 to 7.9 are not required later in this book. *Furthermore, sections 7.5 to 7.9 can be read independently of each other: they presuppose a knowledge only of sections 7.1 to 7.4.*

7.1 THE DEFINITION OF THE PERFECT CLASSICAL GAS

A gas consists of molecules moving about fairly freely in space. The molecules are separated rather far from each other and interact only very weakly. At any instant only a small fraction are interacting strongly, i.e. are colliding. Typical values for the mean separation between molecules under ordinary conditions and for the molecular diameter are of the order of $(22,400/N_0)^{1/3} \approx 30$ Å and 3 Å respectively. The force between molecules, at separations large compared with the molecular diameter, is the weak van der Waals attraction which decreases rapidly with separation, i.e. like the seventh power of the intermolecular distance.* At sufficiently low density, the molecules of a gas will interact very little with each other. The *perfect gas* represents an idealization in which *the potential energy of interaction between the molecules is negligible compared to their kinetic energy of motion.*

Since the interaction energy is negligible, it is an immediate consequence that the energy of a perfect gas subdivides into a sum of 'private' energies for each molecule. These private energies

$$\varepsilon_1 \leqslant \varepsilon_2 \leqslant \cdots \leqslant \varepsilon_r \leqslant \cdots \tag{7.1}$$

correspond to the complete set of discrete quantum states, labelled

$$1, 2, \ldots, r, \ldots, \tag{7.2}$$

in which a *single* molecule can exist. (That these form a *discrete* set — one can hardly imagine anything more *continuous*, thinking intuitively, i.e. classically — may puzzle a reader not familiar with elementary quantum theory. It is discussed in Appendix B.)

Let us consider an enclosure of volume V, containing N molecules of gas, the enclosure being at temperature T, i.e. immersed in a heat bath at temperature T (see Fig. 7.1). We specify the *state of the gas* by listing the number of molecules in each possible (private) state: n_1 molecules in state 1, n_2 molecules in state 2, and n_r in the general state r. n_r is called the *occupation number* of the rth state. (The reader should guard against confusion between the *state of individual molecules*, labelled $1, 2, \ldots r, \ldots$ in Eq. (7.2), and more explicitly referred to as *single-particle states*, and

---

*For a discussion of intermolecular forces in gases, see Flowers and Mendoza[26] or Present,[11] Chapter 12.

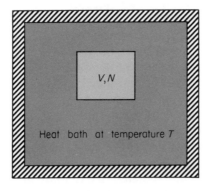

Fig. 7.1. System of fixed volume V containing N molecules of a gas, in a heat bath at temperature T.

states of the system as a whole, i.e. of all the N gas molecules collectively.) The energy of the gas in this state, determined by the occupation numbers $n_1, n_2, \ldots n_r, \ldots$, is

$$E(n_1, n_2, \ldots, n_r, \ldots) = \sum_r n_r \varepsilon_r \qquad (7.3)$$

where the summation is over all single-particle states. We require

$$N = \sum_r n_r \qquad (7.4)$$

since there are N molecules in all.

Let us now derive the partition function for the perfect gas. Since we are neglecting interactions, we may at first consider a single molecule in the container of volume V. For this the partition function is

$$Z_1(T, V) \equiv Z(T, V, 1) = \sum_r \exp(-\beta \varepsilon_r) . \qquad (7.5)$$

It is now tempting BUT WRONG to call

$$\left[\sum_r \exp(-\beta \varepsilon_r) \right]^N = Z(\cancel{T,V,N}) \qquad (7.6)$$

the partition function for the gas consisting of N molecules. We can see this best by considering two molecules ($N = 2$) in the enclosure. In this case expression (7.6) can be written

$$\left[\sum_r \exp(-\beta \varepsilon_r) \right] \left[\sum_s \exp(-\beta \varepsilon_s) \right]$$

$$= \sum_r \exp(-2\beta \varepsilon_r) + \underset{(r \neq s)}{\sum_r \sum_s} \exp[-\beta(\varepsilon_r + \varepsilon_s)]. \qquad (7.7)$$

Here the first term on the right-hand side corresponds to all those terms on the left for which $r = s$, i.e. both molecules are in the same state. The second term on the right corresponds to all remaining terms on the left. For these $r \neq s$: the two molecules are in different states. But looking at the gas (of two molecules) as a whole we have counted each of these states (with $r \neq s$) of the gas twice; e.g. the state with one molecule in state 1 and one molecule in state 2 will occur twice in Eq. (7.7): as $r = 1$, $s = 2$, and as $r = 2$, $s = 1$. Now except for labelling the molecules (calling one of them No. 1 and the other No. 2) these two states of the gas are the same. Since the molecules in a gas are *identical*, as we have stressed previously, we have no business to count this state of the gas twice. What are significant are the *occupation numbers* $n_1, n_2, \ldots n_r, \ldots$ of the various single-particle states. We can *not* attach labels as to *which* molecules we put into *which* of these states. There is no way of distinguishing these two situations experimentally. Hence the second term on the right-hand side of Eq. (7.7) should be divided by 2!, corresponding to the 2! permutations in this sum which represent the *same* state of our (two-molecule) gas. We should, for our two-molecule gas, replace expression (7.7) by

$$\sum_r \exp(-2\beta\varepsilon_r) + \frac{1}{2!} \sum_r \sum_{\substack{s \\ (r \neq s)}} \exp[-\beta(\varepsilon_r + \varepsilon_s)]; \tag{7.8}$$

this counts each state of the gas exactly once.

In general for a gas of N molecules, expression (7.6) will break up into a sum of contributions, the first of which will correspond to *all N* molecules being in the *same* state, and the last of which will correspond to *all N* molecules being in *different* states. The first of these terms is correct as it stands; the last of them must be divided by $N!$ so that each *distinct* state of the N-particle gas is counted *exactly once*. Expression (7.6) should be replaced by

$$\sum_r \exp(-N\beta\varepsilon_r) + \ldots + \frac{1}{N!} \sum_{r_1} \ldots \sum_{\substack{r_N \\ \text{(all } r_i \text{ different)}}} \exp[-\beta(\varepsilon_{r_1} + \varepsilon_{r_2} + \ldots \varepsilon_{r_N})] \tag{7.9}$$

where the terms not written down correspond to states of the gas where some, but not all, molecules are in the same single-particle states. These latter terms contain appropriate factorials in the denominators to ensure that each state of the gas is counted exactly once. We shall not write these terms down as we shall not need them.

We shall restrict ourselves to the *classical* regime. This is the second qualifying word (the first was *perfect*) in the title of this chapter, and we must now explain what it means. We shall restrict ourselves to conditions of the gas where the *probability is very small that any single-particle state is occupied by more than one molecule*. In other words, most single-particle states are empty, a very few contain a single molecule, and an insignificantly small number contain more than one molecule. This situation will occur if the number of single-particle states in any energy region is very large compared with the number of molecules possessing that sort of energy. In this case most states will be empty. This is the *classical regime* which we are considering in this chapter. In section 7.3 we shall discuss the physical conditions under which the classical regime holds. The mathematical consequence of the classical regime holding is that in Eqs. (7.8) and (7.9) *only the last terms are important*. In Eqs. (7.8) and (7.9) *these* have the *correct* statistical weight (each state of the gas counted exactly once); in Eq. (7.6), *these* states of the gas are weighted *wrongly* (i.e. they are overweighted by a factor $N!$). If instead of Eq. (7.6) we consider the expression

$$\frac{1}{N!}\left[\sum_r \exp(-\beta\varepsilon_r)\right]^N \tag{7.10}$$

then *the terms which matter* (all molecules in different states) *are counted correctly* (and agree with the correct Eqs. (7.8) and (7.9)). Only terms which, under classical conditions, are *unimportant* (not all molecules in different states) are *weighted wrongly*; for example, use of (7.10) underweights the first terms in Eqs. (7.8) and (7.9) by a factor $1/N!$. It follows that in the *classical regime the partition function for a perfect gas of N molecules* is given by

$$Z \equiv Z(T,V,N) = \frac{1}{N!}\left[\sum_r \exp(-\beta\varepsilon_r)\right]^N = \frac{1}{N!}\,[Z_1(T,V)]^N\ . \tag{7.11}$$

Eq. (7.11) is our final result from which all the physics of a perfect classical gas follows easily. The factor $(N!)$ in the denominator of this equation is due to the fact that the molecules are *not localized*. This is in contrast to the paramagnetic problem of Chapter 3 where the two situations shown in Figs 2.4(*a*) and (*b*) are physically different: the atoms *are localized*: we can distinguish where the up-spin and down-spin atoms are located.

7.2   THE PARTITION FUNCTION

To evaluate the partition function Z, Eq. (7.11), we need only consider the single-particle partition function Z_1, Eq. (7.5). The motion of one molecule decomposes into two parts, the translational motion of its centre of mass and its internal motions, i.e. rotations and vibrations of the molecule, as well as electronic excitation. Correspondingly the energy of a molecule can be written

$$\varepsilon_r \equiv \varepsilon_{s\alpha} = \varepsilon_s^{\text{tr}} + \varepsilon_\alpha^{\text{int}} \tag{7.12}$$

where $\varepsilon_s^{\text{tr}}$ and $\varepsilon_\alpha^{\text{int}}$ are the energies associated with the translational and internal motions, and the label r of a state of the molecule has been replaced by the pair of labels (s, α): s to specify the state of translational motion, and α the state of internal motion. It follows that the single-particle partition function Z_1 decomposes into a translational and an internal partition function:

$$Z_1 = \sum_r \exp(-\beta \varepsilon_r) = \sum_s \sum_\alpha \exp[-\beta(\varepsilon_s^{tr} + \varepsilon_\alpha^{int})]$$

$$= \left[\sum_s \exp(-\beta \varepsilon_s^{tr})\right]\left[\sum_\alpha \exp(-\beta \varepsilon_\alpha^{int})\right] = Z_1^{tr} Z_{int}. \qquad (7.13)$$

Here

$$Z_1^{tr} \equiv Z^{tr}(T, V, 1) = \sum_s \exp(-\beta \varepsilon_s^{tr}) \qquad (7.14)$$

is the *translational partition function* (and $\sum_s$ denotes the sum over all translational states of the molecule), and

$$Z_{int} \equiv Z_{int}(T) = \sum_\alpha \exp(-\beta \varepsilon_\alpha^{int}) \qquad (7.15)$$

is the *internal partition function* of a molecule (and $\sum_\alpha$ denotes the sum over all states of internal motion; rotations and vibrations of the molecule, etc.).

The fact that the partition function (7.13) factorizes into translational and internal parts is very important. It enables us to consider the two factors Z_1^{tr} and Z_{int} separately. We can evaluate Z_1^{tr} once and for all. The result will apply to any perfect classical gas, irrespective of its internal structure. It will apply to argon (A) as much as to carbon monoxide (CO). We shall evaluate Z_1^{tr} in this section. Z_{int}, on the other hand, depends on the internal properties of the gas molecules and will vary for different gases. We shall consider these internal properties briefly in section 7.5. Here we only note two properties of Z_{int}. Firstly, it necessarily refers to one molecule. Secondly, the energy eigenvalues ε_α^{int} of the internal rotational and vibrational motions of a single molecule do not depend on the volume V of the enclosure containing the gas. Hence Z_{int} *is independent of the volume* V.

We shall now calculate Z_1^{tr}, Eq. (7.14). According to classical mechanics the kinetic energy ε^{tr} of the (translational) centre-of-mass motion of a molecule is related to its momentum $\mathbf{p}$ by

$$\varepsilon^{tr} = \frac{1}{2m} p^2 \qquad (7.16)$$

where m is the mass of the molecule and $p = |\mathbf{p}|$, the magnitude of the momentum. Classically, $\mathbf{p}$ and $\varepsilon^{\mathrm{tr}}$ can assume any values. Quantum-mechanically, Eq. (7.16) still holds, but $\mathbf{p}$ and $\varepsilon^{\mathrm{tr}}$ are restricted to certain discrete values. The problem is analogous to that of finding the modes of a vibrating string with fixed ends. The centre-of-mass motion of the molecule obeys the Schrödinger wave equation and the wave function satisfies certain boundary conditions on the surface of the enclosure containing the gas. The problem of finding the modes of a system of waves or vibrations is a very basic one which occurs in many different contexts in statistical physics. It is solved in Appendix B. For the present problem the result, given by Eq. (B.25), is as follows. For a molecule in an enclosure of volume V, the number of states in which the molecule has a momentum whose magnitude lies in the interval p to $p + \mathrm{d}p$ (and with kinetic energy in a corresponding range $\varepsilon^{\mathrm{tr}}$ to $\varepsilon^{\mathrm{tr}} + \mathrm{d}\varepsilon^{\mathrm{tr}}$, given by Eq. (7.16)) is given by

$$f(p)\,\mathrm{d}p = \frac{V 4\pi p^2 \,\mathrm{d}p}{h^3} \, . \tag{7.17}$$

(Note the denominator contains Planck's constant $h \equiv 2\pi\hbar$.) The reader not familiar with this result is advised to study Appendix B at this stage. Eq. (7.14) is a sum over *all* translational states of motion of the molecule. Eq. (7.17) gives the number of those states for which the magnitude of the momentum lies in the range p to $p + \mathrm{d}p$. Hence we can rewrite Eq. (7.14) by means of Eq. (7.17) as an integral

$$Z_1^{\mathrm{tr}} = \int_0^\infty \frac{V 4\pi p^2 \,\mathrm{d}p}{h^3} \exp(-\beta p^2/2m) \tag{7.18}$$

where Eq. (7.16) was used in the exponent. This integral is of a kind which frequently occurs in the kinetic theory of gases and it is evaluated in Appendix A.3, Eq. (A.31), giving

$$Z_1^{\mathrm{tr}} = V \left(\frac{2\pi m k T}{h^2} \right)^{3/2} , \tag{7.19}$$

where we put $\beta \equiv 1/kT$ again. Note that $(h^2/2\pi m k T)^{1/2}$ has the dimension of length, as it must do to make Z_1^{tr} dimensionless (compare Eq. (7.14)). We shall discuss the physical significance of this length in section 7.3.

Substituting Eqs. (7.19) and (7.13) in the partition function (7.11) gives

$$Z \equiv Z(T, V, N) = \frac{1}{N!} V^N \left(\frac{2\pi m k T}{h^2} \right)^{3N/2} [Z_{\mathrm{int}}(T)]^N . \tag{7.20}$$

To obtain the Helmholtz free energy $F = -kT \ln Z$ we approximate $N!$ by Stirling's formula, for $N \gg 1$,

$$N! = \left(\frac{N}{e}\right)^N \tag{7.21}$$

(see Appendix A.1, Eq. (A.2)). Eq. (7.20) then gives for the Helmholtz free energy of the gas of N molecules in an enclosure of volume V, at temperature T,

$$F \equiv F(T, V, N) = -NkT \ln \left\{ \frac{eV}{N} \left(\frac{2\pi mkT}{h^2}\right)^{3/2} Z_{\text{int}}(T) \right\} . \tag{7.22}$$

This is our final result from which all thermodynamic properties of the perfect classical gas follow. We see from the relation $F = E - TS$ that the Helmholtz free energy F is an extensive quantity, i.e. proportional to the size of the system (see section 2.3). Eq. (7.22) has the correct extensive property: V/N and hence the quantity in curly parentheses are intensive quantities, per molecule, so that F is proportional to N, the number of molecules. This is a consequence of the factor $1/N!$ in the definition (7.11) of the N-particle partition function.

In deriving the entropy of a perfect gas in section 5.4.1 from thermodynamics we had to choose the constant of integration so as to make the entropy extensive. In our present treatment F is automatically extensive and this ensures that the entropy $S = (E - F)/T$ is also an extensive quantity. We shall derive it in section 7.6.

The dependence of the partition function on the external parameter, i.e. the volume V of the enclosure, occurs through the factor $f(p)$ which determines the density of states, i.e. the number of states per unit range of p. $f(p)$ is proportional to V: if we double the volume, the density of states is doubled.

Eq. (7.17) for the density of states has a simple interpretation. Consider a particle moving in one dimension, along the x-axis, say. According to classical mechanics, we can specify its state of motion at any instant by specifying its position coordinate x and its momentum p_x. Its motion will be described by a trajectory such as is shown in Fig. 7.2, with x and p_x as orthogonal axes. We call this two-dimensional space, with x and p_x as axes, the *phase space* of the particle. Any point in the phase space represents a definite state of motion. Although we can no longer draw it so well, it is advantageous to use the same geometrical language to describe the three-dimensional motion of a particle. Phase space is now six-dimensional with space coordinates x, y, z and momentum coordinates p_x, p_y, p_z as orthogonal axes. Again each point in phase space represents a state of motion

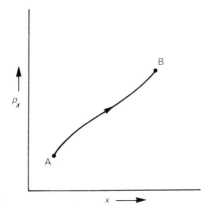

Fig. 7.2. Phase space for a particle moving in one dimension. The trajectory corresponds to a particular motion of the particle, from the state specified by the point A to that specified by the point B.

of the particle. In classical mechanics *every* point represents a possible state of motion, so it is reasonable to say that the number of states in a given region of phase space is proportional to the six-dimensional volume of that region. In particular, the number of states where the particle has position and momentum coordinates within the ranges

$$\left.\begin{array}{l} x, x + \mathrm{d}x;\ y, y + \mathrm{d}y;\ z, z + \mathrm{d}z; \\[4pt] p_x, p_x + \mathrm{d}p_x;\ p_y, p_y + \mathrm{d}p_y;\ p_z, p_z + \mathrm{d}p_z; \end{array}\right\} \tag{7.23}$$

is proportional to the volume of this element of phase space, i.e. to

$$\mathrm{d}^3\mathbf{r}\,\mathrm{d}^3\mathbf{p} \equiv (\mathrm{d}x\,\mathrm{d}y\,\mathrm{d}z)(\mathrm{d}p_x\,\mathrm{d}p_y\,\mathrm{d}p_z)\ . \tag{7.24}$$

This is essentially the description used by Maxwell, Boltzmann and Gibbs in classical statistical mechanics. The constant of proportionality they could not fix, as classically the possible states form a continuum. Eq. (7.17), on the other hand, contains no undetermined constant. The constant of proportionality involves Planck's constant, a consequence of the quantal picture of discrete states from which we started.

We easily find the constant of proportionality by relating Eqs. (7.17) and (7.24). The volume $\mathrm{d}\Gamma$ in phase space for which the particle lies within the enclosure and its momentum has magnitude within the range p to $p + \mathrm{d}p$ is given by

$$\mathrm{d}\Gamma = \int_V \mathrm{d}^3\mathbf{r} \int_{p,\,p+\mathrm{d}p} \mathrm{d}^3\mathbf{p} = V \int_{p,\,p+\mathrm{d}p} \mathrm{d}^3\mathbf{p}\ . \tag{7.25}$$

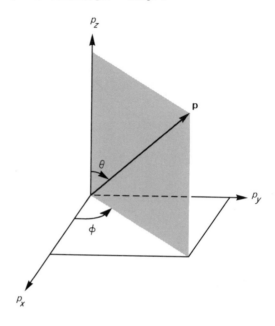

Fig. 7.3. Momentum space, showing the relations between cartesian and polar coordinates:

$$p_x = p \; \sin\theta \; \cos\phi$$
$$p_y = p \; \sin\theta \; \sin\phi$$
$$p_z = p \; \cos\theta.$$

The integral in momentum space (i.e. the three-dimensional space with p_x, p_y and p_z as orthogonal axes — see Fig. 7.3) is simply the volume in momentum space of a spherical shell with radii p and $p + \mathrm{d}p$, i.e.

$$\int_{p,p+\mathrm{d}p} \mathrm{d}^3\mathbf{p} = 4\pi p^2 \, \mathrm{d}p \; . \tag{7.26}$$

(A more complicated way of getting this result is to write Eq. (7.25), not in terms of cartesian coordinates p_x, p_y, p_z in momentum space, but in terms of polar coordinates (p, θ, ϕ), shown in Fig. 7.3, and to perform the integrations with respect to θ and ϕ with p kept constant.) From Eqs. (7.25) and (7.26) we have

$$\mathrm{d}\Gamma = V 4\pi p^2 \, \mathrm{d}p \tag{7.27}$$

for the volume of this region of phase space, while Eq. (7.17) gives the discrete number of states (according to quantum theory) corresponding to this region of phase space. Hence the volume in phase space for each discrete quantum state is h^3. We can think of phase space subdivided into *cells* of volume h^3.

Each such cell represents one quantum state. This result, derived for a particular case, is generally valid. Applied to Eq. (7.24), we have that

$$\frac{1}{h^3} \, d^3 r \, d^3 p \tag{7.28}$$

is the number of states where the particle lies within a volume element $d^3 r$ of (ordinary) coordinate space and within a volume element $d^3 p$ of momentum space. This result, of converting continuous classical phase space into a 'cellular' quantum-mechanical phase space was proposed by Planck in connection with his theory of black-body radiation at the beginning of this century before the existence of quantum mechanics.

In section 7.9 we shall give a somewhat more general discussion of classical statistical mechanics.

7.3   VALIDITY CRITERION FOR THE CLASSICAL REGIME

In the last section we characterized the classical regime by the properties that most of the single-particle states are empty, a very few contain one molecule and an insignificantly small number contain more than one molecule. We now want to formulate this condition mathematically and see what it means physically.

The probability that a molecule is in a particular state s of translational motion (with energy ε_s^{tr}, see Eq. (7.12)) is given, according to the Boltzmann distribution (2.22), by

$$p_s = \frac{1}{Z_1^{tr}} \exp(-\beta \varepsilon_s^{tr}) \; . \tag{7.29}$$

(Do not confuse p_s with a momentum variable!) If the gas consists of N molecules, the mean number $\bar{n}_s$ of molecules which are in the translational state s is given by

$$\bar{n}_s = N p_s \; . \tag{7.30}$$

For each translational state s, the molecule can still be in any one of many different states of internal motion. It follows that if each of the mean occupation numbers $\bar{n}_s$ of the translational states is very small compared to unity, the classical regime holds, i.e. we obtain as a sufficient *condition for the classical regime*

$$\bar{n}_s \ll 1, \quad \text{for } all \ s \; . \tag{7.31}$$

From Eqs. (7.30), (7.29) and (7.19) this condition can be written

$$\bar{n}_s = \left[\frac{N}{V} \left(\frac{h^2}{2\pi mkT} \right)^{3/2} \right] \exp(-\beta \varepsilon_s^{tr}) \ll 1 \tag{7.32}$$

for all s, and this will certainly be the case if

$$\frac{N}{V} \left(\frac{h^2}{2\pi mkT} \right)^{3/2} \ll 1 \ . \tag{7.33}$$

Eq. (7.33) is the *condition for the classical regime* to hold. (It is a sufficient condition.) The inequality (7.33) would certainly hold in the limit $h \to 0$. This is of course the classical limit used by Maxwell and Boltzmann. For this reason the statistics which were derived in sections 7.1 and 7.2 are called *classical* or *Maxwell-Boltzmann statistics*. We see from Eq. (7.33) that classical statistics will hold better at high temperatures and low densities, just as expected, so that the perfect gas laws should hold. Before obtaining quantitative estimates of the inequality (7.33), we shall rewrite it to bring out its physical meaning.

According to quantum mechanics, a particle of momentum p has associated with it a de Broglie wavelength

$$\lambda_{dB} = \frac{h}{p} = \frac{h}{\sqrt{(2m\varepsilon^{tr})}} \tag{7.34}$$

where we used Eq. (7.16) for the translational kinetic energy ε^{tr}. We shall show below (see Eq. (7.42)) that under classical conditions the mean kinetic energy of a gas molecule is

$$\overline{\varepsilon^{tr}} = \tfrac{3}{2} kT$$

where T is the temperature of the gas. Hence the de Broglie wavelength of a gas molecule is of the order of

$$\lambda_{dB} = \frac{h}{\sqrt{(3mkT)}} = \left(\frac{2\pi}{3} \right)^{1/2} \left(\frac{h^2}{2\pi mkT} \right)^{1/2} \ . \tag{7.34a}$$

Thus Eq. (7.33) becomes

$$\left(\frac{3}{2\pi} \right)^{3/2} \frac{N}{V} \lambda_{dB}^3 \ll 1 \ . \tag{7.35}$$

If we introduce the mean distance between molecules

$$l = \left(\frac{V}{N} \right)^{1/3} \tag{7.36}$$

and omit the unimportant factor $(3/2\pi)^{3/2}$, we finally obtain as condition for the classical regime to hold

$$\lambda_{dB}^3 \ll l^3 \ . \tag{7.37}$$

The de Broglie wavelength λ_{dB} *must be small compared to the mean separation l of molecules.* If condition (7.37) holds the wave nature of the molecules is not important. The spacing between molecules is too large for wave-mechanical interference effects between the de Broglie waves of different molecules to be significant. We can treat the molecules as particles obeying Newtonian mechanics. For λ_{dB} comparable to l, interference effects are important. The condition (7.31) no longer holds and our derivation, in sections 7.1 and 7.2, of classical statistics breaks down. We must use quantum statistics. We shall return to this problem in Chapters 9 to 11.

If we estimate the quantities in Eq. (7.37) for a gas under normal conditions, say helium at 273 K and a density of 10^{20} molecules per cubic centimeter, we obtain

$$l \approx 2 \times 10^{-7} \, \text{cm} \ ,$$

and from Eq. (7.34a), with $m = 4/(6 \times 10^{23})$ g,

$$\lambda_{dB} \approx 0.8 \times 10^{-8} \, \text{cm} \ .$$

Under ordinary conditions, quantum-mechanical effects are negligible in a gas. It follows from the above numbers that the same is true for a gas even at very low temperatures.

From the interpretation of Eq. (7.37) one would expect this equation generally to be a criterion for the conditions under which quantum-mechanical effects are important. We give two examples.

For liquid helium the atomic volume is $V/N \approx 5 \times 10^{-23}$ cm^3 per atom, so that $l \approx 4 \times 10^{-8}$ cm. A de Broglie wavelength $\lambda_{dB} \approx 4 \times 10^{-8}$ cm corresponds to a temperature $T \approx 10$ K, so that for liquid helium (normal boiling point 4.2 K) one would expect quantum effects to be important.

Secondly we consider the conduction electrons in a metal. In many respects these behave like a gas being freely mobile within the metal. Because of their small mass compared with the atomic masses we have dealt with so far the de Broglie wavelength, calculated from Eq. (7.34a), will be relatively large. If we assume one conduction electron per atom, then $V/N \approx 10^{-23}$ cm^3 per electron and $l \approx 2 \times 10^{-8}$ cm. With $m = 9.1 \times 10^{-28}$ g for the mass of the electron, a de Broglie wavelength of $\lambda_{dB} \approx 2 \times 10^{-8}$ cm now corresponds to a temperature $T \approx 3 \times 10^5$ K. Hence we expect that quantum statistics are essential to describe the conduction electrons in a metal at all temperatures.

7.4 THE EQUATION OF STATE

In Chapter 1 we characterized a perfect classical gas by two equations, the equation of state

$$PV = NkT \qquad (1.9)$$

and Joule's law

$$E = E(T) \qquad (1.19)$$

which states that the internal energy E is independent of the volume of the gas and depends only on its temperature (and, of course, the number of molecules). We can now easily derive these results from the Helmholtz free energy.

The Helmholtz free energy F, Eq. (7.22), and hence all other thermodynamic properties, divide into two contributions from the translational and from the internal degrees of freedom:

$$F = F_{tr} + F_{int} \qquad (7.38)$$

where

$$F_{tr} = -NkT \ln \left\{ \frac{eV}{N} \left(\frac{2\pi mkT}{h^2} \right)^{3/2} \right\} \qquad (7.39)$$

and

$$\left. \begin{aligned} F_{int} &= -NkT \ln Z_{int}(T) \\ &= -NkT \ln \left\{ \sum_\alpha \exp(-\beta \varepsilon_\alpha^{int}) \right\} \end{aligned} \right\} . \qquad (7.40)$$

The first term, F_{tr}, leads to properties common to all perfect classical gases, while the second term depends on the internal structure of the gas molecules. The equation of state follows directly from

$$P = -\left(\frac{\partial F}{\partial V} \right)_{T,N} . \qquad (4.11)$$

Since F_{int} is independent of the volume, Eqs. (7.38) to (7.40) at once lead to

$$PV = NkT \qquad (7.41)$$

as the equation of state of a perfect classical gas *irrespective of its internal molecular structure*. The derivation of this equation from our statistical premises establishes the identity of the thermodynamic and perfect gas temperature scales. A different proof, based on the Carnot cycle, was given in section 5.2.

From

$$E = -\left(\frac{\partial \ln Z}{\partial \beta}\right)_{V,N} \tag{2.26}$$

and $\ln Z = -F/kT$, Eqs. (7.38) to (7.40) give $E = E_{\text{tr}} + E_{\text{int}}$ with

$$E_{\text{tr}} = (\tfrac{3}{2}kT)N \tag{7.42}$$

and

$$E_{\text{int}} = -N\left[\frac{\mathrm{d}}{\mathrm{d}\beta} \ln Z_{\text{int}}(T)\right] . \tag{7.43}$$

Eq. (7.42) gives the kinetic energy associated with the translational motion of the molecules as a whole. The energy $\tfrac{3}{2}kT$ per molecule or $\tfrac{1}{2}kT$ per translational degree of freedom illustrates the theorem of equipartition of energy, discussed in section 7.9.1.

Eq. (7.43) gives the energy associated with the internal degrees of freedom of the molecules. (The reader should distinguish between this quantity, which we denote by E_{int}, and the energy of the gas, in thermodynamics frequently called the internal energy of the gas, which we denote by E.) Since both E_{tr} and E_{int} are independent of the volume of the gas, it follows that the energy E is independent of the volume of the gas, i.e. we have derived Eq. (1.19).

★ 7.5   THE HEAT CAPACITY

It follows from Eq. (1.19) that the heat capacity at constant volume,

$$C_V = \left(\frac{\partial E}{\partial T}\right)_V , \tag{7.44}$$

is independent of the volume and depends only on the temperature. This conclusion holds for any perfect classical gas, irrespective of its internal molecular structure.

From Eq. (7.42), the contribution to the heat capacity at constant volume from the translational motion is

$$C_V^{\text{tr}} = \tfrac{3}{2}kN . \tag{7.45}$$

In addition there is a contribution to the heat capacity from the internal motion of the molecule which one obtains from Eq. (7.43) and which, unlike the translational part C_V^{tr}, depends on the temperature.

For monatomic gases (i.e. the inert gases) the molar heat capacity at constant pressure C_P is observed to have the value $20.8 \text{ JK}^{-1} \text{mol}^{-1}$ over wide ranges of temperature. Since for a perfect gas

$$C_P - C_V = R \tag{1.20}$$

with $R = N_0 k = 8.31 \text{ J K}^{-1} \text{mol}^{-1}$, it follows that the internal degrees of freedom of a monatomic gas make no contribution to the heat capacity under ordinary conditions: they are not excited. The probability p_α that a molecule should be in the state of internal motion α, with energy $\varepsilon_\alpha^{\text{int}}$, is given by

$$p_\alpha = \frac{1}{Z_{\text{int}}} \exp(-\beta \varepsilon_\alpha^{\text{int}}) \ . \tag{7.46}$$

Let

$$\Delta \varepsilon \equiv \varepsilon_2^{\text{int}} - \varepsilon_1^{\text{int}} \tag{7.47}$$

be the energy difference between the first excited state of the internal motion of a molecule and its ground state. If at the temperature T of the gas

$$\Delta \varepsilon \gg kT \tag{7.48}$$

then it follows from Eqs. (7.46) and (7.47) that all the molecules will be in the ground state. The internal degrees of freedom of the molecule are not excited: they are said to be frozen or dormant. In applying Eq. (7.48) in practice, it is useful to remember that for

$$T = 300 \text{ K}: \quad kT \approx 1/40 \, eV \ . \tag{7.49}$$

(More exactly $1 \, eV \equiv 11{,}605$ K.)

For a monatomic gas the only internal excitations are the electronic excitations of the atoms. From the study of atoms (e.g. from atomic spectra or from Franck–Hertz experiments) one knows that the value of $\Delta \varepsilon$ is typically a few electron volts. Hence only the translational motion contributes to the heat capacity of monatomic gases under ordinary conditions. *The electronic degrees of freedom are not excited.*

For di- or polyatomic gases the situation is more complex and we shall only comment on it briefly. For a detailed discussion see, for example, Hill,[5] or Rushbrooke.[14] (See also problems 7.1, 7.2 and 7.9 to 7.11.)

The electronic degrees of freedom again are usually only excited at very high temperatures. In addition the molecules are now capable of vibrational and rotational motions. The energies of these motions are known from the study of molecular spectra.

For diatomic molecules the energy difference $\Delta \varepsilon_{\text{vib}}$ between the first excited vibrational state and the vibrational ground state is typically a few tenths of an electron volt so that vibrations will not be excited at ordinary temperatures. For example, for N_2 where $\Delta \varepsilon_{\text{vib}} \approx 0.3 \, eV$ the temperature at which vibrational excitation becomes significant is, from Eq. (7.46), of the order of $\Delta \varepsilon_{\text{vib}}/k \approx 3{,}500$ K. At room temperature (300 K) the fraction of molecules vibrationally excited is about e^{-12}.

For polyatomic gases there exist several independent vibrational modes and for some of these the spacings of the vibrational levels $\Delta\varepsilon_{vib}$ may be considerably smaller than for diatomic molecules. (E.g. for BF_3, $\Delta\varepsilon_{vib} = 0.054\,eV$ corresponding to a temperature $\Delta\varepsilon_{vib}/k = 630\,K$.) Hence for polyatomic gases, vibrational degrees of freedom make appreciable contributions to the heat capacity even at room temperature or below.

The spacings of the rotational energy levels of almost all di- and polyatomic molecules are very small. Hence the rotational degrees of freedom are generally fully excited and make their full contribution to the heat capacity: $\frac{1}{2}k$ per molecule for each rotational degree of freedom. This result follows from the quantal treatment of the rotations (see problem 7.1). It also follows from the classical equipartition theorem (section 7.9.1). Linear molecules have only two rotational degrees of freedom corresponding to rotations about two axes normal to the line of the molecule. Nonlinear molecules have three rotational degrees of freedom. Hence the molar rotational heat capacities are R and $3R/2$ for linear and nonlinear molecules respectively.

For most diatomic gases the observed molar heat capacity at constant volume is $5R/2$ in agreement with the above theory. For small light molecules, such as H_2, HD and D_2, the molar heat capacities at constant volume decrease from $5R/2$ to $3R/2$ at comparatively high temperatures. For HD, for example, this decrease sets in fairly sharply as the temperature drops below 40 K, suggesting that at these lower temperatures the rotational degrees of freedom of HD are not fully excited. This dependence on the size of the molecule (actually on its moments of inertia) can be understood quantum mechanically (see problem 7.1).

We mention, in passing, that for molecules containing identical atoms, e.g. H_2 or D_2, additional effects occur which can be understood in terms of the quantum mechanics of identical particles (see problem 7.1).

Very good agreement is obtained between theory and experiment for the temperature variation of the heat capacity of gases, verifying the correctness of the above sort of considerations. For some detailed comparisons, the reader is referred to the books listed earlier in this section.

★ 7.6   THE ENTROPY

The entropy is simply expressed in terms of the Helmholtz free energy

$$F = E - TS \ , \tag{2.37}$$

from which

$$dF = dE - T\,dS - S\,dT \ .$$

If we combine this equation with the fundamental thermodynamic relation

$$T\,dS = dE + P\,dV \tag{4.12}$$

we obtain

$$dF = -S\,dT - P\,dV$$

whence

$$S = -\left(\frac{\partial F}{\partial T}\right)_{V,N}. \tag{7.50}$$

We obtain two contributions to the entropy, one from the translational and one from the internal motions. For the former we obtain, from Eq. (7.39) for F_{tr},

$$S_{\text{tr}} = Nk\left\{\ln\frac{V}{N} + \frac{3}{2}\ln T + \frac{5}{2} + \frac{3}{2}\ln\frac{2\pi mk}{h^2}\right\}. \tag{7.51}$$

The entropy (7.51) tends to infinity as $T \to 0$ (and similarly the heat capacity (7.45) does *not* go to zero as $T \to 0$), in violation of the third law of thermodynamics (section 4.7). This difficulty would not have occurred with the original definition (7.14) of the partition function as a sum over discrete states but results from the replacement of this sum by an integral in Eq. (7.18). This replacement is not justified near the absolute zero of temperature where the ground state will be important. For the ground state (for which the momentum $p = 0$) is excluded altogether by Eq. (7.18). At higher temperatures the ground state is not important, and the density of single-particle levels is high so that the replacement of a sum by an integral is permissible.

In section 5.4.1 we derived the entropy of a perfect gas on the assumption that the heat capacity of the gas is constant. We have seen in section 7.5 that this is a good approximation over large temperature ranges. Under these conditions our earlier result (5.23c) and Eq. (7.51) agree. Eq. (7.51) has the advantage that it determines the entropy completely, unlike Eq. (5.23c) which contained an unspecified constant of integration. We note also that Eq. (7.51) gives a correctly extensive definition for the entropy.

Eq. (7.51) allows us to obtain the temperature dependence of the vapour pressure of a saturated vapour. Putting

$$\frac{V}{N} = \frac{kT}{P}$$

in Eq. (7.51) and solving for $\ln P$ we obtain

$$\ln P = \frac{5}{2}\ln T + \frac{5}{2} + \ln\left[k^{5/2}\left(\frac{2\pi m}{h^2}\right)^{3/2}\right] - \frac{S_{\text{vap}}}{Nk}, \tag{7.52}$$

where S_{vap} shows we are dealing with a vapour phase for which $PV = NkT$. This equation of state usually holds for a saturated vapour in equilibrium with the condensed phase since the vapour pressure falls very rapidly as the temperature decreases below the critical point.* Hence, except close to the critical point, Eq. (7.52) will apply to such an equilibrium state of the liquid or solid with the saturated vapour.

Consider the condensed phase and saturated vapour in equilibrium at temperature T. If L is the molar latent heat of evaporation of the condensate at temperature T, then reversibly adding an amount of heat L to the condensate–vapour system, from a source at temperature T, will evaporate one mole of the condensed phase. If S_{vap} and S_c are the molar entropies of vapour and condensed phase, then

$$S_{vap} - S_c = \frac{L}{T}, \qquad (7.53)$$

and combining this equation with Eq. (7.52) gives

$$\ln P = -\frac{S_c}{R} - \frac{L}{RT} + \frac{5}{2}\ln T + \frac{5}{2} + \ln\left[k^{5/2}\left(\frac{2\pi m}{h^2}\right)^{3/2}\right]. \qquad (7.54)$$

This is the Sackur–Tetrode equation for the vapour pressure of a saturated vapour in equilibrium at temperature T with its condensed state. At sufficiently low temperatures the entropy S_c of the condensate is negligible compared with the entropy change L/T, and knowing L suffices to calculate the vapour pressure P. (If S_c cannot be neglected one can calculate it from measured heat capacities and, if necessary, latent heats.) Comparison of the vapour pressure calculated in this way with the observed vapour pressure then provides a direct test of the constant terms in the entropy expression (7.51). Such comparisons yield excellent agreement (see, for example, R. Fowler and E. A. Guggenheim, *Statistical thermodynamics*, Cambridge University Press, Cambridge, 1965, pp. 200 and 204). One can also compare Eq. (7.51) directly with the entropies obtained from measured heat capacities of gases, and such comparisons also give very good agreement (see, for example, G. N. Lewis and M. Randall, *Thermodynamics*, 2nd edn. (revised by K. S. Pitzer and L. Brewer), McGraw-Hill, New York, 1961, pp. 420–21). See also problems 7.3 and 8.7 which deal with these aspects.

---

*In Chapter 8 we shall consider in detail the equilibrium between different phases of a substance, e.g. a liquid in contact with its vapour. We shall see that there exists a temperature T_c, the critical temperature, such that for $T > T_c$ the liquid and vapour phases cannot be distinguished, while for $T < T_c$ the saturated vapour coexists in equilibrium with the liquid or, at sufficiently low temperatures, with the solid phase.

★ 7.7 THE MAXWELL VELOCITY DISTRIBUTION

We shall now derive the Maxwell velocity distribution for a perfect classical gas, as well as some related distributions and consequences. In sections 7.2 and 7.3 we derived the distribution of single-particle states in momentum space, Eq. (7.17), and the mean occupation number for each of these states given that there are N molecules altogether, Eq. (7.30).

Consider a gas of N molecules in an enclosure of volume V, in thermal equilibrium at temperature T. The mean number of molecules in a state with momentum of magnitude p is given, from Eqs. (7.30), (7.29), (7.16) and (7.19), by

$$\bar{n}(p) = \frac{N}{V}\left(\frac{h^2}{2\pi mkT}\right)^{3/2} \exp(-p^2/2mkT) \ . \tag{7.55}$$

Combining this equation with Eq. (7.17), we obtain the probability that a molecule in the gas possesses momentum of magnitude in the range p to $p+dp$:

$$P(p)dp = \frac{1}{N} f(p)dp\bar{n}(p)$$

$$= \frac{4\pi p^2 \, dp}{(2\pi mkT)^{3/2}} \exp(-p^2/2mkT) \ . \tag{7.56}$$

It follows from the method of derivation that Eq. (7.56) is correctly normalized, i.e.

$$\int_0^\infty P(p)dp = 1 \ , \tag{7.57}$$

as can be verified directly using the integrals of Appendix A.3. (Conversely, one can use Eq. (7.57) to find the correct proportionality factor in Eq. (7.56).)

Eq. (7.56) is the basic result. If we substitute

$$p = mv \tag{7.58}$$

into Eq. (7.56) we obtain the Maxwell speed distribution, i.e. the probability that a molecule should have a speed in the interval v to $v+dv$:

$$P(v)dv = 4\pi v^2 dv \left(\frac{m}{2\pi kT}\right)^{3/2} \exp(-mv^2/2kT) \tag{7.59a}$$

$$= \frac{4}{\sqrt{\pi}} u^2 du \exp(-u^2) \equiv F_1(u) du \tag{7.59b}$$

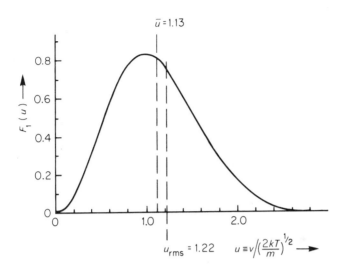

Fig. 7.4. The probability $F_1(u)\,du$, Eq. (7.59b), that a molecule should have a speed in the interval $(2kT/m)^{1/2}u$ to $(2kT/m)^{1/2}(u+du)$. ($\bar{u}$ = mean speed, u_{rms} = root-mean-square speed; both in units of $v_{max} = (2kT/m)^{1/2}$.)

where

$$u \equiv v \Big/ \left(\frac{2kT}{m}\right)^{1/2} \tag{7.60}$$

is the speed measured in units of $(2kT/m)^{1/2}$. This speed distribution is shown in Fig. 7.4. The maximum of the distribution occurs at $u = 1$, i.e. at a speed

$$v_{max} = \left(\frac{2kT}{m}\right)^{1/2} . \tag{7.61}$$

This is the unit of velocity which was used in Eqs. (7.59b) and (7.60).

The calculation of mean values in the kinetic theory involves integrals of the kind

$$I_n(a) = \int_0^\infty dx\, x^n\, e^{-ax^2} .$$

These are evaluated in Appendix A.3. Using the results of this appendix, one obtains from Eq. (7.59b) the mean speed $\bar{v}$ and the root-mean-square speed v_{rms} (see problem 7.5).

$$\bar{v} = \frac{2}{\sqrt{\pi}}\, v_{max}, \qquad v_{rms} = \sqrt{\tfrac{3}{2}}\, v_{max} \tag{7.62}$$

so that v_{max}, $\bar{v}$ and v_{rms} have nearly the same values. The rms-speed for N_2 molecules at 300 K is about 5×10^4 cm/sec.

The probability distribution that a molecule should have translational kinetic energy in the range E to $E + dE$, obtained by substituting $E = p^2/2m$ in Eq. (7.56) is given by

$$P(E)\,dE = \frac{2}{\sqrt{\pi}} \frac{\sqrt{E}\,dE}{(kT)^{3/2}} \exp(-E/kT) \qquad (7.63a)$$

$$= \frac{2}{\sqrt{\pi}} \sqrt{\varepsilon}\,d\varepsilon \exp(-\varepsilon) \equiv F_2(\varepsilon)\,d\varepsilon , \qquad (7.63b)$$

where

$$\varepsilon \equiv \frac{E}{kT} \qquad (7.64)$$

is the energy measured in units of kT. This energy distribution is shown in Fig. 7.5. The mean kinetic energy $\bar{E}$ per molecule is, from Eqs. (7.63) and (7.64), given by

$$\bar{E} = \tfrac{3}{2}kT , \qquad (7.65)$$

in agreement with our earlier result (7.42) for the translational energy of a gas, and with Eq. (7.62) for the rms-speed.

The two 'dashed' curves in Fig. 7.5 show the two factors making up the energy distribution (7.63b):

$$F_2(\varepsilon) = e^{-\varepsilon}[2(\varepsilon/\pi)^{1/2}] ,$$

i.e. the Boltzmann factor $e^{-\varepsilon}$ and the normalized density of states $[2(\varepsilon/\pi)^{1/2}]$. These two factors respectively decrease and increase with energy so that $F_2(\varepsilon)$ has a maximum. This is quite a broad maximum corresponding to the physical interpretation of $F_2(\varepsilon)$: it is the distribution of translational kinetic energy of one molecule in a gas at temperature T, and this is of course not a sharply defined quantity but changes with every collision of that molecule.

We can now easily show that the energy distribution of a gas of N molecules in a heat bath at temperature T has an extremely sharp maximum. We are again not interested in the internal properties of the molecules but only in the total translational kinetic energy. We therefore omit the internal partition function Z_{int} from Eq. (7.13) and obtain, from Eq. (7.11), for the translational partition function of the N-molecule gas:

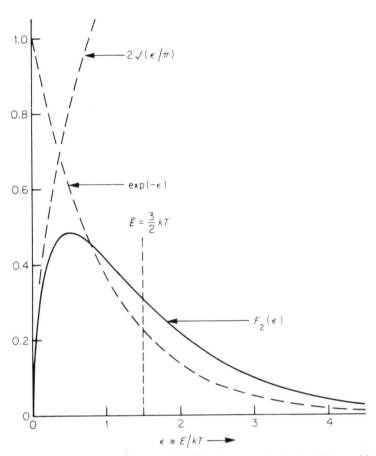

Fig. 7.5. The probability $F_2(\varepsilon)\,\mathrm{d}\varepsilon$, Eq. (7.63b), that a molecule should have kinetic energy in the interval $(kT)\varepsilon$ to $(kT)(\varepsilon + \mathrm{d}\varepsilon)$. ($\bar{E}=$ mean kinetic energy per molecule.) The two dashed curves are the two factors which make up $F_2(\varepsilon)$: the Boltzmann factor $\mathrm{e}^{-\varepsilon}$ and the normalized density of states $2\sqrt{(\varepsilon/\pi)}$.

$$Z_{\mathrm{tr}} \equiv Z_{\mathrm{tr}}(T,V,N) = \frac{1}{N!}\,[Z_1^{\mathrm{tr}}(T,V)]^N \; . \qquad (7.66)$$

We substitute for Z_1^{tr} from Eq. (7.18). For this purpose we use Eq. (7.26) to rewrite (7.18) as a three-dimensional integral with respect to $\mathrm{d}^3\mathbf{p}$. Denoting the variables of integration of the N factors in Eq. (7.66) by $\mathbf{p}_1, \mathbf{p}_2, \ldots \mathbf{p}_N$, we can write that equation

$$Z_{\mathrm{tr}} = \frac{V^N}{N!\,h^{3N}} \int \exp\left[-\beta(p_1{}^2 + p_2{}^2 + \ldots p_N{}^2)/2m\right]\mathrm{d}^3\mathbf{p}_1 \ldots \mathrm{d}^3\mathbf{p}_N \; . \qquad (7.67)$$

Since

$$E = \frac{1}{2m} (p_1^2 + p_2^2 + \ldots + p_N^2) \tag{7.68}$$

is the total translational kinetic energy of the gas, we see that the exponential in Eq. (7.67) is just the Boltzmann factor $\exp(-\beta E)$, which we expect in the energy distribution.

We want to compare Eq. (7.67) with the usual definition of the partition function

$$Z = \sum_{E_r} g(E_r) \exp(-\beta E_r) \ . \tag{2.24}$$

If the energy levels E_r lie very closely together so that there are $f(E) dE$ states in the interval E to $E + dE$, then we can write Eq. (2.24) as

$$Z = \int e^{-\beta E} f(E) dE \ . \tag{7.69}$$

To compare Eqs. (7.67) and (7.69), we introduce new dimensionless variables of integration in Eq. (7.67):

$$\mathbf{p}_i = \mathbf{x}_i \sqrt{(2mE)}, \qquad i = 1, \ldots N \ .$$

Eq. (7.67) then reduces to an integral with respect to E. From purely dimensional considerations (counting factors of E) it follows that Eq. (7.67) becomes

$$Z_{\mathrm{tr}} = A \int e^{-\beta E} E^{3N/2 - 1} dE \ . \tag{7.70}$$

(We shall not require the value of the constant of proportionality A which arises from integrations over variables other than E in Eq. (7.67).)

Eq. (7.70) is our final result: the translational partition function for a perfect classical gas of N molecules. Comparison with Eq. (7.69) shows that the density of states* for this system is given by

$$f(E) dE = A E^{3N/2 - 1} dE. \tag{7.71}$$

This equation gives the dependence of the density of states on the number of molecules N and is seen to be a very steeply increasing function of N. This is a consequence of the fact that the number of ways in which a given energy E can be distributed over N molecules increases enormously rapidly as N increases.

---

*We are of course considering translational states only, and not internal states of motion of the molecules.

From Eq. (7.71) we obtain directly $\Phi(E)$, the number of states with energy less than or equal to E:

$$\Phi(E) = \int_0^E f(E') \, dE' = \text{const.} \, E^{3N/2} \ . \tag{7.71a}$$

(In Eq. (7.71a) we have omitted the constant of integration—it has the value unity since $\Phi(0) = 1$—which in practice is completely negligible.)

From Eq. (7.70) one obtains at once the probability $P(E) \, dE$ that the total translational energy of the gas of N molecules at temperature T should lie in the range E to $E + dE$:

$$P(E) \, dE = A' \, e^{-\beta E} E^{3N/2 - 1} \, dE \ , \tag{7.72}$$

where the constant of proportionality A' (which is of course different from that in Eq. (7.70)) follows from the normalization of probability

$$\int P(E) \, dE = 1 \ . \tag{7.73}$$

(Note for any reader who is puzzled by Eq. (7.72): this equation is related to Eq. (7.70) in the same way in which Eqs. (2.25) and (2.24) are related. Compare also Eq. (2.32).)

In section 2.5 we showed generally that for a macroscopic system in a heat bath the energy is sharply defined, i.e. the energy fluctuations are extremely small (see Chapter 2, pp. 57–60). For the perfect classical gas we can calculate the mean energy $\bar{E}$ and the standard deviation ΔE, Eq. (2.28), explicitly from the energy distribution (7.72). For one molecule we found above a broad energy distribution. As N becomes large, Eq. (7.72) leads to a very sharply defined energy. From Eq. (7.72) one obtains (see problem 7.6) for the mean translational energy

$$\bar{E} = N \cdot \tfrac{3}{2} kT \tag{7.74}$$

and for the corresponding relative fluctuation

$$\frac{\Delta E}{\bar{E}} = \left(\frac{2}{3N} \right)^{1/2} , \tag{7.75}$$

in agreement with Eq. (7.65) (the energy is additive since the molecules are not interacting) and with Eqs. (2.30) and (2.31).

We next consider the Maxwell velocity distribution, i.e. the probability that a molecule possesses a velocity within the range $\mathbf{v}$ to $\mathbf{v} + d^3\mathbf{v}$, i.e. within the velocity space volume element $d^3\mathbf{v}$ which is situated at the velocity $\mathbf{v}$

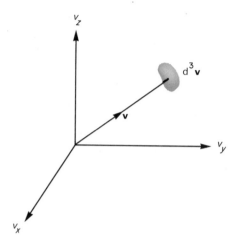

Fig. 7.6. The element $d^3\mathbf{v}$ of velocity space situated at $\mathbf{v}$.

(see Fig. 7.6). To derive this distribution we should go back to Eq. (7.28). However, since the velocity distribution must be isotropic, we can infer it from the speed distribution (7.59a):

$$P(\mathbf{v})d^3\mathbf{v} = d^3\mathbf{v}\left(\frac{m}{2\pi kT}\right)^{3/2}\exp(-mv^2/2kT) \ . \tag{7.76}$$

By integrating Eq. (7.76) over all directions of the velocity $\mathbf{v}$, but keeping its magnitude within the range v to $v+dv$, we get back to Eq. (7.59a).

The probability distribution for each of the cartesian components of $\mathbf{v}$ is obtained from Eq. (7.76) by writing

$$v^2 = v_x{}^2 + v_y{}^2 + v_z{}^2$$

and taking $d^3\mathbf{v} = dv_x dv_y dv_z$. Eq. (7.76) then factorizes into three separate identical distributions for each component; for example, for the x-component v_x one obtains

$$P(v_x)dv_x = dv_x\left(\frac{m}{2\pi kT}\right)^{1/2}\exp(-mv_x{}^2/2kT) \tag{7.77a}$$

$$= \frac{1}{\sqrt{(2\pi)}}\, dw\, \exp(-\tfrac{1}{2}w^2) \equiv F_3(w)dw \ , \tag{7.77b}$$

where

$$w \equiv v_x\left/\left(\frac{kT}{m}\right)^{1/2}\right. \tag{7.78}$$

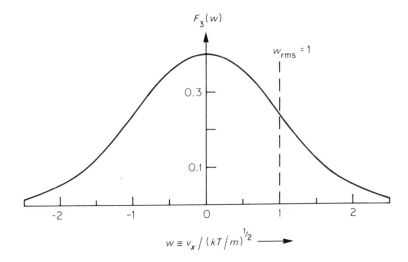

Fig. 7.7. The probability $F_3(w)\,dw$, Eq. (7.77b), that a molecule should have an x-component of velocity in the interval $(kT/m)^{1/2}w$ to $(kT/m)^{1/2}(w+dw)$. ($w_{rms}=$ root-mean-square value of the velocity component, in units of $(kT/m)^{1/2}$.)

is the velocity component measured in units of $(kT/m)^{1/2}$. This distribution is shown in Fig. 7.7. The rms-value of v_x is given by

$$(v_x)_{rms} = \left(\frac{kT}{m}\right)^{1/2}. \tag{7.79}$$

This agrees with the rms-speed, Eqs. (7.62) and (7.61), since

$$\overline{v_x^2} = \overline{v_y^2} = \overline{v_z^2} = \tfrac{1}{3}\overline{v^2}.$$

★ 7.8 REAL GASES

In this section we shall consider real gases. We want to show that at sufficiently low densities a real gas behaves like a perfect gas, and how deviations from perfect gas behaviour arise as the density increases.

The molecules of a real gas are subject to an intermolecular potential whose qualitative features are shown in Fig. 7.8.* The potential has a long-range weak attractive part. This is the van der Waals attraction due to the induced electric dipole moments between molecules. At small distances there is a very

---

*A good semiempirical representation of the intermolecular potential is given by the Lennard–Jones (6, 12)-potential. For further discussion of intermolecular potentials, see Flowers and Mendoza,[26] Hill,[28] and Present.[11]

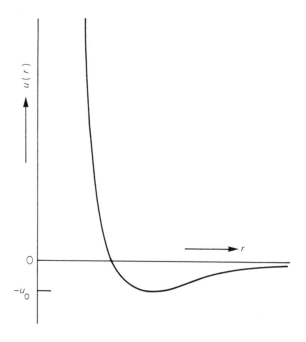

Fig. 7.8. Sketch of the intermolecular potential $u(r)$, i.e. the potential energy of two molecules when a distance r apart.

strong repulsive force between molecules. This is due to the size of the molecules. This force comes into play when two molecules approach sufficiently closely for their electron clouds to begin to overlap. To understand this repulsive force requires quantum mechanics: it is a consequence of the Pauli exclusion principle.

One way of obtaining information about the intermolecular potential is from the transport properties of a gas, for example from measurements of viscosity. Transport properties depend directly on molecular collisions, i.e. on the law of force governing these collisions. For example, in elementary treatments of transport properties one employs the mean free path which depends on the collision cross-section.

A second way of learning about the intermolecular potential is from the equation of state; by studying deviations from the ideal gas equation $PV = NkT$. It is this aspect which we shall consider in this section, generalizing the perfect gas treatment of section 7.2. We can no longer ascribe a 'private' energy to each molecule of a real gas since the molecules of a real gas interact with each other. The fact that the molecules interact makes this a very hard problem. For interacting molecules condensation from the gas to the liquid occurs at appropriate densities and temperatures, and one possesses only a very limited understanding of such phase transitions. We shall not discuss these 'real' problems which occur at high densities, but shall limit ourselves to considering how deviations from perfect gas behaviour first show up as the density of the gas is increased.*

Following the procedure of section 7.2 we shall now obtain the partition function for a real gas. We first rewrite the partition function of a perfect gas. Substituting Eqs. (7.13) and (7.18) in (7.11) gives

$$Z_P \equiv Z_P(T, V, N) = \frac{1}{N!} \left(\int_0^\infty \frac{V 4\pi p^2 \, dp}{h^3} \exp(-\beta p^2/2m) \right)^N (Z_{int}(T))^N \qquad (7.80)$$

where we have written Z_P instead of Z to indicate that we are dealing with a perfect gas. In the last equation we use Eq. (7.26) to replace the integrals over p by ones over $\mathbf{p}$, and write the factor V as a volume integral $\int d^3\mathbf{r}$ over the volume of the gas. Hence Eq. (7.80) becomes

$$Z_P = \frac{1}{N!} \left(\int \frac{d^3\mathbf{r} \, d^3\mathbf{p}}{h^3} \exp(-\beta p^2/2m) \right)^N (Z_{int}(T))^N . \qquad (7.81)$$

Finally, we label the variables of integration in each of the N factors differently: $\mathbf{r}_1, \mathbf{p}_1, \ldots \mathbf{r}_N, \mathbf{p}_N$, and use the multiplicative property of exponentials ($e^A e^B = e^{(A+B)}$) to rewrite Eq. (7.81) as

$$Z_P = \frac{1}{N!} \int \frac{1}{h^{3N}} d^3\mathbf{r}_1 \ldots d^3\mathbf{r}_N d^3\mathbf{p}_1 \ldots d^3\mathbf{p}_N e^{-\beta K} (Z_{int}(T))^N , \qquad (7.82)$$

where

$$K = \frac{1}{2m}(\mathbf{p}_1^2 + \mathbf{p}_2^2 + \ldots + \mathbf{p}_N^2) \qquad (7.83)$$

is the total translational kinetic energy of the gas, when the molecules possess momenta $\mathbf{p}_1, \mathbf{p}_2, \ldots \mathbf{p}_N$.

In order to generalize Eq. (7.82) to the case of a real gas we shall merely use plausibility arguments.[†] We note these features of Eq. (7.82):

(i) By comparison with Eq. (7.28) we interpret

$$\frac{1}{h^{3N}} d^3\mathbf{r}_1 \ldots d^3\mathbf{r}_N d^3\mathbf{p}_1 \ldots d^3\mathbf{p}_N \qquad (7.84)$$

---

*For a more thorough discussion of this latter problem, see Becker,[2] Rushbrooke,[14] Wannier,[20] or Wilson.[21] (Note that many books talk of *imperfect*, rather than *real* gases.)
[†]Some of the points are discussed further in section 7.9.

as the number of states of the gas with molecule No. 1 in a volume element $d^3\mathbf{r}_1$ at $\mathbf{r}_1$ and with momentum within $d^3\mathbf{p}_1$ at $\mathbf{p}_1$, and similarly for the molecules labelled $2, 3, \ldots N$.

(ii) The factor $1/N!$ results, as explained in section 7.1, from the classical approximation. Since the molecules are not localized we cannot label them $1, 2, \ldots N$ in a definite order.

(iii) The internal partition function depends only on the internal properties of a single molecule.

(iv) The exponential $\exp(-\beta K)$ is the Boltzmann factor associated with the total translational kinetic energy K of the gas.

If we consider a real instead of a perfect classical gas, features (i) to (iii) are not altered. For (ii) and (iii) this is trivially obvious. For (i) one must show that the density of states (7.28), which was derived for free particles only, is also valid for interacting particles.* On the other hand (iv) must be modified: for interacting molecules we must replace the Boltzmann factor $\exp(-\beta K)$ by the Boltzmann factor

$$\exp[-\beta H(\mathbf{r}_1, \mathbf{r}_2, \ldots \mathbf{r}_N, \mathbf{p}_1, \mathbf{p}_2, \ldots \mathbf{p}_N)] \ . \tag{7.85}$$

Here

$$H(\mathbf{r}_1, \ldots \mathbf{p}_N) = K + U(\mathbf{r}_1, \ldots \mathbf{r}_N) \tag{7.86}$$

is the sum of the total translational kinetic energy K, Eq. (7.83), and of the total potential energy U due to the molecules interacting with each other. U depends only on the position coordinates $\mathbf{r}_1, \ldots \mathbf{r}_N$ of the molecules, K only on their momenta $\mathbf{p}_1, \ldots \mathbf{p}_N$.

We conclude therefore that the partition function $Z(T, V, N)$ for the real gas is obtained from that for the perfect gas by merely replacing $\exp(-\beta K)$ in Eq. (7.82) by the Boltzmann factor $\exp(-\beta H)$, Eq. (7.85):

$$Z \equiv Z(T, V, N)$$

$$= \frac{1}{N!} \int \frac{1}{h^{3N}} \, d^3\mathbf{r}_1 \ldots d^3\mathbf{r}_N d^3\mathbf{p}_1 \ldots d^3\mathbf{p}_N e^{-\beta H} (Z_{\text{int}}(T))^N \ . \tag{7.87}$$

Now H is the sum of two terms: K which depends only on the momenta, and U which depends only on the position coordinates. Hence the integral in Eq. (7.87), and with it the partition function Z, factorizes, giving

*See, for example, Hill,[5] Chapter 6 and section 22.6, or Rushbrooke,[14] Chapter 4. Both these books contain further references.

$$Z = \left\{ \frac{1}{N!} \int \frac{1}{h^{3N}} \, d^3\mathbf{p}_1 \ldots d^3\mathbf{p}_N e^{-\beta K} (Z_{\text{int}}(T))^N \right\}$$

$$\times \left\{ \int d^3\mathbf{r}_1 \ldots d^3\mathbf{r}_N e^{-\beta U} \right\}$$

$$= Z_P \left\{ \frac{1}{V^N} \int d^3\mathbf{r}_1 \ldots d^3\mathbf{r}_N e^{-\beta U} \right\}$$

$$\equiv Z_P Q_N(T, V) \tag{7.88}$$

where Z_P is the partition function of a perfect gas and

$$Q_N \equiv Q_N(T, V) = \frac{1}{V^N} \int d^3\mathbf{r}_1 \ldots d^3\mathbf{r}_N e^{-\beta U} \tag{7.89}$$

is known as the configurational partition function of the gas.

Eq. (7.88) is our final expression for the partition function Z. We see that Z factorizes into the perfect gas partition function Z_P, which has already been evaluated in Eq. (7.20), and the configurational partition function Q_N. It is a consequence of this factorization that the momentum (i.e. velocity) distribution, which depends only on Z_P, and the spatial distribution, which depends only on Q_N, are independent of each other. Classically one always deals with the Maxwell velocity distribution. The forces which the particles experience determine their spatial configuration only.

We see now that in order to treat a real gas by statistical mechanics, one must calculate the configurational partition function Q_N, Eq. (7.89). A systematic procedure for this very complex problem has been developed by Ursell and by Mayer. We shall only indicate the basic ideas and obtain the lowest approximation, i.e. the onset of deviations from perfect gas behaviour at comparatively low densities.

We shall assume that the potential energy U of the gas results from intermolecular potentials between pairs of molecules which have the qualitative features shown in Fig. 7.8. If we denote this potential energy between molecules i and j, when a distance $r_{ij} \equiv |\mathbf{r}_i - \mathbf{r}_j|$ apart, by $u_{ij} \equiv u(r_{ij})$ then we can write the potential energy U of the gas of N molecules as

$$U = \sum u_{ij} \tag{7.90}$$

where the summation is over all pairs of particles. (In detail: $i = 1, \ldots N$; $j = 1, \ldots N$; with $i < j$. The last inequality ensures that each pair of molecules is counted only once.) Substituting Eq. (7.90) into Eq. (7.89), we obtain

$$Q_N = \frac{1}{V^N} \int d^3\mathbf{r}_1 \ldots d^3\mathbf{r}_N \exp\left(-\beta \sum u_{ij}\right)$$

$$= \frac{1}{V^N} \int d^3\mathbf{r}_1 \ldots d^3\mathbf{r}_N \prod \exp(-\beta u_{ij}) \qquad (7.91)$$

where the product is over all pairs of molecules (with $i < j$).

[For the reader who found the last step difficult we add some explanations and intermediate steps.

The notation $\prod$ denotes a product of factors just like Σ denotes a sum of terms, and the ranges of terms are indicated similarly; e.g.

$$\prod_{k=1}^{n} a_k \equiv a_1 a_2 \ldots a_n . \qquad (7.92)$$

Eq. (7.91) then follows from

$$\exp\left(\sum_{k=1}^{n} a_k\right) = \exp(a_1 + a_2 + \ldots + a_n)$$

$$= (\exp a_1)(\exp a_2) \ldots (\exp a_n)$$

$$= \prod_{k=1}^{n} (\exp a_k) , \qquad (7.93)$$

by taking $a_k = u_{ij}$ and letting k run over all pairs of molecules (with $i < j$). In our case there will be $\frac{1}{2}N(N-1)$ terms, i.e. $n = \frac{1}{2}N(N-1)$.]

The trick in evaluating Eq. (7.91) consists in introducing, instead of the exponentials involving u_{ij}, new quantities

$$\lambda_{ij} \equiv \lambda(r_{ij}) \equiv \exp(-\beta u_{ij}) - 1 , \qquad (7.94)$$

so that Eq. (7.91) becomes

$$Q_N = \frac{1}{V^N} \int d^3\mathbf{r}_1 \ldots d^3\mathbf{r}_N \prod (1 + \lambda_{ij})$$

$$= \frac{1}{V^N} \int d^3\mathbf{r}_1 \ldots d^3\mathbf{r}_N [1 + \sum \lambda_{ij} + \sum \lambda_{ij} \lambda_{kl} + \ldots] , \qquad (7.95)$$

where successive terms in the square parentheses arise as follows by picking terms from

$$\prod (1 + \lambda_{ij}) \equiv (1 + \lambda_{12})(1 + \lambda_{13}) \ldots (1 + \lambda_{N-1,N}) : \qquad (7.96)$$

the first term (1) results from picking the 1 from each factor $(1+\lambda_{ij})$; the second term $(\sum\lambda_{ij})$ results from picking in all possible ways *one* factor λ_{ij} and 1's from *all other* factors $(1+\lambda_{rs})$; the third term $(\sum\lambda_{ij}\lambda_{kl})$ results from picking in all possible ways *two* factors λ_{ij}, λ_{kl}, and 1's from *all other* factors $(1+\lambda_{rs})$; and so on. Thus the substitution (7.94) replaces the very complicated multiple integral (7.91), whose integrand is a product of very many (!) factors, by a sum of multiple integrals. Except for the first two integrals, these integrals are still very difficult to evaluate; (of course, only the first few have been calculated). Nevertheless Eq. (7.95) allows considerable analytic treatment (see the references in the footnote on p. 193) and successive terms in it possess a simple physical interpretation which relates to the deviations from ideal gas behaviour as the density of the gas is increased.

We shall now discuss this interpretation and then consider the first two terms in Eq. (7.95) in detail.

The meaning of the various terms in Eq. (7.95) follows from the way

$$\lambda(r) \equiv \exp\left[-\beta u(r)\right] - 1 \tag{7.94a}$$

depends on r. For the intermolecular potential $u(r)$ shown in Fig. 7.8, $\lambda(r)$ has the form shown in Fig. 7.9. $\lambda(r)$ differs appreciably from zero for small values of the intermolecular separation only, say $r \lesssim 2r_0$, where $2r_0$ is of the order of the diameter of one molecule. Thus λ_{12} is essentially zero unless molecules 1 and 2 are very close together, i.e. they are colliding. Similarly

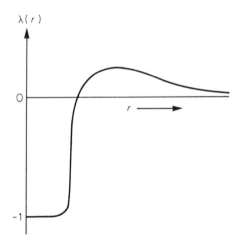

Fig. 7.9. Sketch of

$$\lambda(r) = e^{-\beta u(r)} - 1$$

for the intermolecular potential of Fig. 7.8.

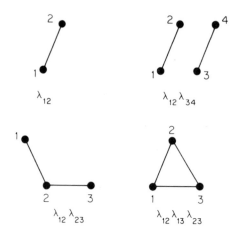

Fig. 7.10. Graphical representation of some of the terms
which occur in the integrand of Eq. (7.95).

$\lambda_{12}\lambda_{34}$ only differs from zero in simultaneous collisions of 1 with 2 and of 3 with 4, but a term such as $\lambda_{12}\lambda_{23}$ only differs from zero if molecules 1, 2 and 3 are all simultaneously close together, i.e. this term represents a triple collision between all three molecules. One obtains similar interpretations for products of three or more factors λ. A graphical representation of such terms is given in Fig. 7.10. Thus the various terms in Eq. (7.95) represent clusters of interacting molecules. The various multiple integrals are known as cluster integrals and the series (7.95) is called a cluster expansion. We shall only consider the first two terms of this expansion.

If we neglect all but the first term (1) in the expansion (7.95), i.e. we neglect all collisions, then Eq. (7.95) reduces to $Q_N = 1$. This is of course just the perfect gas approximation.

Consider next the second term ($\sum \lambda_{ij}$) in Eq. (7.95). There are altogether $\frac{1}{2}N(N-1)$ such terms in the sum $\sum \lambda_{ij}$ corresponding to the number of ways of picking pairs of molecules. These terms are of course all equal since they differ only in the way the variables of integration are labelled. Consequently we can write this second term in Eq. (7.95) as

$$\frac{1}{V^N} \frac{N(N-1)}{2} \int d^3\mathbf{r}_1 \ldots d^3\mathbf{r}_N [\exp\{-\beta u(r_{12})\} - 1] \ . \tag{7.97}$$

Integration with respect to $\mathbf{r}_3, \ldots \mathbf{r}_N$ over the volume of the gas gives a factor V^{N-2}. For the remaining integrations over $\mathbf{r}_1$ and $\mathbf{r}_2$ we introduce new variables of integration

$$\mathbf{r} = \mathbf{r}_1 - \mathbf{r}_2, \qquad \mathbf{R} = \tfrac{1}{2}(\mathbf{r}_1 + \mathbf{r}_2) \ .$$

The integration with respect to $\mathbf{R}$ gives a further factor V and expression (7.97) becomes

$$\frac{N^2}{2V} \int d^3\mathbf{r}(e^{-\beta u(r)} - 1) \equiv \frac{N^2}{2V} I_2 \tag{7.98}$$

where we also replaced $N(N-1)$ by N^2 since $N \gg 1$, and we introduced the abbreviation I_2 for the cluster integral in (7.98).

Using these results, we can write the configurational partition function (7.95) as

$$Q_N = \left\{ 1 + N\left(\frac{N}{2V} I_2\right) + \cdots \right\}. \tag{7.99}$$

The cluster integral I_2, Eq. (7.98), is of the order of magnitude of the volume v_0 of one molecule, since the integrand differs appreciably from zero only within such a volume and is of the order of unity there. Hence the factor $\left(\frac{N}{2V} I_2\right)$ in Eq. (7.99) is independent of N, i.e. of the size of the system, it is an intensive quantity.

Now we know that the Helmholtz free energy is an extensive quantity. From its logarithmic relation to the partition function [Eq. (2.36)], it follows that the latter must depend on N as the Nth power of some intensive quantity. For the perfect gas we had this result explicitly in Eqs (7.20) and (7.22). Since $Z = Z_P Q_N$, Eq. (7.88), it follows that Q_N by itself must be the Nth power of some intensive quantity. If this latter expression for Q_N is to be identical with Eq. (7.99), then it must have the form

$$Q_N = \left(1 + \frac{N}{2V} I_2 + \cdots\right)^N, \tag{7.100}$$

where the later terms, indicated by . . ., must also be intensive.

We have seen that the second term in Eq. (7.100), $\frac{1}{2}NI_2/V$, is of the order of Nv_0/V. This ratio is small for a gas at moderate densities, since Nv_0 is the total volume occupied by the molecules of the gas. To estimate this ratio we take $v_0 = 3 \times 10^{-23}$ cm^3 (corresponding to a molecular radius of about 2 Å). For a gas at S.T.P. this gives $Nv_0/V \approx 10^{-3}$. Analysis of the later terms in Eq. (7.100) shows that these are of the order of higher powers of Nv_0/V. For a gas at low density we may therefore approximate Eq. (7.100) by

$$Q_N = \left(1 + \frac{N}{2V} I_2\right)^N. \tag{7.101}$$

This is our final expression for the configurational partition function. From it the onset of deviations from perfect gas behaviour at low densities follows. To see this, we introduce the Helmholtz free energy $F = -kT \ln Z$. From $Z = Z_P Q_N$ and Eq. (7.101) one obtains

$$F = F_P - kTN \ln \left(1 + \frac{N}{2V} I_2 \right)$$

$$= F_P - \frac{kTN^2}{2V} I_2 \tag{7.102}$$

where we approximated $\ln (1 + NI_2/2V)$ by $(NI_2/2V)$, which is permissible since the latter ratio is small, and we introduced the Helmholtz free energy of the perfect gas $F_P \equiv -kT \ln Z_P$.

We calculate the equation of state of the gas from

$$P = - \left(\frac{\partial F}{\partial V} \right)_{T,N} .$$

The first term (F_P) in (7.102) leads to the ideal gas equation; the second term gives the deviations for a real gas at low density. One obtains directly

$$P = \frac{NkT}{V} \left[1 - \frac{N}{2V} I_2 \right] . \tag{7.103}$$

The correction term is of the order of Nv_0/V. When this term is small, the gas behaves like a perfect gas.

Eq. (7.103) is the beginning of the virial expansion,* which is a representation of the equation of state as a power series in Nv_0/V. The higher-order terms of the virial expansion result from the later terms in Q_N, Eq. (7.100). Writing the virial expansion as

$$P = \frac{NkT}{V} \left[1 + \frac{N}{V} B(T) + \left(\frac{N}{V} \right)^2 C(T) + \cdots \right] , \tag{7.104}$$

we see that the second virial coefficient $B(T)$ is related to the cluster integral I_2, from Eqs. (7.103) and (7.98), through

$$B(T) = - \frac{1}{2} I_2 = \frac{1}{2} \int d^3 \mathbf{r} \, [1 - e^{-\beta u(r)}] . \tag{7.105}$$

---

*Readers of Chapter 5, *which is not presupposed at this stage*, may like to be reminded that the virial expansion was introduced in Eq. (5.34). We have slightly modified the notation in Eq. (7.104).

We conclude this section by calculating the second virial coefficient and the corresponding equation of state for the intermolecular potential shown in Fig. 7.8. We shall assume that the temperature is such that

$$kT \gg u_0 , \tag{7.106}$$

where $(-u_0)$ is the minimum value of the potential $u(r)$; cf. Fig. 7.8. Condition (7.106) states that the kinetic energy of the thermal motion of the molecules is large compared with the cohesive potential energy which tends to bring about condensation. Hence (7.106) is satisfied at temperatures sufficiently high so that the gas cannot liquefy, i.e. well above the critical temperature T_c.*

In order to simplify the calculation we shall not use the Lennard–Jones (6, 12) potential to describe the intermolecular potential of Fig. 7.8. Instead we shall slightly modify the potential of Fig. 7.8 retaining all its essential features. The part of the potential of Fig. 7.8 where $u(r) > 0$, i.e. where there is a very large but finite repulsive force $(-du/dr)$, we shall replace by an infinite repulsive force, as shown in Fig. 7.11(a). At $r = 2r_0$ this potential possesses a vertical slope. This corresponds to thinking of the molecules as possessing a hard core of radius r_0, beyond which they cannot be compressed. The intermolecular potential, shown in Fig. 7.11(a), has a repulsive region (to the left of the minimum in the figure) but it mainly contributes a weak long-range attractive force. Hence we shall refer to it as describing a gas of weakly attracting hard-core molecules.

If the inequality (7.106) holds, we can approximate the integrand in Eq. (7.105) by

$$-\lambda(r) \equiv [1 - e^{-\beta u(r)}] = \begin{cases} 1, & 0 < r < 2r_0, \\ \dfrac{u(r)}{kT}, & 2r_0 < r, \end{cases} \tag{7.107}$$

so that $\lambda(r)$ has the form shown in Fig. 7.11(b), which is to be compared with Fig. 7.9.

To calculate the second virial coefficient we first perform the angular integrations in Eq. (7.105). These are trivial since the integrand depends only on r. One obtains

$$B(T) = 2\pi \int_0^\infty dr\, r^2 [1 - e^{-\beta u(r)}] . \tag{7.108}$$

*In Chapter 8, where liquefaction of gases is considered, we shall see that there exists a critical temperature T_c such that condensation of the gas can only occur at a temperature $T < T_c$.

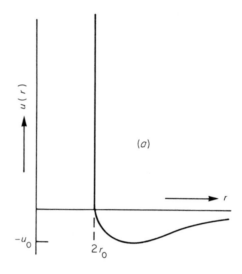

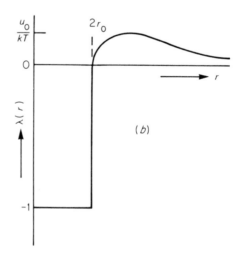

Fig. 7.11. A gas of 'weakly attracting hard-core' molecules. (a) The intermolecular potential $u(r)$. (b) The related function $\lambda(r)$, Eq. (7.94a).

Finally this integral is calculated using the approximate expressions (7.107), giving

$$B(T) = 2\pi \cdot \tfrac{1}{3}(2r_0)^3 + \frac{2\pi}{kT} \int_{2r_0}^{\infty} \mathrm{d}r r^2 u(r) \equiv b - \frac{a}{kT} \tag{7.109}$$

where the positive constant a is defined by

$$a \equiv -2\pi \int_{2r_0}^{\infty} \mathrm{d}r r^2 u(r) \tag{7.110a}$$

and the constant b is defined by

$$b \equiv 4\left(\frac{4\pi}{3} r_0^3\right) \equiv 4 v_0 \ , \tag{7.110b}$$

i.e. b is four times the volume of one molecule.

We substitute Eq. (7.109) for $B(T)$ in the virial expansion (7.104), dropping the third and higher virial coefficients $C(T)$, ... The resulting equation of state is

$$P = \frac{NkT}{V}\left[1 + \frac{N}{V}\left(b - \frac{a}{kT}\right)\right] \tag{7.111}$$

which can be written

$$\left(P + \frac{N^2}{V^2} a\right) = \frac{NkT}{V}\left(1 + \frac{N}{V} b\right) \ . \tag{7.112}$$

If we make the replacement

$$\left(1 + \frac{N}{V} b\right) \to \left(1 - \frac{N}{V} b\right)^{-1} \ , \tag{7.113}$$

then Eq. (7.112) goes over into van der Waals' equation

$$\left(P + \frac{N^2}{V^2} a\right)\left(1 - \frac{N}{V} b\right) = \frac{NkT}{V} \ . \tag{7.114}$$

Furthermore our interpretation of the constants a and b, Eqs. (7.110), are just those of van der Waals: a stems from the long-range weak attractive forces between molecules, and b is four times the volume of one molecule.

The replacement (7.113) is permissible if $Nb/V \ll 1$. However the virtue of van der Waals' equation is that it gives a good qualitative and, to *some* extent even semi-quantitative description of condensation phenomena* where Nb/V is certainly not small. Under these conditions the replacement (7.113) is not justified: for the limit of high densities we have *not* derived van der Waals' equation. This is hardly surprising since in deriving Eq. (7.112) we retained binary clusters only but omitted clusters of three or more molecules, and it will be clusters of very many molecules which are most important when condensation occurs. Furthermore, accurate experiments on substances near their critical point do not agree with van der Waals' equation. (See section 8.6.) So the absence of a theoretical justification of this equation is not to be regretted.

★ 7.9   CLASSICAL STATISTICAL MECHANICS

In sections 7.1 and 7.2 we considered a perfect gas. We started from a quantum viewpoint and derived essentially the results of classical statistical mechanics as formulated by Maxwell, Boltzmann and Gibbs.† We shall now obtain the corresponding classical results for more general systems, again allowing for modifications introduced by quantum aspects. A basic result of classical statistical mechanics is the theorem of equipartition of energy. In section 7.9.1 we shall derive the equipartition theorem from the classical probability distribution, and we shall apply it to a few problems.

In section 2.5 we gave the definition of the partition function as

$$Z = \sum_{E_r} g(E_r) \exp(-\beta E_r) , \qquad (7.115)$$

i.e. as a sum over discrete energy levels. According to classical mechanics a system does not have discrete energy levels. The energy, like all other variables characterizing the state of the system, is a continuous variable. To express Eq. (7.115) in classical terms we must find a classical way of characterizing states and counting them. In section 7.2 we solved this problem for the simple case of a point particle. We shall now generalize the approach of section 7.2 to apply to complicated systems and we must first of all consider the analytic description of such systems in classical mechanics. For reasons of space, we must limit ourselves to a very few remarks relevant for this section.‡

Classically, a system with v degrees of freedom is described by v *generalized coordinates* $q_1, q_2, \ldots q_v$. These coordinates vary with time. At any instant, they completely determine the configuration of the system. The statement

---

*See, for example, Flowers and Mendoza,[26] Chapter 7, or Hill,[5] section 16.1.

†Actually our results go a good deal further. For one thing we obtained a criterion for when the classical result applies. For another, we defined the *absolute* entropy of a system, involving Planck's constant, whereas genuine classical theory was only able to give the entropy up to an additive undetermined constant.

‡For a thorough treatment of classical mechanics see Goldstein[43] or Landau and Lifshitz.[44]

that the system has ν degrees of freedom means that the coordinates $q_1, q_2, \ldots q_\nu$ are independent variables: there are no relations between them expressing constraints. Consider, for example, two point particles constrained to move a fixed distance L apart. We could describe this system by the six cartesian coordinates (x_1, y_1, z_1), (x_2, y_2, z_2) of the two particles, subject to the constraint

$$(x_1 - x_2)^2 + (y_1 - y_2)^2 + (z_1 - z_2)^2 = L^2 .$$

A simpler description for this system with five degrees of freedom is to use, as the five generalized coordinates, the coordinates (x_1, y_1, z_1) of one particle and the polar angles θ, ϕ which define the direction of the vector $\mathbf{r}_2 - \mathbf{r}_1$. ($\theta$, ϕ are defined by: $L \sin \theta \cos \phi = x_2 - x_1$, $L \sin \theta \sin \phi = y_2 - y_1$, $L \cos \theta = z_2 - z_1$.)

To describe the state of *motion* of a system at a given instant of time we must specify both the position coordinates $q_1, q_2, \ldots q_\nu$ and the generalized velocities $\dot{q}_1, \dot{q}_2, \ldots \dot{q}_\nu$ (where $\dot{q} \equiv dq/dt$) at that instant. We shall restrict ourselves to systems for which the potential energy U is a function of the position coordinates $q_1, \ldots q_\nu$ only:

$$U \equiv U(q_1, \ldots q_\nu) . \tag{7.116a}$$

A more general formulation which allows for velocity-dependent forces, i.e. U is a function of the q's and the $\dot{q}$'s, is possible. This generalization is required in the important case of an electrically charged particle moving in a magnetic field. The kinetic energy K of the system at a given time will be a function of the generalized coordinates and velocities:

$$K \equiv K(q_1, \ldots q_\nu, \dot{q}_1, \ldots \dot{q}_\nu) . \tag{7.116b}$$

Expressed in terms of cartesian coordinates (x, y, z) the kinetic energy of a point particle of mass m is

$$K = \tfrac{1}{2} m (\dot{x}^2 + \dot{y}^2 + \dot{z}^2) , \tag{7.117}$$

i.e. it depends only on the velocities $\dot{x}, \dot{y}, \dot{z}$, but not on the position coordinates. But this is a special case and in general the kinetic energy will depend on the generalized position coordinates as well. For example, expressed in terms of spherical polar coordinates (r, θ, ϕ), defined by

$$x = r \sin \theta \cos \phi , \qquad y = r \sin \theta \sin \phi , \qquad z = r \cos \theta , \tag{7.118}$$

the kinetic energy becomes

$$K = \tfrac{1}{2} m (\dot{r}^2 + r^2 \dot{\theta}^2 + r^2 \sin^2 \theta \, \dot{\phi}^2) . \tag{7.119}$$

Instead of using the generalized coordinates and velocities one may alternatively specify the state of motion of a system as follows. Starting from the generalized position coordinates $q_1 \ldots q_v$ and velocities $\dot{q}_1, \ldots \dot{q}_v$ we define the Lagrangian L of the system from Eqs. (7.116) by

$$L(q_1, \ldots q_v, \dot{q}_1, \ldots \dot{q}_v) \equiv K - U \ . \tag{7.120}$$

We now define v *generalized momenta* $p_1, \ldots p_v$ by

$$p_i \equiv \frac{\partial L}{\partial \dot{q}_i} \ , \qquad i = 1, \ldots v \ , \tag{7.121}$$

where in the partial differentiation the remaining $(2v - 1)$ arguments of L, Eq. (7.120), are kept constant. *The v generalized coordinates $q_1, \ldots q_v$ together with the v generalized momenta $p_1, \ldots p_v$ then afford an alternative way of specifying the state of motion of the system.* For a given system one can choose the generalized coordinates in many different ways. For each choice of position coordinates $q_1, \ldots q_v$, Eqs. (7.121) define a particular set of generalized momenta $p_1, \ldots p_v$, and position and momentum coordinates related through Eqs. (7.121) are called *conjugate coordinates*.

The variables p_i are called generalized momenta since they reduce to the usual expressions for the momentum variables if for the q_i one takes cartesian coordinates. Consider a point particle of mass m moving in a potential field $U(x, y, z)$. Treating the cartesian coordinates (x, y, z) as the generalized position coordinates, one obtains from Eqs. (7.117) and (7.120)

$$L = \tfrac{1}{2} m(\dot{x}^2 + \dot{y}^2 + \dot{z}^2) - U(x, y, z) \ .$$

Hence the momentum p_x, conjugate to the position coordinate x, follows from Eq. (7.121):

$$p_x \equiv \frac{\partial L}{\partial \dot{x}} = m\dot{x} \ , \tag{7.122}$$

i.e. p_x is just the usual x-component of linear momentum, and similar results hold for the other components. In this example the difference between using the velocities $\dot{x}, \ldots$ or the momenta $p_x, \ldots$ as variables is not very great since they only differ by a constant factor. In general this is not so. If, for example, we use polar coordinates (r, θ, ϕ) to describe the motion of a point particle of mass m in a central potential $U(r)$, then we obtain from Eqs. (7.119) and (7.120)

$$L(r, \theta, \phi, \dot{r}, \dot{\theta}, \dot{\phi}) \equiv \tfrac{1}{2} m(\dot{r}^2 + r^2\dot{\theta}^2 + r^2 \sin^2\theta\, \dot{\phi}^2) - U(r) \ ,$$

and, from Eqs. (7.121), for the momenta p_r, p_θ and p_ϕ, conjugate to r, θ and ϕ:

$$
\left.
\begin{aligned}
p_r &\equiv \frac{\partial L}{\partial \dot{r}} = m\dot{r} \\[2ex]
p_\theta &\equiv \frac{\partial L}{\partial \dot{\theta}} = mr^2\dot{\theta} \\[2ex]
p_\phi &\equiv \frac{\partial L}{\partial \dot{\phi}} = mr^2 \sin^2\theta\, \dot{\phi}
\end{aligned}
\right\} . \qquad (7.123)
$$

We see that whereas p_r is the linear momentum associated with the radial motion of the particle, p_θ and p_ϕ are the components of angular momentum appropriate to changes of θ and ϕ.

We have obtained two modes of describing a system: in terms of position coordinates and velocities, or in terms of position coordinates and conjugate momenta. For the further development we *must choose the description in terms of position coordinates $q_1, \ldots q_v$ and conjugate momenta $p_1, \ldots p_v$; only in terms of these are the following results valid,* as will be explained below.

We proceed by analogy to the treatment of a point particle in section 7.2. There we introduced a six-dimensional phase space with x, y, z, p_x, p_y, p_z as coordinate axes. (Fig. 7.2 showed a graphical representation for one-dimensional motion along the x-axis. In the more general case of motion in three dimensions we can interpret this figure as the projection of the motion onto the $x - p_x$ plane of phase space.) For the system with v degrees of freedom which we are considering now, the state of motion at a given time is specified by $2v$ coordinates: $q_1, \ldots q_v, p_1, \ldots p_v$. We therefore introduce a $2v$-dimensional *phase space* with the v generalized coordinates $q_1, \ldots q_v$ and the v conjugate momenta $p_1, \ldots p_v$ as coordinate axes. Each point in this phase space corresponds to a definite state of motion of the system. Hence the number of states in a given region of phase space will be proportional to the volume of that region of phase space. The constant of proportionality cannot be determined within the framework of classical mechanics since classically the states of a system form a continuum. Quantum-mechanically we adopt the obvious generalization of Eq. (7.28). The number of states of the system with coordinates and momenta lying within the intervals $q_1, q_1 + dq_1; \ldots q_v, q_v + dq_v; p_1, p_1 + dp_1; \ldots p_v, p_v + dp_v;$ is given by

$$
\frac{1}{h^v} \, dq_1 \ldots dq_v \, dp_1 \ldots dp_v \ . \qquad (7.124)
$$

The result (7.124) for the density of states in phase space has two properties which are essential for the consistency of the whole theory. Firstly, the result is independent of the particular choice of generalized coordinates, i.e. a *given region* of phase space must always contain the *same* number of states, irrespective of what coordinates we use to describe the system. Secondly, consider a given region Γ of phase space at time t. In the course of time, each point of this region traces out a trajectory in phase space, the trajectory from each point being completely determined by the equations of motion of the system. (This is merely stating that Newton's equations of motion are second-order differential equations in time. Hence the initial values of the position coordinates and velocities, or of position coordinates and momenta, completely determine the subsequent motion.) Hence at a later time t', the region Γ will have gone over into a new region Γ' of phase space. Now the regions Γ and Γ' contain the same number of states (since each state in Γ has gone over into a state in Γ'). Hence these two regions must have the same volumes in phase space if Eq. (7.124) is to hold consistently. Both these statements can be proved to be true. They are known as Liouville's theorem which is a consequence of the properties of the equations of motion, expressed in terms of generalized position coordinates and their conjugate momenta.*

An analogous formulation of these results in a 2ν-dimensional space with the position coordinates $q_1, \ldots q_\nu$ and the velocities $\dot{q}_1, \ldots \dot{q}_\nu$ as coordinate axes *fails*. If in Eq. (7.124) one replaces the differentials $dp_1 dp_2 \ldots dp_\nu$ by $d\dot{q}_1 d\dot{q}_2 \ldots d\dot{q}_\nu$, then the above two results do *not* hold, and the theory contains internal inconsistencies. This explains why the above ideas *must in general* be formulated in phase space.[†]

Eq. (7.124) gives us the density of states in phase space. Classically each point of phase space represents a definite state. Quantum-mechanically the states have a discrete nature. We must think of phase space as subdivided into cells of volume h^ν each. Each cell represents one state. Using this density of states we can translate expression (7.115) into classical language. Eq. (7.115) is a sum of Boltzmann factors $\exp(-\beta E_r)$ over all energy levels, each counted the correct number of times. We now replace this sum by a sum over states in phase space; with Eq. (7.124) which gives the density of states in phase space we obtain an integral over phase space. To write down this integral we must use the fact that each point of phase space, labelled

---

*For a discussion of the equations of motion in general form and the proof of Liouville's theorem see the books on mechanics referred to earlier in this section.

[†]However, one frequently deals with *particular* choices of coordinates where one can use velocities instead of momenta. For example, if one describes a system of n point masses $m_1, \ldots m_n$ in terms of cartesian coordinates $(x_1, y_1, z_1), \ldots (x_n, y_n, z_n)$, then one has for the momenta: $p_{x1} = m_1 \dot{x}_1$, etc., and expression (7.124) is easily rewritten in terms of $dx_1 \ldots dz_n d\dot{x}_1 \ldots, d\dot{z}_n$.

by the $2v$ coordinates $q_1, \ldots p_v$, has a definite energy associated with it given, from Eqs. (7.116), by

$$H(q,p) \equiv H(q_1, \ldots q_v, p_1, \ldots p_v) = K + U . \qquad (7.125)$$

To avoid writing all the $2v$ arguments, we have introduced the abbreviation $H(q,p)$ for the energy function $H(q_1, \ldots q_v, p_1, \ldots p_v)$, which is known as the *Hamiltonian* of the system. Hence each point in phase space has a Boltzmann factor $\exp[-\beta H(q,p)]$ associated with it, and the partition function (7.115) goes over into the classical partition function

$$Z = \frac{1}{h^v} \int \exp[-\beta H(q,p)] \, dq_1 \ldots dq_v dp_1 \ldots dp_v . \qquad (7.126)$$

Eq. (7.126) represents our final result which has been obtained by plausibility arguments only.* If the system consists of N identical non-localized subsystems, then the right-hand side of Eq. (7.126) must be divided by $N!$, as explained in section 7.1.

In deriving Eq. (7.126) we made two approximations. Firstly, we replaced the degeneracies $g(E_r)$ in the quantal partition function (7.115) by the density of states (7.124). It turns out that for practical purposes this is always a good approximation.† The second approximation was to replace the discrete discontinuously varying Boltzmann factors $\exp(-\beta E_r)$ by the continuous smoothly varying function $\exp[-\beta H(q,p)]$. This will be justified if successive Boltzmann factors in the quantal partition function (7.115) differ sufficiently little from each other. This will be the case if all level spacings $E_{r+1} - E_r$ are sufficiently small, i.e. if

$$|E_{r+1} - E_r| \ll kT .‡ \qquad (7.127)$$

Eq. (7.127) represents the real limitation on the validity of the classical partition function (7.126). If Eq. (7.127) fails, i.e. at 'low' temperatures such that kT is of the order of or small compared to the level spacings, deviations from classical behaviour occur. The temperatures at which such deviations occur vary greatly, since the level spacings vary greatly in different situations.

---

*More proper derivations are possible but require a good deal of quantum mechanics; see Hill,[5] section 22.6, or Wilson,[21] section 5.52.

†Actually, there is an exception: the phenomenon of Bose–Einstein condensation, to be discussed in section 11.6.

‡A slightly weaker condition suffices. Since the partition function (7.115) is an infinite convergent series, Eq. (7.127) need only apply to those terms in the series which contribute significantly to the series.

7.9.1   The equipartition of energy

Historically, the inadequacy of classical theory first showed up in the failure of the equipartition theorem which we shall now derive. For this purpose we note that the partition function (7.126) in the usual way implies a probability distribution. (Compare the relation between Eqs. (2.22) and (2.23).) The probability that the system, when in thermal equilibrium at temperature T, be in a state with position and momentum coordinates in the intervals

$$q_i, q_i + dq_i \ , \qquad p_i, p_i + dp_i \ , \qquad i = 1, \ldots \nu \ ,$$

is given by

$$P(q_1, \ldots p_\nu) dq_1 \ldots dp_\nu = \frac{\exp\left[-\beta H(q,p)\right] dq_1 \ldots dp_\nu}{\int \exp\left[-\beta H(q,p)\right] dq_1 \ldots dp_\nu} \ . \qquad (7.128)$$

The *theorem of equipartition of energy* can be stated as follows. *Consider a system in thermal equilibrium at the temperature T. A generalized position or momentum coordinate which occurs in the Hamiltonian only as a square term contributes an energy $\tfrac{1}{2}kT$ to the mean energy of the system.*

Assume the coordinate ξ—one of the position or momentum coordinates $q_1 \ldots p_\nu$ describing the system—occurs in the Hamiltonian H in the manner stated, i.e. H is of the form

$$H = A\xi^2 + H' \ , \qquad (7.129)$$

where A and H' do *not* depend on ξ but may depend on any of the other coordinates. It follows from Eq. (7.128) that the probability distribution in ξ, for fixed values of the other coordinates, is given by

$$P(\xi) d\xi = \frac{\exp(-\beta A\xi^2) d\xi}{\int \exp(-\beta A\xi^2) d\xi} \ . \qquad (7.130)$$

The Gaussian distribution (7.130) is peaked at $\xi = 0$, decreasing rapidly as $|\xi|$ becomes large. Hence we may take the range of the variable ξ from $-\infty$ to ∞ and this range of integration will be assumed in all the following integrals. From Eq. (7.130) we then obtain

$$\overline{A\xi^2} = \frac{\int A\xi^2 \exp(-\beta A\xi^2) d\xi}{\int \exp(-\beta A\xi^2) d\xi}$$

$$= -\frac{\partial}{\partial \beta} \ln \int \exp(-\beta A\xi^2) d\xi \ .$$

If we introduce the new variable of integration $t = \xi\sqrt{\beta}$, we obtain

$$\overline{A\xi^2} = -\frac{\partial}{\partial\beta} \ln\left[\beta^{-1/2}\int \exp(-At^2)\,dt\right]$$

$$= \frac{1}{2\beta} = \frac{1}{2}kT. \tag{7.131}$$

In deriving this result we assumed that all coordinates other than ξ were kept constant. We made this assumption since A which occurs in the probability distribution (7.130) may depend on these other variables. However, our final result (7.131) is independent of A and therefore of the values of these other variables, so that *Eq. (7.131) represents the mean of $A\xi^2$ over the complete probability distribution* (7.128). This completes the proof of the equipartition theorem. It is necessary to allow A to depend on coordinates other than ξ, rather than restrict A to be a constant, because the more general case occurs in applications. For example, the kinetic energy of a point particle, expressed in polar coordinates, is from Eqs. (7.119) and (7.123) given by

$$K = \frac{1}{2m}p_r^2 + \frac{1}{2mr^2}\left[p_\theta^2 + \frac{1}{\sin^2\theta}p_\phi^2\right]. \tag{7.132}$$

One uses this expression, for example, when studying the vibrational and rotational properties of diatomic molecules. (See problem 7.9.)

It is a direct consequence of the equipartition theorem that each square term in the Hamiltonian of the system contributes an amount $\frac{1}{2}k$ to the heat capacity of the system. We see now that this result is only valid at sufficiently high temperatures so that the condition (7.127) holds. If the latter condition is violated the discrete quantal nature of the energy levels will manifest itself through a temperature-dependent heat capacity. This is of course also required by the third law of thermodynamics according to which the heat capacity must go to zero as the temperature tends to 0 K (section 4.7).

A typical example of the application of the equipartition theorem is Dulong and Petit's law of the heat capacity of crystalline solids. If in the first approximation we treat the N atoms of such a solid as independently performing simple harmonic oscillations about their equilibrium sites, all with the same circular frequency ω, the crystal will be equivalent to $3N$ independent oscillators, i.e. each atom possesses three degrees of freedom resulting in independent vibrations in the x, y, and z-directions. If we denote the $3N$ cartesian position coordinates $x_1, \ldots z_N$ of the atoms by $q_1, \ldots q_{3N}$, and the corresponding momenta by $p_1, \ldots p_{3N}$, then the Hamiltonian of the crystal becomes

$$H(q,p) = \sum_{i=1}^{3N}\left(\frac{1}{2m}p_i^2 + \frac{1}{2}m\omega^2 q_i^2\right). \tag{7.133}$$

(m is the mass of one atom.) This expression contains $6N$ square terms. Each vibrational degree of freedom provides two such terms, one from the kinetic energy and one from the potential energy. From the equipartition theorem we obtain for the mean energy E of the system

$$E = 3NkT \ , \tag{7.134}$$

which is Dulong and Petit's law.

In section 6.2 we obtained the same result (7.134) as the high-temperature limit of a quantum-mechanical treatment of the atomic vibrations. According to quantum theory, a simple harmonic oscillator of circular frequency ω possesses discrete energy levels equally spaced with separation $\hbar\omega$ between neighbouring levels. In section 6.2 the high-temperature limit was defined by $kT \gg \hbar\omega$. This is just our condition (7.127) for the classical approximation to hold.

The result (7.134) also holds if more realistically one describes the crystal vibrations as the vibrations of N *coupled* atoms. By introducing suitable generalized coordinates $q_1, \ldots q_{3N}$ (known as normal coordinates) one can describe the vibrations of the crystal as a superposition of $3N$ independent oscillations with frequencies $\omega_1, \ldots \omega_{3N}$. (This problem was briefly discussed at the beginning of section 6.3.1.) Correspondingly the Hamiltonian is a sum of $6N$ square terms $q_1{}^2, \ldots q_{3N}{}^2, p_1{}^2, \ldots p_{3N}{}^2$, so that Eq. (7.134) again follows from the equipartition theorem (7.131). In section 6.3.1 this result was obtained as the high-temperature limit of the corresponding quantal description of N coupled oscillators.

A vibrational degree of freedom contributes an energy kT to the mean energy of a system, according to the equipartition law. This is so since a vibrational degree of freedom always involves two square terms in the energy. In addition to the kinetic energy there is the potential energy term which results from the restoring force. More generally, the kinetic energy of a system can always be written as a sum of square terms, leading to an energy $\frac{1}{2}kT$ for each such term. In this way one immediately obtains the value $\frac{3}{2}NkT$ for the mean translational energy of a gas of N molecules, in agreement with our earlier result (7.42).

As another example we consider a system rotating about a fixed axis. The kinetic energy is in this case given by

$$K = \tfrac{1}{2}I\dot{\theta}^2 = \frac{1}{2I} p_\theta{}^2 \ , \tag{7.135}$$

where θ is the angle of rotation, I the relevant moment of inertia, and p_θ the momentum coordinate conjugate to θ:

$$p_\theta = \frac{\partial K}{\partial \dot{\theta}} = I\dot{\theta} \ , \tag{7.136}$$

which is just the angular momentum associated with this rotation. According to the equipartition theorem, this rotational degree of freedom contributes a mean energy $\frac{1}{2}kT$. Correspondingly, one obtains mean energies kT and $\frac{3}{2}kT$ from two or three rotational degrees of freedom (see problems 7.9 and 7.10).

As for vibrations, the quantum-mechanical treatment of rotations leads to discrete energy levels whose spacings determine the 'high-temperature' region where the equipartition theorem holds and where the contribution to the heat capacity from the rotational degrees of freedom is constant. Quantum theory of course also gives the temperature dependence of the rotational heat capacity at lower temperatures (see problem 7.1).

These considerations are very important in the study of the vibrational and rotational heat capacities of gases (section 7.5). The equipartition theorem is in many respects very successful in this field but one needs quantum theory to understand the observed deviations from the equipartition law. However, it was realized long before the advent of quantum theory that the observed specific heats of gases presented the equipartition theorem—and hence classical physics—with fundamental difficulties.*

So far we have discussed microscopic systems, but the equipartition theorem is equally applicable to a macroscopically observable coordinate which occurs as a square in the Hamiltonian of the system. The basic requirement is that the system is in thermal equilibrium. Thus the equipartition theorem is applicable to Brownian motion, i.e. the fluctuating motion of tiny particles suspended in a liquid.† Another important application occurs for a system suspended so that it can rotate about a given axis, for example, the suspended coil of a moving coil galvanometer. The suspended system performs minute irregular movements, partly due to collisions with molecules of the surrounding gas, and partly due to the thermal motion of the atoms in the suspension. The Hamiltonian for this system is given by

$$H = \frac{1}{2I}\, p_\theta^2 + \tfrac{1}{2}\alpha\theta^2 \tag{7.137}$$

where the first term is just the kinetic energy (7.135) of such a system, and the second term is the potential energy for a deflection θ, i.e. the suspension exerts a restoring torque α per unit deflection. From the equipartition theorem we at once obtain for the mean kinetic and potential energies of the system when at temperature T

*See Feynmann,[41] section 40-6, who quotes a paper, written in 1859, by Maxwell, or Sommerfeld,[16] pp. 227 and 233–234, who refers to a lecture by Kelvin in 1884.
†For a discussion of Brownian motion, see Flowers and Mendoza,[26] section 4.4.2, Present,[11] Chapter 9, Feynman,[41] Chapter 41.

$$\frac{1}{2I}\,\overline{p_\theta^2} = \tfrac{1}{2}\overline{I\dot{\theta}^2} = \tfrac{1}{2}kT \tag{7.138a}$$

$$\tfrac{1}{2}\alpha\overline{\theta^2} = \tfrac{1}{2}kT \ . \tag{7.138b}$$

These equations determine the root-mean-square fluctuations of the deflection θ and the angular velocity $\dot\theta$. In particular, we obtain for the rms-deflection θ_{rms}

$$\theta_{rms} = \left(\frac{kT}{\alpha}\right)^{1/2} \ . \tag{7.139}$$

This result implies that there is a minimum current that can be detected with the galvanometer. For a steady current i, the deflection θ is proportional to the current:

$$\theta = \text{const.}\, i \ . \tag{7.140}$$

Hence the minimum observable current i_{min} must produce a deflection at least equal to the deflection θ_{rms} due to thermal fluctuations. Hence we have from Eqs. (7.139) and (7.140):

$$i_{min} = \frac{1}{\text{const.}}\left(\frac{kT}{\alpha}\right)^{1/2} \ . \tag{7.141}$$

These ideas can be greatly extended. For example, in a resistor there are thermal fluctuations of the electron density which produce voltage fluctuations. This phenomenon is known as Johnson noise. These and similar considerations are of practical importance in determining the maximum sensitivity of measuring instruments.*

PROBLEMS 7

7.1 According to quantum mechanics, the molecules of a diatomic gas possess rotational energy levels

$$\varepsilon_r = \frac{\hbar^2}{2I}\, r(r+1) \ , \qquad r = 0, 1, 2, \ldots \,,$$

(where I is a constant), the level ε_r being $(2r+1)$-fold degenerate. Write down an expression for the partition function of the rotational motion. Hence find

*See, for example, B. I. Bleaney and B. Bleaney, *Electricity and Magnetism*, 3rd edn., Oxford University Press, Oxford, 1976, Chapter 23, or M. Garbuny, *Optical Physics*, Academic Press, New York, 1965, section 7.2.

the molar rotational heat capacity of the gas: (i) at low temperatures, (ii) at high temperatures.

For carbon monoxide (CO) the constant I has the value 1.3×10^{-39} g cm^2. What is the molar rotational heat capacity of carbon monoxide at room temperature?

7.2 A diatomic molecule may to a good approximation be considered as vibrating in one-dimensional simple harmonic motion with circular frequency ω. According to quantum mechanics such a system possesses an infinite set of nondegenerate energy levels with energy $\hbar\omega(r+\frac{1}{2})$, $r=0,1,2,\ldots\infty$. Obtain expressions for the vibrational partition function and the vibrational heat capacity. What are the high and low temperature limits of this heat capacity?

Obtain the molar vibrational heat capacity of N_2 at 1000 K given that the vibrational level spacing $\hbar\omega$ is 0.3 eV.

7.3 At its normal boiling point (N.B.P.) of 630 K, saturated mercury vapour is monatomic and satisfies the perfect gas laws. Calculate the entropy per mole of saturated mercury vapour at its N.B.P.

The latent heat of vaporization of mercury at its N.B.P. is 5.93×10^4 J mol^{-1}. Hence find the entropy per mole of liquid mercury at 630 K and 1 atm.

(Atomic weight of mercury = 200.6.)

7.4 Consider a perfect gas of particles in the extreme relativistic range (i.e. the kinetic energy E of a particle and its momentum $\mathbf{p}$ are related by $E=c|\mathbf{p}|$). Find the Helmholtz free energy and hence the heat capacity at constant volume for this gas.

7.5 For a gas obeying the Maxwell velocity distribution obtain:
   (i) the most probable speed of the molecules,
   (ii) the most probable kinetic energy of the molecules,
   (iii) the mean speed of the molecules,
   (iv) the rms speed of the molecules,
   (v) the mean kinetic energy of the molecules.

7.6 Derive Eqs. (7.74) and (7.75). (Hint: Use Eqs. (A.33) and (A.36) from Appendix A.)

7.7 An imperfect gas has an intermolecular potential

$$u(r) = \xi \exp(-\alpha r^2).$$

If ξ is small so that the partition function may be expanded in a series in ξ, calculate the change in the equation of state to order ξ.

7.8 For a real gas at low density obtain the internal energy of the gas up to terms proportional to its density, in terms of the intermolecular potential $u(r)$.

The radial distribution function $g(r, \rho, T)$ is defined as follows. Consider a gas at temperature T in a volume V and let $\rho = N/V$ be its mean particle density. Then

$$\rho g(r, \rho, T) 4\pi r^2 \, dr$$

is the number of molecules which lie within a spherical shell of radius r and thickness dr, given that there *is* a molecule at the origin $r=0$. For a uniform distribution we would have $g \equiv 1$. Thus g allows for deviations from uniform density due to the intermolecular forces. (Experimentally one obtains g from x-ray diffraction experiments.) Show that for a real gas the internal energy is related to the radial distribution function by the equation

$$E = \tfrac{3}{2} NkT + \tfrac{1}{2} N\rho \int_0^\infty dr 4\pi r^2 g(r, \rho, T) u(r).$$

Obtain a relation between the radial distribution function in the limit of zero density and the intermolecular potential $u(r)$.

7.9 A diatomic molecule consists of atoms of masses M_1 and M_2, whose position coordinates are $\mathbf{r}_1$ and $\mathbf{r}_2$. The atoms interact through an attractive potential $V(|\mathbf{r}_1 - \mathbf{r}_2|)$. Show how the energy of the molecule decomposes into translational, vibrational and rotational parts.

Apply the equipartition theorem to your result to derive the heat capacity of a perfect gas of these molecules.

7.10 A polyatomic molecule consists of n atoms, each possessing three degrees of freedom (i.e. we treat each atom as a point mass neglecting its internal degrees of freedom). Divide the degrees of freedom of the molecule into translational, rotational and vibrational degrees: (i) for a nonlinear molecule, (ii) for a linear molecule. (Note that a nonlinear molecule, like a rigid body, possesses three rotational degrees of freedom, e.g. two angles to fix an axis of rotation and a third angle to fix the extent of the rotation about this axis relative to some standard position. On the other hand a linear molecule possesses only two rotational degrees of freedom. We need two angles to fix the interatomic axis but do not now have rotations about *this* axis. See the previous problem where the diatomic molecule was considered in detail.)

Use your results to discuss the heat capacity of a polyatomic perfect gas.

7.11 Carbon dioxide (CO_2) is a linear molecule and it possesses 4 vibrational modes. The vibrational temperatures $\Theta_{vib} \equiv \hbar\omega/k$ of these modes are 3360, 1890, 954 and 954 K (i.e. there are two distinct modes with the same frequency). Use Fig. 6.5 to estimate the molar heat capacity at constant volume of CO_2 at 312 K.

CHAPTER

8

Phase equilibria

We now come to discuss systems consisting of two or more phases in equilibrium. A phase is a homogeneous part of a system bounded by surfaces across which the properties change discontinuously. In the most general situation each phase will contain several components i.e. it will contain several different species of molecules or ions. For example a gaseous phase might consist of a mixture of gases; a liquid phase might be a solution. To specify the properties of a phase we must then specify the concentrations of the various components in it. One then has the possibility of transfer of matter between different phases and also of chemical reactions (which change the amounts of different components). Except for a brief discussion of multi-component systems in section 11.9, we shall restrict ourselves to the discussion of systems containing one component only, since a thorough treatment of multi-component systems would take up too much space.* A one-component system can of course still exist in different phases, corresponding to gaseous, liquid and different crystalline forms of matter.

The problem of the equilibrium between different phases is much harder than any we have studied so far. The atomic viewpoint does, of course, provide a qualitative understanding of the observed phenomena. For example, the increasing degree of molecular disorder of crystalline, liquid and gaseous states explains qualitatively the entropy changes which accompany the processes of melting, evaporation and sublimation. But the statistical

*The reader will find good discussions in the books by Denbigh,[4] Landau and Lifshitz,[7] ter Haar and Wergeland,[18] Wilson,[21] and Zemansky and Dittman.[22]

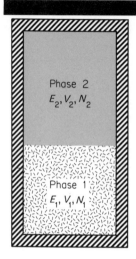

Phase 2
E_2, V_2, N_2

Phase 1
E_1, V_1, N_1

Fig. 8.1. Equilibrium of two phases of a one-component system in total isolation.

approach can deal with phase changes to a very limited degree only, and we shall use the methods of thermodynamics.

We now have the possibility of the transfer of matter between different phases (e.g. in a mixture of ice and water, some or all of the ice may melt). For different phases to be in equilibrium certain conditions must hold so that no mass transfer occurs. These conditions are additional to those we had earlier for temperature and pressure equilibrium. These additional conditions will be derived in sections 8.1 and 8.2, and we shall discuss them in section 8.3. The Clausius–Clapeyron equation and its applications will occupy sections 8.4 and 8.5, while section 8.6 will deal with the critical point and related topics.

8.1 EQUILIBRIUM CONDITIONS

We consider a one-component system and shall derive the conditions for two phases to coexist in equilibrium. The two phases could be solid and liquid (e.g. ice and water), liquid and vapour, solid and vapour, and there are other possibilities, such as different crystalline forms of the same substance, e.g. grey and white tin.

Our approach will be a direct generalization of that used in section 2.3, i.e. we shall assume the system totally isolated, as shown in Fig. 8.1, so that Eqs. (2.4a) to (2.4c) hold, namely

$$E_1 + E_2 = E \tag{8.1a}$$

$$V_1 + V_2 = V \tag{8.1b}$$

$$N_1 + N_2 = N \tag{8.1c}$$

which express the division of constant energy E, volume V and particle number N between the two phases. The entropy of the system is (cf. Eq. (2.6))

$$S(E, V, N, E_1, V_1, N_1) = S_1(E_1, V_1, N_1) + S_2(E_2, V_2, N_2) \qquad (8.2)$$

where, as throughout this chapter, subscripts 1 and 2 label quantities in the two phases 1 and 2 respectively. For equilibrium of the system the entropy must be a maximum, i.e. from Eq. (8.2), treating E_1, V_1, and N_1 as the independent variables and eliminating E_2, V_2 and N_2 by means of Eqs. (8.1),

$$dS = \left[\left(\frac{\partial S_1}{\partial E_1} \right)_{V_1, N_1} - \left(\frac{\partial S_2}{\partial E_2} \right)_{V_2, N_2} \right] dE_1$$

$$+ \left[\left(\frac{\partial S_1}{\partial V_1} \right)_{E_1, N_1} - \left(\frac{\partial S_2}{\partial V_2} \right)_{E_2, N_2} \right] dV_1$$

$$+ \left[\left(\frac{\partial S_1}{\partial N_1} \right)_{E_1, V_1} - \left(\frac{\partial S_2}{\partial N_2} \right)_{E_2, V_2} \right] dN_1 = 0 . \qquad (8.3)$$

Since E_1, V_1 and N_1 are independently variable, each of the square parentheses in Eq. (8.3) must vanish for the system to be in equilibrium. The first two of these conditions (coefficients of dE_1 and dV_1 zero) are the conditions for equal temperatures (cf. Eq. (2.9)) and equal pressures (cf. Eq. (2.12)) for the two phases, as we discussed in section 2.3. The third condition

$$\left(\frac{\partial S_1}{\partial N_1} \right)_{E_1, V_1} = \left(\frac{\partial S_2}{\partial N_2} \right)_{E_2, V_2} , \qquad (8.4)$$

mentioned in passing in Eq. (2.13), is the condition for particle equilibrium, i.e. no transfer of molecules between the two phases. We define the *chemical potential* μ_i for each phase by

$$\mu_i = - T_i \left(\frac{\partial S_i}{\partial N_i} \right)_{E_i, V_i} , \qquad i = 1, 2 . \qquad (8.5)$$

Condition (8.4) then becomes

$$\mu_1 = \mu_2 : \qquad (8.6)$$

in equilibrium the chemical potentials of the two phases must be the same.

Before considering this condition further, we want to extend the thermodynamic relations which were derived in Chapters 2 and 4 to allow for the variable particle number. Considering only a single phase for the moment and so omitting the suffix i, we have for $S = S(E, V, N)$:

$$dS = \left(\frac{\partial S}{\partial E}\right)_{V,N} dE + \left(\frac{\partial S}{\partial V}\right)_{E,N} dV + \left(\frac{\partial S}{\partial N}\right)_{E,V} dN$$

$$= \frac{1}{T} dE + \frac{1}{T} PdV - \frac{1}{T} \mu dN , \tag{8.7}$$

where we used Eqs. (2.9), (2.12) and (8.5), or on slight rearrangement

$$dE = TdS - PdV + \mu dN . \tag{8.8}$$

Eq. (8.8) is the generalization of the fundamental thermodynamic relation (4.12) for a one-component system whose size, i.e. particle number, is not kept fixed.

From Eq. (8.8) the chemical potential can be expressed in various ways depending on which of the variables E, S and V are held constant. We can also introduce different independent variables and thermodynamic potentials, as in sections 4.4 and 4.5. It is left as an easy exercise for the reader (compare the derivations of Eqs. (4.43) and (4.49)) to show that, expressed in terms of the Helmholtz and Gibbs free energies, Eq. (8.8) becomes

$$dF = -SdT - PdV + \mu dN , \tag{8.9}$$

$$dG = -SdT + VdP + \mu dN , \tag{8.10}$$

whence

$$\mu = \left(\frac{\partial F}{\partial N}\right)_{T,V} = \left(\frac{\partial G}{\partial N}\right)_{T,P} . \tag{8.11}$$

We see from Eqs. (8.8) to (8.10) that the chemical potential is always the derivative with respect to the particle number N.

In the case of a single one-component phase, the Gibbs free energy $G(T,P,N)$ is an extensive variable, i.e. it is proportional to the particle number N:

$$G(T,P,N) = NG(T,P,1) \equiv Ng(T,P) \tag{8.12a}$$

where

$$g(T,P) = \frac{1}{N} G(T,P,N) \tag{8.12b}$$

is the corresponding *Gibbs free energy per particle*. It follows from Eqs. (8.11) and (8.12) that

$$\mu = g(T,P) : \tag{8.13}$$

the chemical potential μ is simply the Gibbs free energy per particle. But this result (8.13), like Eq. (8.12), is *only true* for a *homogeneous one-component* system. It is *not valid generally*.

Reverting now to our system consisting of one component in two phases, each phase by itself is, of course, a homogeneous one-component system so that Eq. (8.13) applies to it. Hence we can write the equilibrium condition (8.6) in the form:

$$g_1(T,P) = g_2(T,P) \ : \tag{8.14}$$

the Gibbs free energies per particle for the two phases must be the same.

So far one of our equilibrium conditions has always been that the system be at a uniform pressure. This condition frequently applies, but it depends on the neglect of surface effects. If, for example, we consider drops of liquid suspended in their vapour then surface tension effects will be important if the drops are small, and the pressures within and outside them are no longer the same.*

In the next section we shall give an alternative derivation of the equilibrium conditions between phases.

★ 8.2   ALTERNATIVE DERIVATION OF THE EQUILIBRIUM CONDITIONS

In the last section we considered a one-component system in two phases, which was completely isolated. The more usual experimental conditions are that the temperature and pressure of the system are held constant, for example the system is immersed in a heat bath and is subject to atmospheric pressure. An idealized arrangement of such a system existing in two phases is shown in Fig. 8.2. We shall obtain the equilibrium conditions for this system. We know from section 4.5 that for a system at fixed temperature T and pressure P, the Gibbs free energy G is a minimum in equilibrium. We can write the Gibbs free energy as

$$G \equiv G(T,P,N_1,N_2) = N_1 g_1(T,P) + N_2 g_2(T,P) \tag{8.15}$$

where we used the fact that each phase by itself is a homogeneous one-component system (cf. Eqs. (8.12)); g_i and N_i are the Gibbs free energies per particle and the particle numbers of the two phases. From Eq. (8.15) we have

*This problem is treated, for example, by Pippard[9] and Wannier.[20]

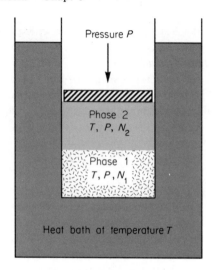

Fig. 8.2. Equilibrium of two phases of a one-component system in a heat bath and at constant pressure.

$$dG = \left[N_1 \frac{\partial g_1}{\partial T} + N_2 \frac{\partial g_2}{\partial T} \right] dT$$

$$+ \left[N_1 \frac{\partial g_1}{\partial P} + N_2 \frac{\partial g_2}{\partial P} \right] dP$$

$$+ g_1(T, P) dN_1 + g_2(T, P) dN_2 \ . \tag{8.16}$$

For equilibrium we require G to be a minimum, i.e.

$$dG = 0 \tag{8.17}$$

subject to the constraints

$$dT = 0 \tag{8.18a}$$
$$dP = 0 \tag{8.18b}$$
$$N_1 + N_2 = N \tag{8.18c}$$

which express constant temperature, constant pressure and constant total number of particles for the system. From Eqs. (8.16) to (8.18) we obtain at once the equilibrium condition

$$dG = [g_1(T, P) - g_2(T, P)] dN_1 = 0 \ , \tag{8.19}$$

whence the original equilibrium condition Eq. (8.14), $g_1 = g_2$, follows since N_1 is freely variable. These equilibrium conditions for a system at constant pressure and temperature are the same as those for an isolated system. This is to be expected. For equilibrium between two phases they must be at the same pressure, temperature and chemical potential, irrespective of the applied constraints. Condition (8.14) ensures equilibrium with respect to particle transfer between the two phases.

8.3   DISCUSSION OF THE EQUILIBRIUM CONDITIONS

The condition (8.14),

$$g_1(T,P) = g_2(T,P) \; , \tag{8.14}$$

defines a curve in the (T,P) plane, as shown in Fig. 8.3. At a point on the curve the chemical potentials of the two phases are the same; at a point off the curve they differ. Let us denote by 1 and 2 the regions in which we have $g_1 < g_2$ and $g_2 < g_1$ respectively. If N_1 and N_2 are the number of molecules in phases 1 and 2 respectively (with $N_1 + N_2 = N = \text{constant}$) the Gibbs free energy of the system is

$$G = N_1 g_1(T,P) + N_2 g_2(T,P) \; . \tag{8.20}$$

For a system at a pressure and temperature given by the point A of Fig. 8.3, where $g_1 < g_2$, the Gibbs free energy (8.20) is a minimum if all the substance

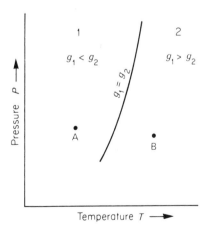

Fig. 8.3. The phase equilibrium curve $(g_1 = g_2)$ separating regions in which phases 1 and 2 respectively are the stable phases.

is in phase 1 (i.e. $N_1 = N$, $N_2 = 0$, $G = Ng_1(T,P)$). Similarly corresponding to point B, G is a minimum if all the substance is in phase 2. Thus the curve (8.14) divides the (P,T) plane into regions where the one or other phase represents the stable equilibrium state. It is only on the curve (where $g_1 = g_2$ so that $G = Ng_1$ irrespective of the values of N_1 and $N_2 = N - N_1$) that the two phases can coexist in equilibrium. This curve is called a *phase equilibrium curve*. If the two phases are liquid and vapour it is called the vapour pressure curve, if solid and vapour the sublimation curve, and if solid and liquid the melting curve. For two coexisting phases of a one-component system we cannot choose the temperature *and* pressure arbitrarily. Give the temperature, the pressure is determined, and vice versa. For example, water vapour in contact with water at 100 °C always has a pressure of one atmosphere.

Let us next consider the equilibrium of three different phases of a one-component system, solid, liquid and vapour, say, and let us label these 1, 2 and 3 respectively. If we repeated the derivations of sections 8.1 or 8.2 we would obtain the equilibrium conditions

$$g_1(T,P) = g_2(T,P) = g_3(T,P) \ . \tag{8.21}$$

But we can also see this directly, since equilibrium between all three phases demands equilibrium between any pair of them. So we obtain a condition like Eq. (8.14) for each pair of phases, and hence Eqs. (8.21) follow.

We have seen that an equation such as Eq. (8.14) defines a phase equilibrium curve (Fig. 8.3). For points on this curve the two phases are in equilibrium. Eqs. (8.21) represent the intersection of the two curves $g_1 = g_2$ and $g_2 = g_3$ in the (T,P) diagram. This intersection defines a point in the (T,P) diagram: the *triple point*. This is shown in Fig. 8.4 which is known as the *phase diagram* of the system. (In general, the phase diagram even of a one-component system will be more complex. See, for example, Fig. 8.5 below.) Of course, at the triple point we also have $g_1 = g_3$. The triple point is the intersection of three phase equilibrium curves: the vapour pressure curve, the sublimation curve and the melting curve. At the triple point all three phases are in equilibrium with each other. The triple point of water, for example, occurs at $P = 4.58$ mmHg and $T = 0.01$ °C. Because ice, water and water vapour coexist at a unique temperature, the triple point of water is used as the fixed point of the absolute temperature scale (see section 1.2). It follows from Eqs. (8.21) (two equations in two unknowns) that the maximum number of coexisting phases of a one-component system is three.

On the other hand a pure substance may be capable of existing in more than one allotropic form and it will then have several triple points. Fig. 8.4 is a particularly simple phase diagram. Fig. 8.5 shows schematically the phase diagram for sulphur which can exist in two different crystalline forms,

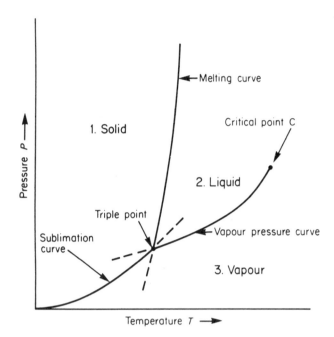

Fig. 8.4. Phase diagram of a one-component system possessing one triple point only.

rhombic and monoclinic. It has three triple points which occur at: (A) 10^{-6} atmospheres, 95.5 °C; (B) 4×10^{-6} atmospheres, 119 °C; (C) 1,400 atmospheres, 153.7 °C.

Reverting to Fig. 8.4, the three phase equilibrium curves divide the (T, P) plane into three regions in which the solid, liquid and gaseous phases respectively are the stable state (i.e. have the lowest chemical potential). Although not true equilibrium states (i.e. G is not a minimum), metastable states can exist for quite long times under suitable conditions. For example, it is possible for a supersaturated vapour to exist in region 2 (of Fig. 8.4), well above the saturation vapour pressure. Similarly one can have supercooled liquids in region 1, or superheated liquids in region 3. In each case the persistence of the metastable state depends on the absence of nuclei which can start condensation, etc.

We may also imagine the phase equilibrium curves continued beyond the triple point as shown by the dashed curves in Fig. 8.4. For example, along the dashed continuation of the vapour pressure curve supercooled liquid and vapour may coexist in a metastable state (along this curve $g_2 = g_3 > g_1$). For sulphur, Fig. 8.5, the existence of metastable phase equilibrium curves leads to a metastable triple point at which rhombic sulphur coexists with the liquid and vapour.

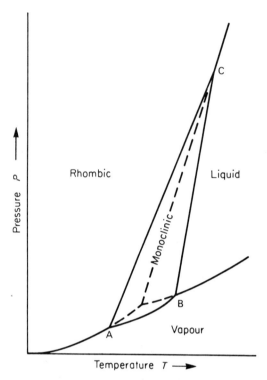

Fig. 8.5. Schematic phase diagram for sulphur.

In Fig. 8.4 the vapour pressure curve comes to an abrupt end at the critical point, the significance of which will be discussed in section 8.6.

8.4   THE CLAUSIUS–CLAPEYRON EQUATION

The Clausius–Clapeyron equation which we shall now derive is the differential equation for a phase equilibrium curve. The equation specifies the slope dP/dT at each point of the curve. It thus gives information about the dependence on pressure and temperature of the phase equilibrium.

For two points A and B on a phase equilibrium curve, Fig. 8.6, we have

$$g_1(T,P) = g_2(T,P) \tag{8.22a}$$

$$g_1(T+dT,P+dP) = g_2(T+dT,P+dP) , \tag{8.22b}$$

whence by subtraction

$$\left(\frac{\partial g_1}{\partial T}\right)_P dT + \left(\frac{\partial g_1}{\partial P}\right)_T dP = \left(\frac{\partial g_2}{\partial T}\right)_P dT + \left(\frac{\partial g_2}{\partial P}\right)_T dP \tag{8.23}$$

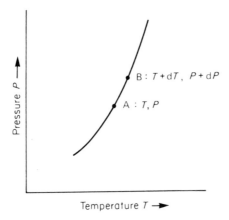

Fig. 8.6. The phase equilibrium curve $g_1(T,P) = g_2(T,P)$.

so that

$$\frac{dP}{dT} = - \frac{\Delta\left(\dfrac{\partial g}{\partial T}\right)_P}{\Delta\left(\dfrac{\partial g}{\partial P}\right)_T} \ . \tag{8.24}$$

Here we introduced the notation

$$\Delta f \equiv f_2 - f_1 \tag{8.25}$$

for the difference of the quantity f in the two phases.

We can write Eq. (8.10) for each phase separately as

$$dG_i = - S_i dT + V_i dP + \mu_i dN_i \ . \tag{8.26}$$

For a one-component phase, we have from Eqs. (8.12) and (8.13) that $\mu_i = g_i$ and $G_i = N_i g_i$. Hence Eq. (8.26) becomes

$$N_i dg_i = - S_i dT + V_i dP \ , \tag{8.27}$$

so that

$$\left(\frac{\partial g_i}{\partial T}\right)_P = - \frac{S_i}{N_i} \ , \qquad \left(\frac{\partial g_i}{\partial P}\right)_T = \frac{V_i}{N_i} \ . \tag{8.28}$$

Substituting expressions (8.28) into Eq. (8.24) gives

$$\frac{dP}{dT} = \frac{\Delta\left(\dfrac{S}{N}\right)}{\Delta\left(\dfrac{V}{N}\right)} = \frac{\Delta S}{\Delta V} \ , \tag{8.29}$$

where in the last expression on the right-hand side ΔS and ΔV *must* refer to the *same* quantity of substance: it could be one molecule or one mole. ΔS is the change in entropy, ΔV the change in volume of this same quantity of substance as it undergoes a phase transformation from phase 1 to phase 2. Eq. (8.29), the *Clausius–Clapeyron equation*, relates these changes occurring at a temperature T and pressure P corresponding to a definite point on the equilibrium curve to the slope of the curve at that point.

A phase change, which is accompanied by a change in entropy ΔS, involves the absorption or emission of heat. If L_{12} is the latent heat of transformation when a given quantity of the substance is transformed from phase 1 to phase 2, at the temperature T, then the entropy change is

$$\Delta S = S_2 - S_1 = \frac{L_{12}}{T} \ . \tag{8.30}$$

Combining Eqs. (8.29) and (8.30), we obtain as an alternative form of the Clausius–Clapeyron equation

$$\frac{dP}{dT} = \frac{L_{12}}{T\Delta V} \ . \tag{8.31}$$

Again, the extensive quantities L_{12} and ΔV *must* refer to the same amount of substance.

The phase change we have considered is characterized by the discontinuities (as one goes from one phase to the other) of the numerator and denominator of Eq. (8.24). These discontinuities imply differences in entropy and molar volume (i.e. density) between the phases. The entropy difference implies that the phase change involves a latent heat. This type of phase change is called a *first-order transition*. Solid-liquid-vapour phase changes and many allotropic transitions, e.g. grey to white tin, are of this kind. There are phase changes in which the numerator and denominator of Eq. (8.24) are continuous, i.e. entropy and density are continuous (so that no latent heat is involved) but some higher-order derivatives of g_i and other thermodynamic quantities, such as the specific heat, are discontinuous or may become infinite. We shall not consider these kinds of phase changes.

From the Clausius–Clapeyron equation several interesting deductions are possible. For the processes of melting, evaporation and sublimation, the entropy changes ΔS (and hence the corresponding latent heats) are

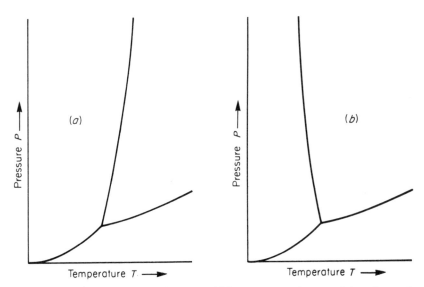

Fig. 8.7. Phase diagram for a substance which possesses only one triple point, and which: (a) expands on melting, (b) contracts on melting.

positive*. That ΔS is positive for these phase changes is a consequence of the fact that these are changes from more to less ordered states of the system. The atoms in a crystal are arranged regularly at the lattice sites so that one here has long-range order over the whole size of the crystal. Except near the boiling point, liquids show a considerable degree of short-range order, i.e. liquids possess localized structure, as in a crystal except that there are many vacant sites; about one in ten. In a gas there is almost no structure left. These different degrees of order show up in x-ray diffraction experiments (see Hill,[28] section 6.3). For a transformation from solid or liquid to gas the density decreases so that ΔV is positive (i.e. the molar volume increases). It follows from the Clausius–Clapeyron equation that the slope dP/dT of the vapour-pressure and sublimation curves is positive; as we drew it in Fig. 8.4. For the process of melting ΔV may be positive or negative depending on whether the solid expands or contracts on melting. Most substances expand upon melting. This is a consequence of the large number of vacant sites in a liquid, so that the density of the liquid is lower than that of the solid. (These vacancies also allow the molecules in the liquid to migrate easily and this leads to their characteristic property of fluidity.) For a substance which

*The behaviour of the isotope helium 3 (^{3}He) at temperatures below 0.3 K and at pressures in the range of about 29 to 34 atm is an exception. If we heat liquid ^{3}He at a constant pressure in this range, starting at a sufficiently low temperature, the liquid will first solidify and then melt again. See, for example, McClintock et al.,[34] p. 171.

expands on melting the slope dP/dT of the melting curve is positive, as drawn in Fig. 8.7(a). But there are exceptional substances such as water (ice floats on water!) which contract on melting and for which the melting curve has a negative slope as shown in Fig. 8.7(b).* This behaviour in ice is due to its very open tetrahedral crystal structure. On melting, both the nearest-neighbour distance and the number of nearest neighbours increases, and the latter effect outweighs the former.

8.5   APPLICATIONS OF THE CLAUSIUS–CLAPEYRON EQUATION

We give some simple illustrative applications of the Clausius–Clapeyron equation in this section.

8.5.1   Pressure dependence of the melting point

We consider the equilibrium between ice and water as an example of the pressure dependence of the melting point. To calculate this effect from the Clausius–Clapeyron equation (8.31) we require: (i) the latent heat of fusion L_{12} of water: at $0\,°C$

$$L_{12} = 3.35 \times 10^5 \text{ J/kg ;}$$

(ii) the volumes per gram in the solid and liquid phase are, respectively,

$$V_1 = 1.09070 \text{ cm}^3/\text{g}, \qquad V_2 = 1.00013 \text{ cm}^3/\text{g} ,$$

so that $\Delta V = V_2 - V_1 = -0.0906 \times 10^{-3} \text{ m}^3/\text{kg}$. Hence from Eq. (8.31)

$$\frac{dP}{dT} = -\frac{3.35 \times 10^5}{273.2 \times 0.0906 \times 10^{-3}} = -1.35 \times 10^7 \text{ N m}^{-2}\,(°C)^{-1}$$

$$= -134 \text{ atm/}°C .$$

Thus an increase in pressure of 134 atmospheres lowers the melting point by $1\,°C$. An increase in pressure of 1,000 atmospheres lowers the melting point by $7.5\,°C$, in excellent agreement with the observed value of $7.4\,°C$. Similarly good agreement is obtained for other substances. The lowering of the melting point of ice under pressure is responsible for the motion of glaciers. The deeper parts of a glacier melt under the weight of ice on top of them allowing them to flow. They freeze again when the pressure decreases. In this way ice

*Actually water has a much more complicated phase diagram since ice can exist in several modifications.

against a rock on the bed of the glacier can flow round the obstructing rock. Ice skaters similarly depend for lubrication on the melting of the ice in contact with the skate.

8.5.2   Pressure dependence of the boiling point

Since ΔV is always positive for the transformation of liquid to vapour, increasing the pressure on a liquid always increases the boiling point. We again take water at the normal boiling point as an example. In this case we have: (i) the latent heat of vaporization

$$L_{12} = 2.257 \times 10^6 \text{ J/kg} \; ;$$

(ii) the volume per gram in the liquid and vapour phase at $T = 373.15$ K and $P = 1$ atm are, respectively,

$$V_1 = 1.043 \text{ cm}^3/\text{g}, \qquad V_2 = 1{,}673 \text{ cm}^3/\text{g} \; . \tag{8.32}$$

Hence Eq. (8.31) gives

$$\frac{dP}{dT} = \frac{2.257 \times 10^6}{373.15 \times 1.672} = 3.62 \times 10^3 \text{ N m}^{-2} \, (°\text{C})^{-1}$$

$$= 27 \text{ mmHg/}°\text{C} \; . \tag{8.33}$$

A pressure of 1 atmosphere is 1.0132×10^5 N/m². Thus at a pressure of 3.6×10^4 N/m² (approximately atmospheric pressure at an altitude of 8 km, the height of Mount Everest)

$$\Delta P = -6.5 \times 10^4 \text{ N/m}^2 \; ,$$

whence

$$\Delta T \approx -\frac{65}{3.6} = -18 \, °\text{C}$$

i.e. water boils at about 80 °C at this height.

8.5.3   The vapour pressure curve

The Clausius–Clapeyron equation for the vapour pressure curve admits the following approximate treatments.

Firstly, we may neglect the volume of the liquid compared with that of the gas: $V_1 \ll V_2$ so that

$$\Delta V = V_2 - V_1 \approx V_2 \; . \tag{8.34}$$

For water at its normal boiling point, for example, we see from Eq. (8.32) that this introduces an error of less than 0.1 per cent. Secondly, we shall assume that the vapour behaves like a perfect gas. For one gram of vapour we then have

$$PV_2 = RT/M \; , \qquad (8.35)$$

where M is the gram-molecular weight. Substitution of Eqs. (8.34) and (8.35) in Eq. (8.31) leads to

$$\frac{\mathrm{d}}{\mathrm{d}T} \ln P = \frac{1}{P} \frac{\mathrm{d}P}{\mathrm{d}T} = \frac{ML_{12}}{RT^2} \; . \qquad (8.36)$$

For water at its normal boiling point, the data in section 8.5.2 give $\mathrm{d}P/\mathrm{d}T = 3.56 \times 10^3 \, \mathrm{N\,m^{-2}\,K^{-1}}$, in very good agreement with the value 3.62×10^3, Eq. (8.33), obtained in the exact calculation.

If we treat the latent heat of vaporization L_{12} as a constant, Eq. (8.36) can be integrated analytically to give

$$P = \mathrm{const.} \exp \left[-\frac{ML_{12}}{RT} \right] \; . \qquad (8.37)$$

Eq. (8.37) will always hold for small temperature changes. For larger variations of T, Eq. (8.37) gives only a very rough value for the vapour pressure, as the latent heat varies appreciably for large temperature changes. Fig. 8.8 illustrates this for water.

Eq. (8.37) should be compared with the Sackur–Tetrode vapour pressure formula (7.54). The latter, derived from statistical mechanics, contains *no* undetermined constant. Furthermore in deriving Eq. (7.54) the latent heat was *not* assumed independent of temperature (the most objectionable feature in the derivation of Eq. (8.37)). The only approximation made in obtaining the Sackur–Tetrode equation was to treat the vapour as a perfect gas, and this is generally a good approximation.

★ 8.6   THE CRITICAL POINT

At the end of section 8.3 we mentioned that the vapour pressure curve which separates the liquid and gaseous phases in the phase diagram comes to an abrupt end at the critical point, labelled C in Fig. 8.9. We now want to discuss the significance of the critical point.

As one follows the vapour pressure curve towards the critical point (see Fig. 8.9) the latent heat and the volume change ΔV associated with the phase transformation decrease until they become zero at the critical point. Fig. 8.8

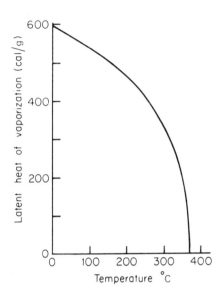

Fig. 8.8. The temperature dependence of the latent heat of vaporization of water.

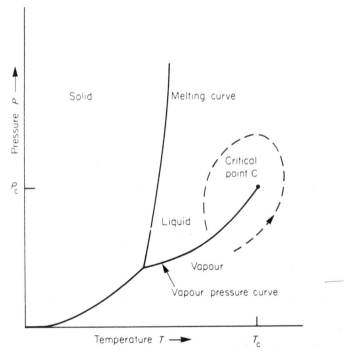

Fig. 8.9. The vapour pressure curve and critical point C. The dashed path represents a change from vapour to liquid without abrupt phase transition.

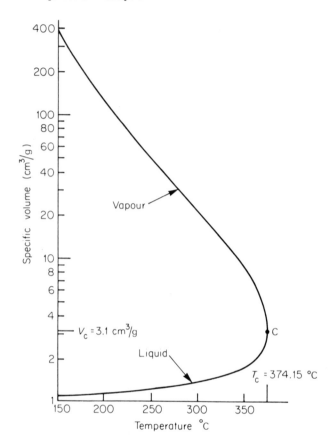

Fig. 8.10. The temperature dependence of the specific volume (in cm³/g) of liquid water and water vapour in equilibrium. C is the critical point.

showed the temperature dependence of the latent heat of vaporization of water. The latent heat vanishes at 374.15 °C, the critical temperature of water. In Fig. 8.10 the temperature dependence of the specific volumes of liquid water and water vapour in equilibrium are shown. At 374.15 °C, the specific volumes of the liquid and vapour become equal; their difference ΔV vanishes. The phase diagram 8.10 is typical. At temperatures below the critical temperature T_c the fluid can co-exist in two states with different specific volumes; the liquid phase with specific volume $V_1 < V_c$, and the vapour phase with specific volume $V_2 > V_c$. At the critical temperature T_c both specific volumes become equal to V_c. Above the critical temperature the substance exists in one fluid phase only.

To understand this behaviour better, let us consider the isotherms of one gram of fluid in a (P, V) diagram, shown schematically in Fig. 8.11. Consider one gram of fluid enclosed in a cylinder with a piston at one end, and immersed in a heat bath, as shown in Fig. 8.2. Suppose initially the cylinder is filled with vapour at a temperature T_1, corresponding to the point A in Fig. 8.11. If the system is compressed isothermally, the vapour pressure and density of the vapour increase as its volume decreases. At a certain point (A$_2$ on the diagram) the vapour has just reached saturation at the given temperature. As the volume is reduced further, condensation occurs at constant pressure. We now have two phases present, liquid and vapour, and the system is at the saturation vapour pressure P_1 corresponding to the temperature T_1. Eventually (point A$_1$ on the figure) all the vapour is condensed; the cylinder is filled with liquid. Further reduction of the volume now requires very large pressures, due to the comparatively low compressibility of liquids, and results in a steeply rising isotherm (the portion A$_1$A$'$ of it). During condensation (i.e., the portion A$_2$A$_1$ of the isotherm) the fraction α of fluid in the liquid phase and the volume V of the system are related by

$$V = \alpha V_1 + (1 - \alpha) V_2 \tag{8.38}$$

where V_1 and V_2 are the specific volumes, at temperature T_1, of liquid and vapour respectively.

Fig. 8.11 shows two isotherms similar to that at temperature T_1. They correspond to temperatures $T_3 > T_2 > T_1$. We see that as the temperature increases, the horizontal portions of the isotherms (A$_1$A$_2$,B$_1$B$_2$,D$_1$D$_2$), along which liquid and vapour are in equilibrium, become shorter. For the isotherm at the *critical temperature* T_c, this horizontal portion has shrunk to a horizontal point of inflection at the *critical point* C. The mathematical definition of the critical point is

$$\left(\frac{\partial P}{\partial V} \right)_{T = T_c} = \left(\frac{\partial^2 P}{\partial V^2} \right)_{T = T_c} = 0 \ . \tag{8.39}$$

The corresponding critical pressure P_c and critical specific volume V_c are shown on Fig. 8.11.

The isotherms at temperatures above T_c are monotonic decreasing continuous curves, that is, following one of these curves in an isothermal compression the properties of the system change continuously; at no point is there a discontinuous change like a phase transition. Starting from very low density, where the fluid behaves like a gas, one can compress the fluid to high densities where it behaves like a liquid. These sort of considerations can also be carried out for one of the alternative forms of the phase diagram, Figs. 8.9 and 8.10. In Fig. 8.9, for example, the dashed curve represents a continuous transition from a gas-like to a liquid-like state without any

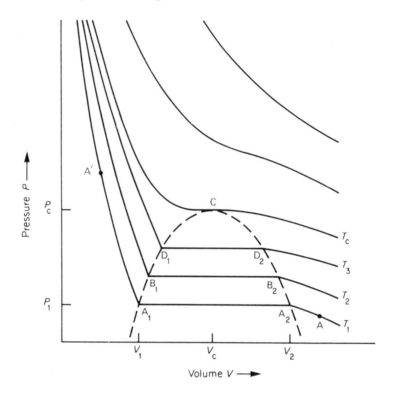

Fig. 8.11. Schematic diagram of the isotherms of a fluid. V is the specific volume (i.e., volume per gram), V_1 and V_2 are the specific volumes of the liquid and vapour coexisting in equilibrium at temperature T_1. C is the critical point, and T_c, P_c and V_c are the critical constants.

abrupt phase transition having occurred. The dashed curve in Fig. 8.11 shows the specific volumes of the liquid and vapour phases as a function of pressure. It is very similar to Fig. 8.10 showing these specific volumes as functions of the temperature.

Experimentally one finds that the critical isotherm of a substance possesses a *very flat* point of inflection at the critical point, as indicated in the schematic diagram 8.11. This suggests that not only the first two derivatives (8.39) vanish at the critical point but also higher derivatives. This is in contrast to the predictions from van der Waals' equation:* the critical isotherm which results from van der Waals' equation possesses a very sharp point of inflection. Thus van der Waals' equation gives at best a qualitative description. In particular, it always leads to a uniform density, whereas we know that under suitable

---

*For a derivation and discussion of van der Waals' equation see the end of section 7.8.

conditions (underneath the dashed curve A_1CA_2 in Fig. 8.11) two separate phases of different densities coexist. It seems that no simple equation of state, like van der Waals', which allows a power series expansion near the critical point can provide a correct description near the critical point. As mentioned previously (section 2.5, see also Flowers and Mendoza,[26] section 7.7, and Present,[11] section 9.2), a substance displays very peculiar properties near the critical point which result from large fluctuations in density, and an adequate description of these properties is very difficult.

The question naturally arises whether there also exists a critical point on the melting curve, Fig. 8.9. Experimental evidence strongly suggests that no such critical point exists (see Pippard,[9] pp. 122–24). This conclusion is supported by the following theoretical argument (which is not accepted by everyone). Liquids and crystalline solids are *in principle* very different. A crystalline solid has a geometrical structure which gives it preferred directions (lines of atoms, etc.). A liquid has no such preferred directions; its properties are isotropic. Now a system either possesses certain geometrical symmetries or it does not. No *continuous* transition from the one to the other state of affairs is possible. On the other hand we have seen that the existence of a critical point implies the occurrence of such continuous transitions. As one approaches the critical point along the phase equilibrium curve the properties of the two phases gradually become more and more similar until at the critical point they are identical. Since no such continuous transition seems possible for the symmetries of a system, a solid–liquid critical point cannot exist.

PROBLEMS 8

8.1 Derive Eqs. (8.9) and (8.10) from Eq. (8.8).

8.2 The transition temperature of grey and white tin at a pressure of one atmosphere is 291 K, grey tin being the stable modification below this temperature. The change in enthalpy for this transition is 2238 J/mol. The densities of grey and white tin are 5.75 and 7.30 g/cm³ respectively, and the atomic weight of tin is 118.7. What is the change in the transition temperature if the system is at a pressure of 100 atm?

8.3 In the temperature range 700 to 730 K, the vapour pressure of magnesium can be approximately represented by

$$\log_{10}P = -\frac{7,500}{T} + 8.6$$

(P in mmHg, T in K). What is the molar latent heat of sublimation of magnesium in this temperature range?

8.4 The latent heat of fusion and the latent heat of vaporization of neon at the triple point have the values 335 J/mol and 1803 J/mol respectively. The temperature of the triple point is 24.57 K. What is the latent heat of sublimation of neon at the triple point? What is the change in entropy when one mole of liquid neon at the triple point is vaporized?

8.5 The vapour pressure of water at 298.15 K is 23.75 mmHg. What is the vapour pressure of water at 273.16 K, given that the latent heat of evaporation of water at 298.15 and 273.16 K is 43,991 and 45,049 J/mol respectively?

8.6 The liquid and solid phases of helium 4 may coexist in equilibrium at 0 K, the density of the solid being greater than that of the liquid. What is the slope dP/dT of the phase equilibrium curve at the absolute zero of temperature?

8.7 Derive an expression for the vapour pressure of a solid monatomic substance given that the vapour may be treated as a perfect monatomic gas and that the solid can be described by the Einstein model, the temperature of the system being large compared with the Einstein temperature Θ_E of the solid.
   (Hint: Look at the 'hints' to problem 6.2.)

8.8 Obtain the critical constants for Dieterici's equation of state

$$P(V-b) = RT \exp\left(-\frac{a}{RTV}\right) \cdot$$

CHAPTER

The perfect quantal gas

9.1 INTRODUCTORY REMARKS

In Chapter 7 we considered a perfect classical gas, that is a gas of non-interacting particles under conditions such that for all single-particle states $r(=1,2,\ldots)$ the mean occupation numbers $\bar{n}_r$ are very small compared to unity:

$$\bar{n}_r \ll 1 , \quad \text{(all } r) . \tag{9.1}$$

We know from section 7.3 that if this condition does not hold quantum effects may become important; we saw that this may be the case for electrons in metals or for the atoms in liquid helium. In this and the two following chapters we shall deal with the modifications of the classical statistics of Chapter 7 when quantum effects are important, i.e. with *quantum statistics*. The basic theory will be developed in this chapter. In Chapter 10 it is applied to black-body radiation, culminating in Planck's radiation law. In Chapter 11 we shall return to the perfect quantum-mechanical gas consisting of material particles.

In Chapter 7, states of the gas with some $n_r > 1$ were treated wrongly. Such states were allowed to occur, but they were given the wrong weight. In the classical regime (9.1) this did not matter: these states were not important. This treatment led to the simple mathematical result (7.11) but very careful argument was required to derive it. These difficulties disappear in the following account of quantum statistics which is conceptually more straight-forward: it does not require the sort of contortions which took us to Eq. (7.11).

9.2   QUANTUM STATISTICS

We consider the same problem as in section 7.1, i.e. we have a gas of non-interacting identical particles, with discrete single-particle states $1, 2, \ldots,$ $r, \ldots$, possessing energies $\varepsilon_1 \leqslant \varepsilon_2 \leqslant \ldots \varepsilon_r \leqslant \ldots$. For a gas of N particles, a state of the gas as a whole is specified by the set of occupation numbers

$$n_1, n_2, \ldots n_r, \ldots \tag{9.2}$$

where n_r is the number of particles in the single-particle state labelled r.

The basic question now is the following: what values can the occupation numbers $n_1, n_2, \ldots$ assume? According to quantum mechanics, which gives an unambiguous answer to this question, the occupation numbers $n_1, n_2, \ldots$ cannot arbitrarily assume all values: they are subject to certain restrictions. A full answer requires a detailed quantum-mechanical analysis. We shall be content to state the conclusions of such an analysis, which is adequate for our purposes.

Two kinds of situation arise which divide particles into two mutually exclusive classes.

In the first case there is *no restriction on the occupation numbers n_r*. They can and do assume all integer values

$$n_r = 0, 1, 2, \ldots \quad (\text{all } r) \ . \tag{9.3}$$

This kind of statistics is known as *Bose–Einstein statistics* (BE statistics for short). It was first introduced by Bose in 1924 in order to derive the Planck radiation law, and Einstein, recognizing its full importance, applied it in the same year to a perfect gas of particles with mass. Examples of *bosons*, i.e. particles obeying BE statistics, are photons, π mesons and K mesons.

In the second kind of situation the occupation numbers are restricted: *at most one particle can be in any state*, i.e. the occupation numbers n_r are *restricted* to the values

$$n_r = 0, 1 \quad (\text{all } r) \ . \tag{9.4}$$

This kind of statistics was first considered in 1926 by Fermi and, independently, by Dirac. It is known as *Fermi–Dirac statistics* (FD statistics for short). Examples of *fermions*, that is particles obeying FD statistics, are electrons, positrons, protons and neutrons. The restriction (9.4) states the *exclusion principle* for non-interacting particles (the case we are considering): *two (identical) fermions cannot be in the same single-particle state.* The exclusion principle was first proposed by Pauli (1925) to explain the observed atomic spectra and the structure of the periodic table.* We see that particles which obey the Pauli exclusion principle are fermions.

There is an important difference in the physical properties of fermions and bosons which we must now discuss. Most particles possess an *intrinsic angular momentum.* It is not connected with the motion of the particle as seen in some frame of reference; the particle possesses this angular momentum even when its centre of mass is at rest. It is like the angular momentum of a spinning top whose centre of mass is at rest. For this reason, this intrinsic angular momentum of the particle is called its *spin.* According to quantum mechanics angular momentum is quantized in multiples of $\frac{1}{2}\hbar$ which would appear to be the natural unit of angular momentum. (However, it is usual to call this angular momentum $\frac{1}{2}$, measuring angular momentum in units of $\hbar$.) The angular momentum of a system can only assume values from amongst the following.[†]

$$0, \tfrac{1}{2}\hbar, \hbar, \tfrac{3}{2}\hbar, 2\hbar, \ldots . \qquad (9.5)$$

The spin of a particle, being an angular momentum, can also assume only one of these values.

The intrinsic spin of each elementary particle, such as electron, photon, proton, etc., has a definite value. From a study of the physical properties of these particles, one finds *bosons always have integral spin*, i.e. their intrinsic angular momentum is restricted to one of the values $0, \hbar, 2\hbar, \ldots$. For example a π meson has spin 0, and a photon has spin 1 (in units of $\hbar$). It is similarly found that *fermions always have half-integral spin*, i.e. their intrinsic angular momentum is restricted to one of the values $\frac{1}{2}\hbar, \frac{3}{2}\hbar, \ldots$. Electrons, positrons, protons, neutrons, etc. all possess spin $\frac{1}{2}$.

We thus have *a most remarkable connection between spin and statistics*: *bosons possess integral spin, fermions half-integral spin.* This connection, to which no exception is known, is very well established since the spin of a particle decisively affects its behaviour in collision processes and in spectroscopy. The different statistics which fermions and bosons obey lead to very different and therefore easily distinguishable properties. (For example, the doublet D lines of sodium and the Zeeman effect provide strong evidence that the electron has spin $\frac{1}{2}$. The periodic table and the properties of

---

*For a discussion of the Pauli principle see, for example, Eisberg,[39] French and Taylor,[42] Mandl,[45] Rae[47] or Willmott[48].

[†]For a discussion of these and other properties of angular momentum in quantum mechanics, see, for example, Mandl,[45] Chapter 6, or Rae,[47] Chapters 5 and 6.

electrons in metals require that electrons obey Fermi–Dirac statistics.) It was first shown by Pauli (1940) that there are very fundamental reasons for this connection between spin and statistics. It is a consequence of demanding that the quantum theory describing these particles is consistent with the special theory of relativity. Pauli's proof, which was later much generalized, represents impressive evidence in support of relativistic quantum field theory and must be regarded as one of the triumphs of theoretical physics.*

The conditions (9.3) and (9.4) which bosons and fermions must satisfy are related to the symmetry properties of the wave functions describing systems of such particles. *The wave functions for identical bosons must be symmetric* if two bosons are interchanged. Suppose we have two identical bosons in the single-particle states r and s, and let ϕ_r and ϕ_s be the wave functions of these states. If we label the two bosons 1 and 2, then the state of the combined two-particle system is described by the wave function $\phi_r(1)\phi_s(2)$. But since the particles are indistinguishable we could equally well write the wave function of this state as $\phi_r(2)\phi_s(1)$. We cannot insist on calling the bosons in state r and state s by the labels 1 and 2 respectively. We could interchange these labels. It follows from quantum mechanics that for *bosons* the correct wave function is the *symmetric* linear combination

$$\Psi_{BE}(1,2) = \phi_r(1)\phi_s(2) + \phi_r(2)\phi_s(1) \ . \tag{9.6}$$

Interchanging the labels 1 and 2 we obtain

$$\Psi_{BE}(2,1) = +\Psi_{BE}(1,2) \ . \tag{9.7}$$

This is the basic symmetry requirement for wave functions describing bosons.

In contrast to this behaviour, *the wave functions for identical fermions must be antisymmetric.* Eq. (9.6) is now replaced by

$$\Psi_{FD}(1,2) = \phi_r(1)\phi_s(2) - \phi_r(2)\phi_s(1) \tag{9.8}$$

whence

$$\Psi_{FD}(2,1) = -\Psi_{FD}(1,2) : \tag{9.9}$$

the wave function of a system of identical *fermions* is *antisymmetric* if two fermions are interchanged. If in Eq. (9.8) we take $r = s$, i.e. both fermions in the *same* single-particle state so that $n_r = 2$, the wave function (9.8) vanishes identically. We see that for an antisymmetric wave function we regain the Pauli principle (9.4). For bosons, on the other hand, we can have $r = s$ in Eq. (9.6), and $n_r = 2$.

*For a simple discussion see Mandl and Shaw,[46] pp. 72–73.

In Eqs. (9.6) and (9.8) the wave functions of the two-particle systems are written as the sum and difference of the products of two single-particle wave functions ϕ_r and ϕ_s. This is only possible if the two particles do not interact with each other or their interaction is neglected. In general, if there is interaction, the wave functions are more complicated than Eqs. (9.6) and (9.8). However the more general symmetry requirements (9.7) and (9.9) still apply. In particular Eq. (9.9), that the wave function for fermions must be antisymmetric, is the general statement of the Pauli principle.

All these considerations can be generalized to more than two particles without additional new concepts.

Above we talked of protons, electrons, neutrons, etc., as elementary particles. This was meant in contrast to such obviously composite particles as an alpha particle (consisting of two protons and two neutrons), a tritium nucleus (consisting of one proton and two neutrons) or a hydrogen atom (consisting of one proton and one electron). But no division into elementary and composite particles is possible. (Is a neutron elementary? It is not even stable, decaying into proton plus electron plus antineutrino with a half-life of about 11 minutes.) Nor is such a division needed. The above discussion about the statistics of identical particles applies equally well to composite particles, provided the excitation energies of the internal degrees of freedom of the composite particles are large compared to the interaction energies of the system, i.e. provided the composite particles remain in their ground state throughout. Only in this case can we apply the statistics of identical particles. Under these conditions the spin-statistics relation also holds for composite particles.*

A composite particle has a spin, i.e. an angular momentum about its centre of mass. This spin is the resultant of the angular momenta of its constituents and is obtained by vector addition of these angular momenta. To perform this operation requires a detailed knowledge of the addition of angular momenta in quantum mechanics. For our purposes the following result, which will only be quoted, suffices.

If a composite particle contains an odd number of fermions (i.e. particles of spin $\frac{1}{2}, \frac{3}{2}, \ldots$) and any number of bosons (i.e. particles of spin $0, 1, \ldots$), then its spin is half-integral and it behaves like a fermion. If it contains an even number of fermions and any number of bosons, then its spin is integral and it behaves like a boson. Thus of the above examples the alpha particle and hydrogen atom have integral spin and behave as bosons, whereas the tritium nucleus has half-integral spin and behaves like a fermion. These conclusions affect, for example, the scattering of deuterons by deuterons. (A deuteron consists of a proton and a neutron, and is therefore a boson.)

*For further discussion of this problem, and references, see ter Haar,[17] p. 136 and pp. 146–147.

An interesting example occurs in low temperature physics. In liquid helium consisting of the usual isotope ^{4}He (i.e. alpha particle nuclei) the atoms obey BE statistics, while in liquid ^{3}He (in which the nuclei consist of two protons and one neutron) they obey FD statistics. This difference in statistics is responsible for the very different properties of liquid ^{4}He and liquid ^{3}He. (See section 11.6.)

9.3 THE PARTITION FUNCTION

In this section we shall obtain the expression for the partition function of a perfect quantal gas. We shall largely be able to treat fermions and bosons simultaneously.

Consider a gas of N non-interacting identical particles. In the state specified by the occupation numbers (9.2), i.e.

$$n_1, n_2, \ldots n_r, \ldots , \tag{9.2}$$

it has the energy

$$E(n_1, n_2, \ldots n_r, \ldots) = \sum_r n_r \varepsilon_r , \tag{9.10}$$

where, as throughout the following, $\sum_r$ stands for summation over all single-particle states, $r = 1, 2, \ldots$. For fermions and bosons, the occupation numbers n_r are given by Eqs. (9.4) and (9.3) respectively:

$$n_r = 0, 1 \qquad \text{all } r \quad \text{FD} \tag{9.4}$$
$$n_r = 0, 1, 2, \ldots \quad \text{all } r \quad \text{BE} . \tag{9.3}$$

In addition we must have, for a gas of N particles, that

$$\sum_r n_r = N . \tag{9.11}$$

Each set of occupation numbers (9.2) satisfying Eq. (9.11), and either Eq. (9.4) or Eq. (9.3)—depending on whether we are dealing with fermions or bosons—defines a state of the gas. Its energy is given by Eq. (9.10). Furthermore the totality of such sets of occupation numbers defines all possible states of the gas. Hence the partition function of such a gas, in equilibrium at temperature T in an enclosure of volume V, is given by

$$Z(T, V, N) = \sum_{n_1 n_2 \ldots} \exp \left\{ -\beta \sum_r n_r \varepsilon_r \right\}$$
$$= \sum_{n_1 n_2 \ldots} \exp\{ -\beta (n_1 \varepsilon_1 + n_2 \varepsilon_2 + \cdots) \} . \tag{9.12}$$

Here, as in the following, $\sum_{n_1 n_2 \ldots}$ stands for summations over all sets of occupation numbers satisfying Eq. (9.11), and either Eq. (9.4) or Eq. (9.3).

From Eq. (9.12) we obtain an expression for the mean occupation number $\bar{n}_i$ of the single-particle state i, for the gas in equilibrium. We differentiate $\ln Z$ with respect to ε_i, keeping T and all other energy levels ε_r, $r \neq i$, fixed. We obtain on dividing by $(-\beta)$:

$$-\frac{1}{\beta} \left(\frac{\partial \ln Z}{\partial \varepsilon_i} \right)_{T, \varepsilon_r (r \neq i)} = \frac{\displaystyle\sum_{n_1 n_2 \ldots} n_i \exp\{-\beta(n_1 \varepsilon_1 + n_2 \varepsilon_2 + \cdots)\}}{\displaystyle\sum_{n_1 n_2 \ldots} \exp\{-\beta(n_1 \varepsilon_1 + n_2 \varepsilon_2 + \ldots)\}}$$

$$= \bar{n}_i \, , \tag{9.13}$$

since $\exp\{-\beta(n_1 \varepsilon_1 + n_2 \varepsilon_2 + \ldots)\}$ is the relative probability that the gas should be in the state specified by the occupation numbers $n_1, n_2, \ldots$, and the summation is over all states of the gas.

From Eq. (9.13) one easily obtains the root-mean-square fluctuation Δn_i of the occupation number of the ith state, defined by

$$(\Delta n_i)^2 \equiv \overline{(n_i - \bar{n}_i)^2} = \overline{n_i^2} - \bar{n}_i^2 \, . \tag{9.14}$$

Differentiation of Eq. (9.13) with respect to ε_i and again dividing by $(-\beta)$ leads to (remember ε_i occurs in both numerator and denominator)

$$\frac{1}{\beta^2} \left(\frac{\partial^2 \ln Z}{\partial \varepsilon_i^2} \right) = -\frac{1}{\beta} \left(\frac{\partial \bar{n}_i}{\partial \varepsilon_i} \right) = (\Delta n_i)^2 \, . \tag{9.15}$$

(In section 2.5 very similar procedures—differentiation with respect to β—led to the mean value and the root-mean-square fluctuation of the energy, Eqs. (2.26) and (2.29).)

Eqs. (9.12) and (9.13) for the partition function and the mean occupation numbers are our basic results. From these all the physics of a given system follows, once one has evaluated the partition function. In Chapters 10 and 11 we shall consider these problems for photons and for material particles respectively. We shall find that for photons the direct evaluation of the partition function (9.12) is easy, while for material particles it is awkward. This leads one to an important generalization of the formalism which greatly simplifies calculations and allows applications to a wide range of new situations.

PROBLEMS 9

9.1 Classify the following particles according to the statistics which they obey
    (i) ^{12}C atom   (ii) $^{12}C^+$ ion   (iii) $^4He^+$ ion
    (iv) H^- ion   (v) ^{13}C atom   (vi) positronium atom.

Black-body radiation

10.1 INTRODUCTORY REMARKS

In the last chapter we began to consider the quantum statistics of a perfect gas of identical particles. We shall now apply these ideas to thermal radiation which is a simple and very interesting example of a perfect quantal gas.

All bodies emit electromagnetic radiation by virtue of their temperature but usually this radiation is not in thermal equilibrium. If we consider the radiation within an opaque enclosure whose walls are maintained at a uniform temperature T then radiation and walls reach thermal equilibrium, and in this state the radiation has quite definite properties. To study this equilibrium radiation one cuts a small hole in the walls of the enclosure. Such a hole, if sufficiently small, will not disturb the equilibrium in the cavity and the emitted radiation will have the same properties as the cavity radiation. The emitted radiation also has the same properties as the radiation emitted by a perfectly black body at the same temperature T as the enclosure.* For this reason cavity radiation is also called black-body radiation.

Kirchhoff and Boltzmann applied thermodynamic methods to the problem of black-body radiation. This approach had some outstanding successes but,

*A perfectly black body absorbs all radiation falling on it, and the very small hole in the walls of our enclosure will also have this property to a very high degree: radiation entering the cavity through the hole is in the course of repeated reflections nearly all absorbed by the inner surfaces of the enclosure. The radiation emitted from the hole and by a perfectly black body then also have the same properties, as follows from Kirchhoff's laws of radiation; cf. Adkins[1] or Sommerfeld.[16]

typically, could not reveal the whole story. We shall discuss it briefly in section 10.5, but we shall first continue the statistical approach of Chapter 9.

The wave-particle duality of electromagnetic radiation leads to two alternative statistical treatments of black-body radiation. On the wave picture, one considers the normal modes of electromagnetic waves in a cavity (a problem similar to that of the harmonics of a vibrating string). Calculating the mean energy of each mode from quantum mechanics, rather than from the classical theorem of equipartition of energy, leads to Planck's radiation law. We consider this method briefly at the end of section 10.3, but shall mainly follow the alternative approach which treats radiation as a gas of radiation quanta, i.e. of photons, obeying BE statistics. In section 10.2 we derive the partition function for a photon gas and from it, in section 10.3, Planck's law. The properties of cavity radiation will be discussed in section 10.4.

10.2   THE PARTITION FUNCTION FOR PHOTONS

In this section we shall derive the partition function for a photon gas. Three specific properties of photons are important for us. Firstly, photons behave in many respects like particles of spin 1; hence *photons obey* BE *statistics*. Secondly, photons do not interact with each other. This follows, for example, from the linearity of Maxwell's equations which allows one to superpose electromagnetic fields.* The fact that photons do not interact with each other means that *a photon gas is a perfect gas*. The process which produces thermal equilibrium between the radiation in the cavity and the enclosure is the continual emission and absorption of photons by the atoms of the surrounding wall maintained at temperature T. From this follows the third property of the photon gas which we require: *the number of photons in the cavity is not constant*. This number fluctuates about a mean which is determined by the equilibrium conditions themselves. This means that for photons the partition function is given by Eq. (9.12), with the occupation numbers n_r given by Eq. (9.3), viz.

$$n_r = 0, 1, 2, \ldots \quad \text{all } r , \qquad (9.3)$$

but the occupation numbers are *not* subject to the constraint (9.11) which specifies the total number of particles in the gas. In other words, each of the occupation numbers $n_1, n_2, \ldots$ in Eq. (9.12) assumes all possible values $0, 1, 2, \ldots$, *independently of the values of the other occupation numbers*.

---

*This neglects an extremely small quantum-mechanical effect which leads to the scattering of photons by photons. This photon–photon interaction is under almost all circumstances much too weak to be observable, let alone to produce thermal equilibrium.

This greatly simplifies the summations in Eq. (9.12) which we can now at once write (omitting the argument N from Z, since N is now *not* fixed, and adding a subscript ph to indicate that we are dealing with photons) in the form

$$Z_{\mathrm{ph}}(T, V) = \sum_{n_1=0}^{\infty} \sum_{n_2=0}^{\infty} \ldots \exp\{-\beta(n_1\varepsilon_1 + n_2\varepsilon_2 + \cdots)\} . \qquad (10.1)$$

To evaluate this expression consider for simplicity first the double sum

$$X \equiv \sum_{n_1} \sum_{n_2} e^{(a_1 n_1 + a_2 n_2)} . \qquad (10.2a)$$

n_1 and n_2 cover the range $0, 1, 2, \ldots$ to ∞, as in Eq. (10.1), but indeed any other ranges would do. We can rewrite expression (10.2a) as

$$X = \sum_{n_1} \sum_{n_2} e^{a_1 n_1} \cdot e^{a_2 n_2} = \left\{\sum_{n_1} e^{a_1 n_1}\right\}\left\{\sum_{n_2} e^{a_2 n_2}\right\} . \qquad (10.2b)$$

(A reader puzzled by the last step should write out explicitly the two series on the right-hand side of Eq. (10.2b). On multiplying the two series together, term by term, all the terms on the left-hand side of Eq. (10.2b) are obtained.) Rewriting the last expression in terms of the conventional product notation* we obtain

$$X \equiv \sum_{n_1} \sum_{n_2} e^{(a_1 n_1 + a_2 n_2)} = \prod_{r=1}^{2} \left\{\sum_{n_r} e^{a_r n_r}\right\} . \qquad (10.2c)$$

This equation is at once generalized from two factors ($r = 1, 2$) to any number of factors, including infinitely many,

$$\sum_{n_1} \sum_{n_2} \ldots \ldots e^{(a_1 n_1 + a_2 n_2 + \ldots)} = \prod_{r=1}^{\infty} \left\{\sum_{n_r} e^{a_r n_r}\right\} . \qquad (10.4)$$

Since n_r takes on all values $0, 1, 2, \ldots$ to ∞, each factor in curly parentheses in the last equation is a geometrical series with the sum

$$\left\{\sum_{n_r=0}^{\infty} e^{a_r n_r}\right\} = \frac{1}{1 - \exp a_r} . \qquad (10.5)$$

The geometrical series (10.5) always converges since $\exp a_r < 1$, i.e. $a_r < 0$. This last inequality will be satisfied since we shall take

$$a_r \equiv -\beta\varepsilon_r ,$$

*Π is used to denote a product of factors just as Σ denotes a sum of terms, and ranges of terms are indicated similarly; e.g.

$$\prod_{r=1}^{p} a_r \equiv a_1 a_2 \ldots a_p . \qquad (10.3)$$

We could have an infinite product with $p = \infty$.

and the photon energy ε_r in the state r is related to the circular frequency ω_r of the photon in this state by $\varepsilon_r = \hbar\omega_r$, so that we have $\varepsilon_r > 0$ and $a_r < 0$. With this choice of a_r the left-hand side of Eq. (10.4) becomes identical with the photon partition function Z_{ph}, Eq. (10.1), and substituting Eq. (10.5) on the right-hand of Eq. (10.4), we finally obtain

$$Z_{ph}(T, V) = \prod_{r=1}^{\infty} \frac{1}{1 - \exp(-\beta\varepsilon_r)} \ . \tag{10.6}$$

We can calculate the mean occupation number $\bar{n}_i$, i.e. the mean number of photons in the state i. From Eq. (10.6)

$$\ln Z_{ph}(T, V) = - \sum_{r=1}^{\infty} \ln\left[1 - \exp(-\beta\varepsilon_r)\right] \ , \tag{10.7}$$

whence Eq. (9.13) gives

$$\left. \begin{aligned} \bar{n}_i &= - \frac{1}{\beta} \frac{\partial}{\partial\varepsilon_i} (\ln Z_{ph}) \\[2mm] &= \frac{1}{\exp(\beta\varepsilon_i) - 1} \end{aligned} \right\} \ . \tag{10.8}$$

To obtain the Planck radiation law, and with it a full understanding of black-body radiation, we require the distribution of the single-photon states which are involved in the summation in Eq. (10.7). Once we know this distribution we can find Planck's formula in a variety of ways. We shall turn to this problem in the next section.

10.3 PLANCK'S LAW: DERIVATION

We shall derive the Planck law for black-body radiation by treating the radiation as a photon gas, as in the last section.

A photon of energy ε has a (circular) frequency ω given by

$$\varepsilon = \hbar\omega \ . \tag{10.9}$$

The momentum of a photon with this energy has magnitude

$$p = \hbar\omega/c \ , \tag{10.10}$$

since photons have zero rest mass.

Secondly we need to know the number of states, for a photon in an enclosure of volume V, such that the magnitude of the photon momentum lies in the interval p to $p+dp$. In Chapter 7 we had the same problem for a gas of material particles, and we can take over the result (7.17) from there but it requires one modification. A photon in addition to a momentum **p** possesses an internal degree of freedom: its polarization. A photon of given momentum **p** can still exist in two different states of polarization. There are various ways of describing these states: one can choose two mutually perpendicular directions of linear polarization, or right- and left-circular polarizations, for example. The net effect is always that the expression (7.17) for the density of states is multiplied by a factor 2. If we also express the momentum p in terms of the frequency ω, from Eq. (10.10), Eq. (7.17) gives for the number of photon states in which the photon has a frequency in the range ω to $\omega+d\omega$

$$f(\omega)\,d\omega = 2\,\frac{V4\pi(\hbar\omega/c)^2\hbar\,d\omega/c}{h^3}$$

$$= \frac{V\omega^2\,d\omega}{\pi^2c^3}\,. \tag{10.11}$$

Combining Eqs. (10.8) and (10.11), and using Eq. (10.9), we obtain for the number of photons in the frequency range ω to $\omega+d\omega$

$$dN_\omega = \frac{V}{\pi^2c^3}\,\frac{\omega^2\,d\omega}{\exp(\beta\hbar\omega)-1}\,, \tag{10.12}$$

and hence for the energy of the radiation in this frequency range:

$$dE_\omega = \hbar\omega\,dN_\omega = \frac{V\hbar}{\pi^2c^3}\,\frac{\omega^3\,d\omega}{\exp(\beta\hbar\omega)-1}\,. \tag{10.13}$$

As discussed in Appendix B, the result (7.17) on which Eq. (10.11) is based depends only on the volume V of the enclosure but not its shape. The same is therefore also true of the results (10.12) and (10.13), so that the photon density (number of photons per unit volume) and energy density are uniform throughout the enclosure. From Eq. (10.13) we obtain for the latter quantity

$$u(\omega,T)\,d\omega = \frac{\hbar\omega^3\,d\omega}{\pi^2c^3\,[\exp(\beta\hbar\omega)-1]}\,. \tag{10.14}$$

This is *Planck's radiation law*: it gives the distribution of energy density as a function of frequency for radiation in thermal equilibrium. It is a function of only the temperature T of the system. *Planck's law gives a complete*

description of black-body radiation, in excellent agreement with observations.
We shall discuss the properties of black-body radiation in the next section.

The alternative approach treats cavity radiation as waves instead of photons. We now need the normal modes of the electromagnetic waves in the enclosure. The frequency distribution of these modes, obtained directly from the wave picture, is of course again given by Eq. (10.11). (See Appendix B Eq. (B.34).) The equation of motion of each normal mode is that of a harmonic oscillator of the same frequency. We obtained the mean energy of a harmonic oscillator in Eq. (6.7). If we omit the zero-point energy term* from Eq. (6.7) (and write ω for ω_E), Eq. (6.7) is identical with the mean energy, obtained from Eq. (10.8), of photons in a state of energy $\hbar\omega$. So we have again derived Eqs. (10.8) and (10.11), from which Planck's law (10.14) follows, but this time in terms of a wave picture.

This treatment of cavity radiation is closely analogous to that of elastic waves in Debye's theory (section 6.3). In both theories we have normal modes of waves whose frequency distributions, Eqs. (6.19b) and (10.11), are essentially the same. (The extra factor $\frac{3}{2}$ in (6.19b) occurs since elastic waves can be longitudinally as well as transversely polarized.) In both theories the mean energy of each mode is that of a harmonic oscillator of the same frequency, Eqs. (6.7) and (10.8). Hence the two theories are mathematically very similar.

This suggests that for elastic waves too an alternative description in terms of quanta is possible. These quanta are called phonons. It follows from the fact that phonons and photons have essentially the same statistical distributions that phonons obey BE statistics and that the number of phonons is not constant, i.e. phonons can be emitted and absorbed.[†]

10.4   THE PROPERTIES OF BLACK-BODY RADIATION

We shall now discuss some properties of black-body radiation which follow from Planck's law, Eq. (10.14). The energy density $2\pi u(\omega, T)$ of the radiation, per unit volume per unit range of the frequency $v = (\omega/2\pi)$, is plotted in Fig. 10.1 as a function of v for three temperatures: 2000, 4000 and 6000 K. Also shown in this figure is a wave-number scale, $1/\lambda = v/c$, and the visible part of the spectrum is marked. At 6000 K, the maximum of the distribution lies at the edge of the visible part of the spectrum. 6000 K is the temperature of the surface of the sun, which approximates a black body fairly closely.

We see from Fig. 10.1 that the maximum of the distribution shifts to higher frequencies with increasing temperature. This maximum is obtained from

---

*This simply corresponds to changing the zero of the energy scale. We can make this change since only energy *differences* are physically significant.
[†]See Hall[27] or Kittel.[29]

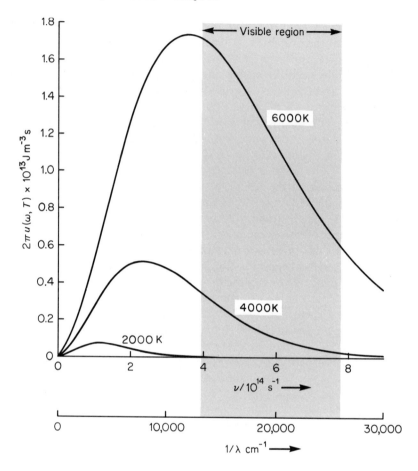

Fig. 10.1. Planck's law, Eq. (10.14). The energy density per unit interval of frequency ν as a function of ν or $1/\lambda$.

Eq. (10.14) from the condition $[\partial u(\omega, T)/\partial \omega]_T = 0$. This differentiation leads to the transcendental equation

$$(3 - x)e^x = 3 \qquad (10.15)$$

where $x \equiv \beta \hbar \omega_{max} \equiv \beta h \nu_{max}$. The solution of Eq. (10.15), which must be found graphically or numerically, is

$$\frac{\hbar \omega_{max}}{kT} = \frac{h\nu_{max}}{kT} = 2.822 \ . \qquad (10.16)$$

(One easily sees that Eq. (10.15) is nearly satisfied by $x = 3$.) Eq. (10.16), known as Wien's displacement law, is extremely well confirmed by experiment.

If we integrate Eq. (10.14) over all frequencies ω we obtain the total energy density of black-body radiation at a given temperature T. If in Eq. (10.14) we put $x = \hbar\omega/kT$, we obtain at once

$$u(T) = \int_0^\infty u(\omega, T)\,d\omega = aT^4 , \qquad (10.17)$$

with the constant of proportionality a given by

$$a \equiv \frac{\hbar}{\pi^2 c^3} \left(\frac{k}{\hbar}\right)^4 \int_0^\infty \frac{x^3\,dx}{(e^x - 1)} = \frac{\pi^2 k^4}{15\hbar^3 c^3} . \qquad (10.18)$$

(The integral in the last equation is evaluated in Appendix A, Eq. (A.19).) Eq. (10.17) is called the Stefan–Boltzmann law and the constant

$$\sigma \equiv \frac{c}{4} a = \frac{\pi^2 k^4}{60\hbar^3 c^2} = 5.67 \times 10^{-8}\,\text{J m}^{-2}\,\text{s}^{-1}\,\text{K}^{-4} \qquad (10.19)$$

is called the Stefan–Boltzmann constant.* The Stefan–Boltzmann law (10.17) was first obtained experimentally by Stefan (1879). Boltzmann in 1884 derived the T^4-dependence theoretically using thermodynamic methods (see section 10.5) but he could not of course calculate the constant of proportionality which involves Planck's constant.

In the *limit of low frequencies*, i.e.

$$\hbar\omega \ll kT , \qquad (10.20)$$

Planck's law (10.14) becomes, approximately,

$$u(\omega, T)\,d\omega = \frac{\omega^2\,d\omega}{\pi^2 c^3}\,kT, \qquad (\hbar\omega \ll kT) . \qquad (10.21)$$

Comparison with Eq. (10.11) shows that in this limit the mean energy per photon, or per normal mode if we think of the radiation as a superposition of normal modes, is kT. We see that this result only holds for *low* frequencies,

---

*As in the kinetic theory of gases, one shows that $cu(T)/4 = \sigma T^4$ is the energy incident per unit area per unit time on the wall of the enclosure containing the radiation. If one cuts a *small* hole of area A in the wall, then $\sigma T^4 A$ is the energy radiated per unit time from the hole, i.e. σT^4 is the energy emitted per unit time per unit area by a black body. This brings out the significance of the Stefan–Boltzmann constant.

but it is the energy distribution predicted *for all frequencies* by classical physics. (Note that Planck's constant no longer appears in Eq. (10.21)!) For, according to the classical theorem of equipartition of energy (section 7.9) the mean energy of a harmonic oscillator is kT, and the equation of motion describing a normal mode is just that of a harmonic oscillator. Eq. (10.21) was first derived in this classical manner (without the proviso $\hbar\omega \ll kT$) by Rayleigh in 1900. It was straightaway realized that it must be wrong: if we calculate the total energy density by integrating Eq. (10.21) over all frequencies (see Eq. (10.17)) the result is divergent because of the contributions from high frequencies. This was given the rather picturesque name of the ultra-violet catastrophe. All the same Eq. (10.21) is partially correct, and for low frequencies agrees with observations.

In the opposite *limit of high frequencies*, i.e.

$$\hbar\omega \gg kT \,, \tag{10.22}$$

Planck's law (10.14) approximates to

$$u(\omega, T)\,d\omega = \frac{\hbar\omega^3}{\pi^2 c^3}\,e^{-\beta\hbar\omega}\,d\omega \,, \qquad (\hbar\omega \gg kT) \,. \tag{10.23}$$

Eq. (10.23) is known as Wien's law; it was proposed by Wien (1896) empirically from experimental data as the energy spectrum for all frequencies, but it agrees with observations in the high frequency limit only.

All thermodynamic properties of black-body radiation follow from the partition function (10.7), and we shall derive them in this way at the end of this section. But some properties follow more directly from the picture of a perfect photon gas. For example, the equilibrium radiation in a cavity exerts a pressure on the walls of the enclosure given by

$$P = \tfrac{1}{3}u(T) \tag{10.24}$$

where $u(T)$ is the total energy density (10.17). This result follows directly from Maxwell's electromagnetic theory, and below we shall derive it thermodynamically. But it also follows very easily from the picture of a perfect photon gas. For a gas of particles of mass m the pressure P is given by

$$P = \tfrac{1}{3}\rho\overline{v^2} \tag{10.25}$$

where ρ is the density of the gas and $(\overline{v^2})^{1/2}$ the rms-speed of its molecules.

This result follows from elementary kinetic theory.* For the photon gas, all the photons have the same velocity c, the velocity of light, so that ρc^2 can be interpreted, from Einstein's mass-energy relation, as the energy density $u(T)$. Hence for the photon gas Eq. (10.25) reduces to Eq (10.24).

The pressure P and other thermodynamic properties of the photon gas are easily obtained from the Helmholtz free energy. Combining Eqs. (10.7) and (10.11) leads to

$$\ln Z_{ph}(T, V) = - \int_0^\infty \frac{V\omega^2 \, d\omega}{\pi^2 c^3} \ln\left[1 - \exp(-\beta\hbar\omega)\right] \, , \tag{10.26}$$

whence we obtain for the Helmholtz free energy

$$F(T, V) = -kT \ln Z_{ph}(T, V)$$
$$= \frac{VkT}{\pi^2 c^3} \int_0^\infty \omega^2 \, d\omega \ln\left[1 - \exp(-\beta\hbar\omega)\right] \, . \tag{10.27}$$

If we introduce the integration variable

$$x = \beta\hbar\omega \, ,$$

then Eq. (10.27) becomes

$$F(T, V) = VT^4 \frac{k^4}{\pi^2 c^3 \hbar^3} \int_0^\infty x^2 \, dx \ln(1 - e^{-x}) \, . \tag{10.28}$$

The last integral is easily evaluated by integration by parts: it equals

$$\frac{1}{3} \int_0^\infty \ln(1 - e^{-x}) \, d(x^3)$$
$$= \left[\tfrac{1}{3} x^3 \ln(1 - e^{-x})\right]_0^\infty - \frac{1}{3} \int_0^\infty \frac{x^3 \, dx}{e^x - 1} = -\frac{\pi^4}{45} \, , \tag{10.29}$$

since in the middle expression in Eq. (10.29) the first term vanishes[†] and

---

*This result holds for any velocity distribution. It does not presuppose a Maxwellian distribution. For a proof see, for example, Present,[11] section 2.3.

[†](i) For $x \to \infty$

$$e^{-x} \to 0 \quad \text{so} \quad x^3 \ln(1 - e^{-x}) \to -x^3 e^{-x} \to 0.$$

(ii) For $x \to 0$

$$e^{-x} \to 1 - x, \quad \text{so} \quad x^3 \ln(1 - e^{-x}) \to x^3 \ln x \to 0.$$

the second term is evaluated in Appendix A, Eq. (A.19). Combining Eqs. (10.28) and (10.29) gives

$$F(T,V) = -\tfrac{1}{3}aVT^4 \qquad (10.30)$$

where the constant a is defined in Eq. (10.18).

Eq. (10.30) is our final result. From it we obtain at once

$$S = -\left(\frac{\partial F}{\partial T}\right)_V = \tfrac{4}{3}aVT^3 , \qquad (10.31)$$

$$P = -\left(\frac{\partial F}{\partial V}\right)_T = \tfrac{1}{3}aT^4 , \qquad (10.32)$$

and

$$E = F + TS = aVT^4 . \qquad (10.33)$$

Eliminating (aT^4) between Eqs. (10.32) and (10.33) gives

$$P = \frac{1}{3}\frac{E}{V} = \frac{1}{3}u(T) , \qquad (10.34)$$

in agreement with Eq. (10.24). Eq. (10.33) for the total energy E could also have been obtained directly by integrating Eq. (10.13) over all frequencies, using Eq. (A.19) from Appendix A for the integral and the definition (10.18) of the constant a. We see that all the extensive thermodynamic properties of black-body radiation are proportional to the volume, so that the intensive properties, such as P, F/V, S/V, E/V, are functions of the temperature only.

★ 10.5 THE THERMODYNAMICS OF BLACK-BODY RADIATION

We shall now briefly discuss the thermodynamic approach to black-body radiation. We shall obtain no new results in this way but the fact that radiation also can be treated in this way shows the great generality of thermodynamic arguments. We shall derive the Stefan–Boltzmann law and shall briefly discuss how much further thermodynamics can go in the study of radiation. As stated before, it cannot give us the spectral distribution function of the energy density, but it does provide some information about the functional form of this distribution function.

For radiation at equilibrium in an enclosure whose walls are at a uniform temperature T, Kirchhoff proved that the qualitative properties of the radiation depend only on the temperature T and are independent of the nature of the walls of the enclosure or of its shape or size.* In particular, this implies that the energy density per unit volume of radiation with circular frequencies within a given range ω to $\omega + d\omega$ is a function of T only. We shall denote this energy density, as before, by $u(\omega, T)d\omega$. It also follows from Kirchhoff's results that extensive properties of the black-body radiation in a cavity, such as its energy or entropy, are proportional to the volume of the cavity.

To apply thermodynamics to cavity radiation, we also need to know that the radiation exerts a pressure P on the walls of the cavity given by

$$P = \tfrac{1}{3}u(T) \tag{10.24}$$

where $u(T)$ is the total energy density per unit volume of the black-body radiation at temperature T. As mentioned in section 10.4, this expression for the pressure follows from electromagnetic theory or from the general picture of a photon gas.

We shall derive the Stefan–Boltzmann law from the fundamental thermodynamic relation (4.12)

$$T dS = dE + P dV \ . \tag{4.12}$$

Our system now consists of the equilibrium radiation in the cavity of volume V and at temperature T. With

$$E = V u(T) \tag{10.35}$$

and Eq. (10.24) for the radiation pressure, Eq. (4.12) becomes

$$dS = \frac{4u(T)}{3T} dV + \frac{V}{T} \frac{du(T)}{dT} dT \ . \tag{10.36}$$

Hence

$$\left(\frac{\partial S}{\partial V}\right)_T = \frac{4u(T)}{3T} \ , \qquad \left(\frac{\partial S}{\partial T}\right)_V = \frac{V}{T} \frac{du(T)}{dT} \ , \tag{10.37}$$

and by forming the second-order derivative $\partial^2 S / \partial V \partial T$ from these equations in two ways

$$\frac{\partial}{\partial T} \left[\frac{4u(T)}{3T} \right] = \frac{\partial}{\partial V} \left[\frac{V}{T} \frac{du(T)}{dT} \right] \ . \tag{10.38}$$

*For a discussion of these results, and of the thermodynamics of black-body radiation generally, see Adkins,[1] Pippard,[9] Sommerfeld[16] or Wannier.[20]

The last equation simplifies to

$$\frac{4}{3}\left(\frac{1}{T}\frac{du(T)}{dT} - \frac{u(T)}{T^2}\right) = \frac{1}{T}\frac{du(T)}{dT}$$

which reduces to

$$\frac{du(T)}{dT} = \frac{4u(T)}{T} \ ;$$

on integration this equation gives

$$u(T) = aT^4 \ . \tag{10.39}$$

This is Boltzmann's classic derivation of the Stefan–Boltzmann law. The proportionality factor a is a constant of integration which remains undetermined. Our earlier statistical derivation determined this factor. It was given by Eq. (10.18).

We obtain the radiation entropy by integrating the first of Eqs. (10.37):

$$S = \frac{4}{3T}u(T)V + \text{const.}$$

The constant of integration must be put equal to zero to make S proportional to V. On substituting for $u(T)$ from Eq. (10.39) we obtain

$$S = \tfrac{4}{3}aT^3 V \ . \tag{10.40}$$

It is not possible to obtain the spectral distribution function $u(\omega, T)$ by thermodynamic methods. Wien studied the way the frequencies of the black-body radiation change when the volume of the cavity is expanded adiabatically. In this way he obtained the following result:[*]

$$\frac{u(\omega, T)}{\omega^3} = \phi\left(\frac{\omega}{T}\right) \tag{10.41}$$

where ϕ is an arbitrary function which thermodynamics is not capable of determining. Eq. (10.41) is of course consistent with Planck's law (10.14) which gives the function ϕ. Eq. (10.41) is known as Wien's displacement law. From it one can predict how the frequency ω_{max}, at which the spectral

---

[*]For details see, for example, Adkins,[1] ter Haar and Wergeland,[18] or Wannier.[20]

distribution $u(\omega, T)$ has its maximum, is displaced to higher frequencies as the temperature of the radiation increases. The result is of the form

$$\frac{\omega_{max}}{T} = \text{const.} \tag{10.42}$$

where the constant is determined by the function ϕ. For Planck's law we obtained this result in Eq. (10.16). Finally we note that Eq. (10.41) of course leads to the Stefan–Boltzmann law (10.39), for any arbitrary function ϕ. From Eq. (10.41) one has

$$u(T) = \int u(\omega, T)\,d\omega = \int \omega^3 \phi\left(\frac{\omega}{T}\right) d\omega = T^4 \int x^3 \phi(x)\,dx \tag{10.43}$$

but the constant of proportionality (i.e. the integral in the last expression) is only known if one knows the function ϕ.

PROBLEMS 10

10.1 Let $\varrho(\lambda, T)\,d\lambda$ be the energy density per unit volume in the wavelength range λ to $\lambda + d\lambda$ of black-body radiation at temperature T. At what wavelength does the wavelength spectrum $\varrho(\lambda, T)$ have its maximum value?
  The radiation emitted by the sun has very nearly the spectral distribution of black-body radiation. If the maximum intensity per unit wavelength occurs at a wavelength of 4800 Å what is the temperature of the surface of the sun?
10.2 Assume the sun emits radiation with the properties of a black-body at 6000 K. What is the power radiated by the sun per megacycle bandwidth at a wavelength of 2 cm? (Radius of sun $= 7 \times 10^{10}$ cm.)
10.3 According to the Rayleigh–Jeans law (10.21), valid at low frequencies, the mean energy U per normal mode of black-body radiation at temperature T is

$$U = kT \ .$$

According to Wien's law (10.23), valid at high frequencies,

$$U = \hbar\omega\, e^{-\beta\hbar\omega} \ .$$

Calculate $d(1/T)/dU$ for these two cases. Invent an interpolation formula for $d(1/T)/dU$ between these two limits. By integrating this interpolation formula obtain an expression for U for any frequency. With the right choice of the interpolation formula one derives Planck's law in this way.

Systems with variable particle numbers

In Chapter 9 we obtained the general expression for the partition function of a perfect quantal gas: Eq. (9.12). In the last chapter we evaluated this expression for the photon gas. That case is particularly simple to handle since photons are not conserved, so that we do not have the subsidiary condition (9.11), i.e.

$$\sum_r n_r = N \ , \qquad (11.1)$$

which expresses the fact that we have a gas with a definite number, N, of particles.

For a gas of electrons or helium atoms in an enclosure the number of particles is fixed. Hence the subsidiary condition (11.1) holds in such cases and must be allowed for in evaluating the partition function, which makes this a harder calculation. There are several ways of dealing with this problem. By far the simplest is to modify the constraints on the system so that the subsidiary condition is dropped altogether, i.e. so that the number of particles in the system is not kept fixed but may vary. We achieve this by considering a system in contact with a particle reservoir with which the system can exchange particles, in the same way in which a system in contact with a heat bath can exchange thermal energy with the latter. This approach, due to

Gibbs, simplifies the mathematics greatly. It represents a powerful generalization of the theory, with a wide range of new applications, e.g. to chemical or nuclear reactions where the numbers of particles of individual species are not conserved. This is the method we advocate. Its basic formalism is developed in section 11.1. In section 11.2 we apply it at once to a perfect quantal gas to derive the BE and FD distribution functions.

In section 11.3 we give an *alternative* derivation of these distribution functions by calculating the partition function of a perfect quantal gas when the number of particles is conserved. This derivation is mathematically more involved and the method is restricted to a system of a fixed number of non-interacting particles. However, it demonstrates one way of allowing for the subsidiary condition (11.1). *Sections 11.1–11.2 and section 11.3 are independent of each other. A knowledge of either approach suffices for understanding the applications in sections 11.4–11.6*: the classical limit of the quantal gas, the free electron model of metals, and BE condensation.

The remaining three final sections of the chapter deal with systems with variable particle numbers and presuppose section 11.1 only. In section 11.7 we develop the basic thermodynamics of systems with variable particle numbers. In section 11.8 we consider once more the perfect classical gas, and in section 11.9 we briefly illustrate the application of the Gibbs formalism to chemical reactions.*

11.1   THE GIBBS DISTRIBUTION

We have so far developed two approaches to statistical physics and we shall now give a third. In Chapter 2 we considered an isolated system (Fig. 11.1(a)) for which energy E, volume V and the number of particles N are fixed. The statistical description of the system is in terms of its statistical weight $\Omega(E, V, N)$ which leads to the thermodynamic description in terms of the entropy S through Boltzmann's equation (2.34):

$$S = k \ln \Omega(E, V, N) \ . \tag{11.2}$$

The second formulation of statistical physics, in section 2.5, was for a system of fixed volume V and given number of particles N at temperature T (Fig. 11.1(b)), i.e. immersed in a heat bath at this temperature. The statistical and thermodynamic descriptions are now given by the partition function $Z(T, V, N)$ and the Helmholtz free energy F, these two quantities being related through Eq. (2.36):

$$F = -kT \ln Z(T, V, N) \ . \tag{11.3}$$

---

*For other applications of the Gibbs approach see, for example, Hill,[5] Rushbrooke,[14] Chapter 17, Kittel and Kroemer,[6] or ter Haar.[17]

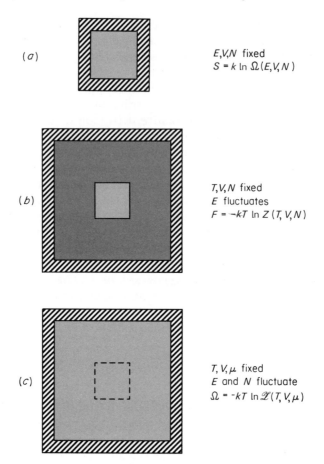

(a)

E, V, N fixed
$S = k \ln \Omega(E, V, N)$

(b)

T, V, N fixed
E fluctuates
$F = -kT \ln Z(T, V, N)$

(c)

T, V, μ fixed
E and N fluctuate
$\Omega = -kT \ln \mathscr{Z}(T, V, \mu)$

Fig. 11.1. A system subject to different conditions:
    (a) isolated system
    (b) system in contact with heat bath
    (c) system in contact with heat bath/particle reservoir.

For the system in the heat bath the energy is not constant; it fluctuates. For a macroscopic system these fluctuations are generally extremely small so that the energy of the system has after all a sharply defined value. In considering the properties of the system, it is then immaterial whether we consider the system at fixed energy (isolated) or at fixed temperature (heat bath).

We shall now obtain a third formulation of the laws of statistical physics, by considering a system of given volume V immersed in a heat bath at temperature T. But the number of particles N of the system is now *not* constant. The system can exchange particles with the heat bath which thus also acts as a particle reservoir. This situation is shown in Fig. 11.1(*c*). The system consists of the fluid within the net-like cage (shown as dashed lines); the heat bath consists of the *same* fluid surrounding the cage. The cage permits transfer both of energy and of particles. It is clear that the cage serves no concrete purpose at all. It was merely introduced to focus our attention on a given region of space: one particular cubic centimetre of the fluid, say. *This* particular volume element — whatever goes on inside it — forms our system, the rest of the fluid the heat bath/particle reservoir. We can therefore dispense with the cage altogether. For example, in studying the properties of a gas, the system would consist of a given region of the gas, of definite shape and volume, and the rest of the gas would act as heat bath. In this sort of situation both the energy E and the particle number N of the system will fluctuate, but the fluctuations will in general be negligibly small for a macroscopic system. For example, for a gas in equilibrium the mean density within a volume of macroscopic dimensions is very sharply defined.

For the system shown in Fig. 11.1(*c*) the temperature T and the volume V are given. We shall see that for a complete description of the system we must also specify a third parameter, the chemical potential μ.* We shall find that the statistical and thermodynamic descriptions of the system lead to two new quantities: the grand partition function $\mathscr{Z}(T, V, \mu)$ and the grand potential[†] Ω which are related by

$$\Omega = -kT \ln \mathscr{Z}(T, V, \mu) \ . \tag{11.4}$$

In this section we shall obtain the general expression for the grand partition function $\mathscr{Z}$ and the corresponding probability distribution which is known as the Gibbs distribution. The thermodynamics of a system with a variable number of particles will be discussed in section 11.7.

We now develop the theory for the system shown in Fig. 11.1(*c*). Our method is a straightforward generalization of that of section 2.5. The only difference is that the system can now exchange particles as well as energy with the heat bath. The composite system, consisting of system plus heat bath, is isolated. It contains N_0 particles in a volume V_0 and possesses energy E_0, and these quantities are divided between system and heat bath. The volume V of the system is fixed, but the number of particles N of the system can vary: $N = 0, 1, 2, \ldots$. For any given value of N, the system possesses a sequence of states which we arrange in increasing energy

$$E_{N1} \leqslant E_{N2} \leqslant \ldots \leqslant E_{Nr} \leqslant \ldots \tag{11.5}$$

*The chemical potential has already been introduced earlier, in section 8.1. The present chapter does NOT presuppose a knowledge of Chapter 8 and is independent of it.

[†]We shall use Ω for the grand potential since it is most frequently denoted by this symbol. No confusion should result from the fact that we also used Ω for the statistical weight, since the two quantities will never occur simultaneously.

i.e. E_{Nr} is the energy of the rth state of N particles of the system; we shall call it the state Nr, for short. If the system is in the state Nr, the heat bath possesses energy $E_0 - E_{Nr}$, consists of $N_0 - N$ particles, and occupies a volume $V_0 - V$. We denote the statistical weight of this macrostate of the heat bath by $\Omega_2(E_0 - E_{Nr}, V_0 - V, N_0 - N)$. The probability p_{Nr} of finding the system in the state Nr follows from the postulate of equal *a priori* probabilities (section 2.3):

$$p_{Nr} = \text{const.}\, \Omega_2(E_0 - E_{Nr}, V_0 - V, N_0 - N) \ , \tag{11.6}$$

or in terms of the entropy of the heat bath:

$$p_{Nr} = \text{const.} \exp\left[S_2(E_0 - E_{Nr}, V_0 - V, N_0 - N)/k \right] \ . \tag{11.7}$$

We now use the fact that the system is small compared with the heat bath and expand S_2 in a Taylor series*

$$S_2(E_0 - E_{Nr}, V_0 - V, N_0 - N)$$

$$= \left\{ S_2 - \frac{\partial S_2}{\partial V_0} V \right\} - \frac{\partial S_2}{\partial E_0} E_{Nr} - \frac{\partial S_2}{\partial N_0} N \ , \tag{11.8}$$

where S_2 stands for $S_2(E_0, V_0, N_0)$, and in each of the differentiations the 'other two' variables are held constant. We introduce the temperature of the heat bath through Eq. (2.9)

$$\frac{1}{T} = \frac{\partial S_2}{\partial E_0} \tag{11.9}$$

and we define the chemical potential μ of the heat bath by

$$\mu \equiv - T \frac{\partial S_2}{\partial N_0} \ . \tag{11.10}$$

We know from Eq. (2.8) in section 2.3 that in equilibrium the system and the heat bath have the same temperature. Eq. (2.13) similarly is the equilibrium condition for particle transfer between the system and the heat bath. It follows from Eqs. (2.13) and (11.10) that in equilibrium the system and the heat bath have the same chemical potential. This is the significance of the chemical potential. Thus Eqs. (11.9) and (11.10) also give the

---

*The reason why we must expand S_2 and not Ω_2 is discussed in section 2.5.

temperature and the chemical potential of the system. On account of Eqs. (11.9) and (11.10), Eq. (11.8) becomes

$$S_2(E_0 - E_{Nr}, V_0 - V, N_0 - N) = \left(S_2 - \frac{\partial S_2}{\partial V_0} V \right) - \frac{E_{Nr}}{T} + \frac{\mu N}{T} \ . \qquad (11.11)$$

We substitute this expression in Eq. (11.7). Since V is kept constant, and only E_{Nr} and N vary, we can write Eq. (11.7) as

$$p_{Nr} = \text{const.} \exp\left[\beta(\mu N - E_{Nr})\right] \ . \qquad (11.12)$$

Eq. (11.12) gives for the correctly normalized probability distribution

$$p_{Nr} = \frac{\exp\left[\beta(\mu N - E_{Nr})\right]}{\mathcal{Z}} \qquad (11.13)$$

where we have introduced the *grand partition function* $\mathcal{Z}$ of the system

$$\mathcal{Z} \equiv \mathcal{Z}(T, V, \mu) \equiv \sum_{Nr} \exp\left[\beta(\mu N - E_{Nr})\right] \ .* \qquad (11.14)$$

(The grand partition function and the partition function are different entities and must not be confused with each other!)

Eq. (11.13) is known as the *Gibbs distribution* or *the grand canonical distribution*. It is our basic result from which the statistical properties of a system with a variable number of particles follow. These properties can be expressed in terms of the grand partition function (11.14) which plays a central role in this theory, analogous to that of the partition function Z for a system with a fixed number of particles in a heat bath. We see from Eq. (11.14) that $\mathcal{Z}$ is a function of T, V and μ, corresponding to the fact that N is now not constant.

11.2   THE FD AND BE DISTRIBUTIONS

We continue with the discussion, begun in Chapter 9, of the perfect quantal gas, i.e. of a perfect gas of identical particles obeying Fermi–Dirac (FD) statistics or Bose–Einstein (BE) statistics. We shall now apply the Gibbs formalism, developed in the last section, to such a gas. In this way, we avoid the subsidiary condition (11.1)—that the number of particles in the system is fixed—which makes the evaluation of the partition function (9.12)

---

*The summation $\sum\limits_{Nr}$ in Eq. (11.14) is short for

$$\sum_{N=0}^{\infty} \left\{ \sum_{r=1}^{\infty} \exp\left[\beta(\mu N - E_{Nr})\right] \right\} \ . \qquad (11.14a)$$

somewhat awkward. (The partition function will be calculated in section 11.3.)

A state of the gas is specified by a set of occupation numbers $n_1, n_2, \ldots$ $n_i, \ldots$ of single-particle states with energies $\varepsilon_1 \leqslant \varepsilon_2 \leqslant \ldots \leqslant \varepsilon_i \leqslant \ldots$. In this state, the system consists of N particles and possesses energy E_{Nr}, given by

$$N = \sum_i n_i , \qquad E_{Nr} = \sum_i n_i \varepsilon_i . \qquad (11.15)$$

(We use the notation of the last section. E_{Nr} is the energy of the rth N-particle state.) The grand partition function (11.14) becomes

$$\mathcal{Z} = \sum_{n_1 n_2 \ldots} \exp\{\beta [\mu(n_1 + n_2 + \ldots) - (n_1 \varepsilon_1 + n_2 \varepsilon_2 + \ldots)]\} \qquad (11.16)$$

where each occupation number n_i is summed from zero to infinity in the case of bosons, and only over the values 0 and 1 for fermions. [See Eqs. (9.3) and (9.4).] The probability of finding n_1 particles in the single-particle state 1, n_2 particles in the single-particle state 2, etc, is given by the Gibbs distribution (11.13) which becomes

$$p_{Nr} = p(n_1, n_2, \ldots) = \frac{\exp\{\beta [\mu(n_1 + n_2 + \cdots) - (n_1 \varepsilon_1 + n_2 \varepsilon_2 + \cdots)]\}}{\mathcal{Z}} . \qquad (11.17)$$

We shall now rewrite Eq. (11.17) in a way which leads to a remarkable simplification. We can write the numerator in Eq. (11.17) in the form

$$\begin{aligned} \exp\{\beta [\mu(n_1 + n_2 + \ldots) &- (n_1 \varepsilon_1 + n_2 \varepsilon_2 + \ldots)]\} \\ &= \exp[\beta(\mu - \varepsilon_1)n_1 + \beta(\mu - \varepsilon_2)n_2 + \ldots] \\ &= e^{\beta(\mu - \varepsilon_1)n_1} \cdot e^{\beta(\mu - \varepsilon_2)n_2} \cdots \end{aligned} \qquad (11.18)$$

where the dots in the last line stand for further similar exponential factors with n and ε having suffixes $3, 4, \ldots$. The expression in the last line of Eq. (11.18) consists of a product of factors, each of which refers to one single-particle state only. The denominator of Eq. (11.17), i.e. the grand partition function (11.16), factorizes similarly. We see from Eqs. (11.16) and (11.18) that

$$\begin{aligned} \mathcal{Z} &= \sum_{n_1 n_2 \ldots} \exp\{\beta [\mu(n_1 + n_2 + \ldots) - (n_1 \varepsilon_1 + n_2 \varepsilon_2 + \ldots)]\} \\ &= \sum_{n_1 n_2 \ldots} \{ e^{\beta(\mu - \varepsilon_1)n_1} \cdot e^{\beta(\mu - \varepsilon_2)n_2} \cdots \} \\ &= \left\{ \sum_{n_1} e^{\beta(\mu - \varepsilon_1)n_1} \right\} \left\{ \sum_{n_2} e^{\beta(\mu - \varepsilon_2)n_2} \right\} \cdots \end{aligned} \qquad (11.19)$$

where the dots in the last two lines stand for similar factors with n and ε having suffixes $3, 4, \ldots$ [A reader who has difficulty following the last step in Eq. (11.19) should note the identity

$$\left\{\sum_r A_r\right\}\left\{\sum_s B_s\right\} = (A_1 + A_2 + \ldots)(B_1 + B_2 + \ldots) = \sum_{r,s} A_r B_s \ .$$

Eq. (11.19) is an obvious generalization of this identity to an infinite number of factors.] Using the conventional product notation,* we can write the grand partition function (11.19)

$$\mathcal{Z} = \prod_{i=1}^{\infty} \mathcal{Z}_i \tag{11.20}$$

with

$$\mathcal{Z}_i = \sum_{n_i} e^{\beta(\mu - \varepsilon_i)n_i} \ . \tag{11.21}$$

Substituting Eqs. (11.18) and (11.20)–(11.21) into (11.17), we obtain our final result

$$p(n_1, n_2, \ldots) = \prod_{i=1}^{\infty} p_i(n_i) \tag{11.22}$$

with

$$p_i(n_i) = \frac{e^{\beta(\mu - \varepsilon_i)n_i}}{\mathcal{Z}_i} \ . \tag{11.23}$$

Eqs. (11.22)–(11.23) represent a significant simplification compared with the probability distribution (11.17) from which we started. They show that the Gibbs probability distribution $p(n_1, n_2, \ldots)$ factorizes, with each factor $p_i(n_i)$ on the right-hand side of Eq. (11.22) depending on one single-particle state i only. This factorization means that the probability of finding n_i particles in the single-particle state i is independent of the occupation numbers $n_j, j \neq i$, of all other single-particle states and is given by Eq. (11.23). [This interpretation of Eq. (11.23) as the probability of finding n_i particles in the single-particle state i also follows by summing the probability distribution (11.22) over all occupation numbers $n_1, n_2, \ldots$, other than n_i. Summing Eq. (11.23) over n_i gives

$$\sum_{n_i} p_i(n_i) = 1 \ ,$$

confirming that the probability distribution $p_i(n_i)$ is correctly normalized.] Physically, the factorization of the probability distribution (11.22) is due to the fact that the state of a system of non-interacting particles can be specified in terms of single-particle states, so that the energy of the system, Eq. (11.15), is just the sum of the energies of the individual particles.

---

*This notation is explained in Eq. (10.3).

Hereafter we need only consider one single-particle state at a time. For FD statistics (i.e. $n_i = 0, 1$), Eq. (11.21) gives at once

$$\mathcal{Z}_i = 1 + e^{\beta(\mu - \varepsilon_i)} \ . \qquad \text{FD} \qquad (11.24a)$$

For BE statistics (i.e. $n_i = 0, 1, 2, \ldots$), Eq. (11.21) is a geometric series which converges only if $\exp[\beta(\mu - \varepsilon_i)] < 1$, i.e. if $\mu < \varepsilon_i$. This condition must hold for all single-particle states and will do so if it holds for the single-particle ground-state: $\mu < \varepsilon_1$.* In this case the geometric series (11.21) converges for all i and has the sum

$$\mathcal{Z}_i = \frac{1}{1 - e^{\beta(\mu - \varepsilon_i)}} \ . \qquad \text{BE} \qquad (11.24b)$$

In the following it will often be convenient to consider the FD and BE cases together and to combine Eqs. (11.24a) and (11.24b) into

$$\mathcal{Z}_i = [1 \pm e^{\beta(\mu - \varepsilon_i)}]^{\pm 1} \qquad \left\{ \begin{array}{l} \text{FD} \\ \text{BE} \end{array} \right. \qquad (11.24)$$

where the upper or lower signs have to be taken together throughout and refer to FD and BE statistics respectively, as indicated by the abbreviations FD and BE to the right of the equation. Equations without any marking hold for both statistics. These conventions will be used throughout this section.
Next, we calculate the mean occupation number $\bar{n}_i$ of the ith single-particle state. From Eqs. (11.23) and (11.21), $\bar{n}_i$ is given by

$$\bar{n}_i = \sum_{n_i} n_i p_i(n_i) = kT \left(\frac{\partial \ln \mathcal{Z}_i}{\partial \mu} \right)_{T,V} \ . \qquad (11.25)$$

Substituting Eq. (11.24) for $\mathcal{Z}_i$ in Eq. (11.25), we obtain

$$\bar{n}_i = \frac{1}{e^{\beta(\varepsilon_i - \mu)} \pm 1} \ . \qquad \left\{ \begin{array}{l} \text{FD} \\ \text{BE} \end{array} \right. \qquad (11.26)$$

Eq. (11.26) gives the distribution functions for a perfect quantal gas; with the *plus* sign, it is the Fermi–Dirac (FD) distribution; with the *minus* sign, the Bose–Einstein (BE) distribution. The mean occupation number $\bar{n}_i$ must of course be non-negative. For the FD case, this is so for all values of μ.

---

*One frequently takes the single-particle ground-state energy as the zero of the energy scale. We then have $\varepsilon_1 = 0$, and the convergence criterion becomes $\mu < 0$: the chemical potential of a BE gas must always be negative.

For the BE case, it implies $\mu < \varepsilon_i$ for all single-particle states, which will be ensured if $\mu < \varepsilon_1$, the single-particle ground-state energy. This is just our earlier condition for the convergence of the geometric series (11.21) for $\mathscr{Z}_i$. Some of the properties of a perfect quantal gas which follow from these distribution functions will be discussed in sections 11.4 to 11.6.

Above we obtained the probability distribution (11.23) for the occupancy of the ith single-particle state. The denominator, $\mathscr{Z}_i$, of this distribution is given by Eq. (11.21) which has a simple interpretation. Comparing Eq. (11.21) with the general definition (11.14) of the grand partition function, we see that $\mathscr{Z}_i$ has the structure of the grand partition function of a system of non-interacting particles, all confined to the same single-particle state i. (The state of the system which contains n_i particles has energy $n_i \varepsilon_i$.) Hence we can think of $\mathscr{Z}_i$ as the grand partition function for the ith single-particle state. These interpretations of Eqs. (11.21) and (11.23) show that we can think of the ith single-particle state as forming our system and of all other single-particle states as forming our heat bath/particle reservoir. If one adopts this rather abstract definition of a system and of a heat bath/particle reservoir from the start, one can write down the single-particle-state grand partition function (11.21) and the corresponding probability distribution (11.23) directly from Eqs. (11.14) and Eqs. (11.13). The justification for this device (of treating one single-particle state as system and all others as heat bath/particle reservoir) is of course that for non-interacting particles the Gibbs distribution factorizes, i.e. our derivation of Eqs. (11.22–23) from Eq. (11.17).*

Summing Eq. (11.26) over all single-particle states, we obtain the mean total number of particles in the system:

$$\bar{N} = \sum_{i=1}^{\infty} \bar{n}_i = \sum_{i=1}^{\infty} \frac{1}{e^{\beta(\varepsilon_i - \mu)} \pm 1} \cdot \qquad \left\{ \begin{array}{l} \text{FD} \\ \text{BE} \end{array} \right. \qquad (11.27)$$

For given temperature T and volume V, Eq. (11.27) relates $\bar{N}$ to the chemical potential μ. The actual number of particles in the system fluctuates, due to exchange of particles between the system and the particle reservoir. In section 11.7.1 we shall show that, as expected, the fluctuations in $\bar{N}$ are usually completely negligible for a macroscopic system. We can then think of $\bar{N}$ as the actual number of particles in the system, and Eq. (11.27) as determining the chemical potential μ for a system of $\bar{N}$ particles in a volume V in equilibrium at temperature T. This situation is similar to that encountered

---

*Perhaps we should remind the reader that, when we speak of non-interacting particles, we always mean particles with very weak interactions between them. Some interaction is required, so that equilibrium can be established. Provided this interaction is sufficiently weak, we can describe the state of the system in terms of single-particle states.

in section 2.5: for a system in a heat bath, the energy E fluctuates; but for a macroscopic system, these fluctuations are negligible and the energy of the system is sharply defined and given by the mean energy $\bar{E}$.

★ 11.2.1   Fluctuations in a perfect gas

We shall now calculate the fluctuations of the occupation numbers n_i of the single-particle states. Differentiating the equation

$$\bar{n}_i = \sum_{n_i} p_i(n_i) n_i$$

with respect to μ gives

$$\left(\frac{\partial \bar{n}_i}{\partial \mu}\right)_{T,V} = \sum_{n_i} \left(\frac{\partial p_i(n_i)}{\partial \mu}\right)_{T,V} n_i = \sum_{n_i} p_i(n_i) \left(\frac{\partial \ln p_i(n_i)}{\partial \mu}\right)_{T,V} n_i \ . \tag{11.28}$$

(The motivation for this approach is that just as differentiating $\ln \mathscr{Z}_i$ with respect to μ brings down a factor n_i from the exponents in Eq. (11.21), leading to Eq. (11.25) for $\bar{n}_i$, so differentiating $\ln \mathscr{Z}_i$ twice, i.e. $\bar{n}_i$ once, will bring down a second n_i, leading to an expression for $\overline{n_i^2}$.) From Eq. (11.23)

$$\ln p_i(n_i) = \beta(\mu - \varepsilon_i) n_i - \ln \mathscr{Z}_i$$

whence

$$\left(\frac{\partial \ln p_i(n_i)}{\partial \mu}\right)_{T,V} = \beta n_i - \left(\frac{\partial \ln \mathscr{Z}_i}{\partial \mu}\right)_{T,V} = \beta(n_i - \bar{n}_i) \ , \tag{11.29}$$

where the last step follows from Eq. (11.25). Substituting Eq. (11.29) in (11.28), we obtain

$$\left(\frac{\partial \bar{n}_i}{\partial \mu}\right)_{T,V} = \sum_{n_i} p_i(n_i) \beta(n_i - \bar{n}_i) n_i = \beta(\overline{n_i^2} - \bar{n}_i^2)$$

and for the standard deviation Δn_i:

$$(\Delta n_i)^2 = \overline{n_i^2} - \bar{n}_i^2 = kT \left(\frac{\partial \bar{n}_i}{\partial \mu}\right)_{T,V} \ . \tag{11.30}$$

Substituting the distribution functions (11.26) for $\bar{n}_i$ in Eq. (11.30) leads to our final result

$$(\Delta n_i)^2 = \bar{n}_i(1 \mp \bar{n}_i) \ . \qquad \left\{ \begin{array}{l} \text{FD} \\ \text{BE} \end{array} \right. \qquad (11.31a)$$

If $\bar{n}_i \ll 1$, the quantal distribution functions (11.26) become

$$\bar{n}_i = e^{\beta(\mu - \varepsilon_i)} \qquad \text{MB}$$

which is just the classical or Maxwell-Boltzmann distribution (MB distribution for short). This is not surprising, since we know from section 7.3 that $\bar{n}_i \ll 1$, for all i, is the condition for MB statistics to be valid. (We shall discuss the classical limit further in section 11.4.) Taking $\bar{n}_i \ll 1$ in Eq. (11.31a), or substituting the MB distribution in Eq. (11.30), we obtain the standard deviation Δn_i for a classical perfect gas:

$$(\Delta n_i)^2 = \bar{n}_i \ . \qquad \text{MB} \quad (11.31b)$$

We see from Eqs. (11.31a) and (11.31b) that the relative fluctuation $\Delta n_i / \bar{n}_i$ is large for all three statistics, except in the case of an extremely degenerate FD gas for which the energetically lowest lying single-particle states are occupied: $\bar{n}_i \approx 1$. (See section 11.5.1.)

Instead of considering one single-particle state, let us take a group of G states whose energies lie within a narrow energy band $\Delta \varepsilon \ll kT$. It follows that all the single-particle states within this group have the same mean occupation number $\bar{n}$. Hence we have for the total number $\bar{N}$ of particles in the group

$$\bar{N} = G\bar{n} \ .$$

It follows from the statistical independence of the probability distributions of the different single-particle states, i.e. from Eq. (11.22) that

$$(\Delta N)^2 = G(\Delta n)^2 \ .$$

Hence combining the last two equations with Eq. (11.31a) or (11.31b) we obtain

$$(\Delta N)^2 = G\bar{n}(1 \mp \bar{n}) = \bar{N}\left(1 \mp \frac{\bar{N}}{G}\right) \qquad \left\{ \begin{array}{l} \text{FD} \\ \text{BE} \end{array} \right. \qquad (11.32a)$$

and

$$(\Delta N)^2 = \bar{N} \ . \qquad \text{MB} \quad (11.32b)$$

We see from these equations that for FD and MB statistics the relative fluctuation $\Delta N/\overline{N}$ is small provided $\overline{N}$ is large. For BE statistics we require in addition that G is large for $\Delta N/\overline{N}$ to be small. (For FD statistics, we have in any case $G \geqslant \overline{N}$.)

We apply Eq. (11.32a) to calculate the energy fluctuations of black-body radiation. For radiation within a narrow frequency band ω to $\omega + \Delta\omega$ we have from Eqs. (10.8) and (10.11)

$$\frac{\overline{N}}{G} = \bar{n} = \frac{1}{e^{\beta\hbar\omega} - 1} , \qquad G = \frac{V\omega^2\Delta\omega}{\pi^2 c^3} .$$

Eq. (11.32a) gives for the energy fluctuations

$$(\Delta E)^2 \equiv (\hbar\omega\Delta N)^2 = (\hbar\omega)^2\overline{N}\left(1 + \frac{\overline{N}}{G}\right) . \qquad (11.33)$$

The two terms in this last expression admit the following interpretations. If $\hbar\omega \gg kT$, then $\overline{N}/G \ll 1$ and Eq. (11.33) becomes

$$(\Delta E)^2 = (\hbar\omega)^2\overline{N}.$$

This result is typical of the corpuscular nature of radiation. The fluctuations are proportional to the square root of the mean number of particles, each of which carries an energy $\hbar\omega$. On the other hand, if $\hbar\omega \ll kT$ then the second term in Eq. (11.33) dominates. Since now $\overline{N}/G = kT/\hbar\omega$, Eq. (11.33) reduces to

$$(\Delta E)^2 = (\hbar\omega)^2 G\left(\frac{\overline{N}}{G}\right)^2 = \left(\frac{V\omega^2\Delta\omega}{\pi^2 c^3}\right)(kT)^2 .$$

This equation no longer contains Planck's constant and we could have derived it from Eq. (11.33) by taking the limit $h \to 0$, showing that it represents the classical result for the energy fluctuations of black-body radiation.

★11.3   THE FD AND BE DISTRIBUTIONS: ALTERNATIVE APPROACH

We next consider a perfect gas of a *fixed* number $\overline{N}$ of particles* obeying FD or BE statistics. We must now impose the subsidiary condition (11.1), and we shall see one way of coping with the complications which arise from

---

*We use $\overline{N}$, instead of N as previously, for the number of particles to conform to the notation of the last section.

this constraint on $\overline{N}$. The gas is contained in a vessel of volume V immersed in a heat bath at temperature T. What are the equilibrium properties of the gas?

We start from the fact that the energy levels

$$\varepsilon_1 \leqslant \varepsilon_2 \leqslant \ldots \leqslant \varepsilon_r \leqslant \ldots \tag{11.34}$$

of the single-particle states, which describe one molecule of the gas, are spaced *extremely* close together. Therefore we shall not now consider individual single-particle states. Instead we shall consider *groups* of neighbouring states. The energies of each group of single-particle states lie very close together so that we can to sufficient accuracy consider them all to have the *same* energy. We shall label these groups $i = 1, 2, \ldots$. The ith group contains G_i single-particle states whose energy we take to be E_i. The exact way this grouping is done is not important, as long as the energy variation of the exact single-particle energies within each group is small and the number of states within each group is very large:

$$G_i \gg 1 \ . \tag{11.35}$$

This situation is schematically indicated in Fig. 11.2. A state of the gas is now specified by the set of occupation numbers

$$N_1, N_2, \ldots N_i, \ldots , \tag{11.36}$$

N_i being the total number of particles occupying single-particle states in the ith group. Since the gas consists of $\overline{N}$ particles we have the subsidiary condition

$$\sum_i N_i = \overline{N} \ , \tag{11.37}$$

where the summation is, of course, over *groups* of states.

To find the equilibrium properties of the gas we shall derive an expression for the Helmholtz free energy F of the gas as a function of the occupation numbers (11.36). For a system of constant volume at a constant temperature the Helmholtz free energy is a minimum.* Hence we shall minimize F with respect to the occupation numbers (11.36), which are restricted by the subsidiary condition (11.37).

We find the Helmholtz free energy F from

$$F = E - TS \ . \tag{2.37}$$

---

*This result is stated near the end of section 2.5 and is derived in sections 4.2 and 4.4, Eqs. (4.29) and (4.41) respectively.

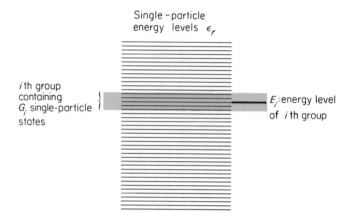

Fig. 11.2. Schematic diagram of single-particle energy levels and of a group of G_i neighbouring single-particle states possessing very nearly the energy E_i.

For the state of the gas specified by the occupation numbers (11.36) the energy E is given by

$$E = \sum_i N_i E_i \ . \qquad (11.38)$$

The entropy S of this state is related to its statistical weight $\Omega \equiv \Omega(N_1, N_2, \ldots)$ through Boltzmann's definition

$$S = k \ln \Omega \ . \qquad (2.34)$$

We must therefore find Ω. Let $\Omega_i(N_i)$ be the number of ways of putting N_i particles into the G_i states of the ith group. We shall derive explicit expressions for $\Omega_i(N_i)$ below. It follows that

$$\Omega = \prod_{i=1}^{\infty} \Omega_i(N_i) \qquad (11.39)$$

since the arrangements of particles in the different groups are independent of each other. It follows from the last four equations that

$$F \equiv F(N_1, N_2, \ldots) = \sum_i [N_i E_i - kT \ln \Omega_i(N_i)] \ . \qquad (11.40)$$

We now minimize F with respect to the occupation numbers $N_1, N_2, \ldots$. But these are *not independent of each other*. They are restricted by the condition (11.37). Hence the occupation numbers cannot all be varied independently of each other. We shall single out one particular occupation number, say N_j, which we shall assume given by Eq. (11.37) in terms of all other occupation numbers, i.e.

$$N_j = \bar{N} - \sum_i' N_i \qquad (11.41)$$

where, as throughout the following, the prime on the summation sign means that the term with $i = j$ is excluded from the summation, i.e. the summation is over $i = 1, 2, \ldots j-1, j+1, \ldots$ only. Accordingly, we write Eq. (11.40)

$$F = \sum_i' [N_i E_i - kT \ln \Omega_i(N_i)] + [N_j E_j - kT \ln \Omega_j(N_j)] , \qquad (11.40a)$$

and the minimum of F is now determined by

$$\frac{\partial F}{\partial N_i} = 0, \qquad i = 1, 2, \ldots j-1, j+1, \ldots . \qquad (11.42)$$

From Eqs. (11.40a) and (11.42) we obtain

$$\left. \begin{array}{l} \dfrac{\partial F}{\partial N_i} = \left[E_i - kT \dfrac{\partial \ln \Omega_i(N_i)}{\partial N_i} \right] + \left[E_j - kT \dfrac{\partial \ln \Omega_j(N_j)}{\partial N_j} \right] \dfrac{\partial N_j}{\partial N_i} = 0, \\[4mm] i = 1, 2, \ldots j-1, j+1, \ldots . \end{array} \right\} \qquad (11.43)$$

From Eq. (11.41)

$$\frac{\partial N_j}{\partial N_i} = -1, \qquad i = 1, 2, \ldots j-1, j+1, \ldots , \qquad (11.44)$$

hence Eqs. (11.43) reduce to

$$\left. \begin{array}{l} \left[E_i - kT \dfrac{\partial \ln \Omega_i(N_i)}{\partial N_i} \right] = \left[E_j - kT \dfrac{\partial \ln \Omega_j(N_j)}{\partial N_j} \right] , \\[4mm] i = 1, 2, \ldots j-1, j+1, \ldots . \end{array} \right\} \qquad (11.45)$$

It follows from Eqs. (11.45) that the expression in square parentheses has the same value for *all* the groups $i = 1, 2, \ldots$, i.e. including the jth group:

$$E_i - kT \frac{\partial \ln \Omega_i(N_i)}{\partial N_i} = \mu , \qquad i = 1, 2, \ldots , \qquad (11.46a)$$

where the quantity μ is the same for all groups. In the following, equations will hold again for *all* groups, i.e. all values of the index i, and we shall not write $i = 1, 2, \ldots$ after each equation. We rewrite Eq. (11.46a) as

$$\frac{\partial \ln \Omega_i(N_i)}{\partial N_i} = \beta(E_i - \mu) = \alpha + \beta E_i \ , \tag{11.46b}$$

with

$$\alpha \equiv -\mu\beta \equiv -\mu/kT \ . \tag{11.47}$$

To proceed further we require the statistical weights Ω_i. These are different for fermions and bosons so that we must now consider these two cases separately.

For FD statistics, we can place at most one particle into any single-particle state. For the ith group, we must therefore choose N_i *different* single-particle states from amongst the G_i single-particle states which comprise this group. The number of ways of doing this is given by

$$\Omega_i(N_i) = \frac{G_i!}{N_i!(G_i - N_i)!} \ . \qquad \text{FD} \qquad (11.48a)$$

We have labelled this equation FD to show that it holds for fermions only, and we shall label equations BE which hold for bosons only.

For BE statistics, any number of particles may be in the same single-particle state. Consider the ith group and imagine the N_i particles of this group arranged in a straight line. By means of $(G_i - 1)$ partitions we can then divide these particles up into G_i sets of particles. Each set represents the particles in one single-particle state. The number of ways of doing this is given by

$$\Omega_i(N_i) = \frac{(N_i + G_i - 1)!}{N_i!(G_i - 1)!} \ . \qquad \text{BE} \qquad (11.49)$$

The numerator of this expression gives the total number of ways of permuting particles and partitions, while the two factors in the denominator allow for the fact that permuting the particles amongst themselves or the partitions amongst themselves produces no new arrangement. Fig. 11.3 shows one such arrangement for a group of five single-particle states ($G_i = 5$) containing 7 particles ($N_i = 7$). On account of the condition $G_i \gg 1$, Eq. (11.35), we can write Eq. (11.49) as

$$\Omega_i(N_i) = \frac{(N_i + G_i)!}{N_i! G_i!} \ . \qquad \text{BE} \qquad (11.48b)$$

In order to calculate $\ln \Omega_i$ from Eqs. (11.48) we shall want to use Stirling's formula

$$\ln N! = N \ln N - N \tag{11.50}$$

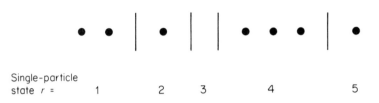

Single-particle
state $r =$ 1   2   3   4   5

Fig. 11.3. One way of arranging 7 bosons amongst a group of 5 single-particle states,
i.e. $G_i = 5$, $N_i = 7$.

which is valid for $N \gg 1$ (see Appendix A, Eq. (A.2)), and we shall therefore
assume

$$G_i \gg 1 \ , \qquad N_i \gg 1 \ , \qquad G_i - N_i \gg 1 \ . \tag{11.51}$$

The last of these conditions is required for the Fermi-Dirac formula (11.48a)
only.* Using Stirling's formula (11.50) we obtain from Eqs. (11.48):

$$\ln \Omega_i(N_i) = \pm \ln G_i! \mp \ln (G_i \mp N_i)! - \ln N_i! \qquad \left\{ \begin{array}{l} \text{FD} \\ \text{BE} \end{array} \right.$$

$$= \pm G_i \ln G_i \mp (G_i \mp N_i) \ln (G_i \mp N_i) - N_i \ln N_i \qquad \left\{ \begin{array}{l} \text{FD} \\ \text{BE} \end{array} \right. \tag{11.52}$$

*where the upper or the lower signs have to be taken together throughout and
refer to FD and BE statistics respectively, as indicated.* From Eq. (11.52)
one obtains directly

$$\frac{\partial \ln \Omega_i(N_i)}{\partial N_i} = \ln \frac{G_i \mp N_i}{N_i} \ , \qquad \left\{ \begin{array}{l} \text{FD} \\ \text{BE} \end{array} \right. \tag{11.53}$$

and hence from Eq. (11.46b):

$$\frac{N_i}{G_i} = \frac{1}{e^{\alpha + \beta E_i} \pm 1} \ . \qquad \left\{ \begin{array}{l} \text{FD} \\ \text{BE} \end{array} \right. \tag{11.54}$$

*One can always arrange for the condition $G_i \gg 1$ to hold. Unfortunately, the two other
conditions (11.51) do not always hold, and the present approach then leads to difficulties. This
occurs, for example, for a gas of fermions very close to or at the absolute zero of temperature.
In this case the first $\bar{N}$ single-particle states, ordered in increasing energy as in Eq. (11.34), are
occupied and so are the corresponding groups of states. Hence for these groups $G_i = N_i$. In the
following treatment the conditions (11.51) will be presupposed, but we shall assume that the
results we obtain hold generally. The same results can be obtained by other methods, for example
that of section 11.1, without these restrictive conditions.

This equation represents our result. It tells us how in thermal equilibrium at temperature T the fermions or bosons of a perfect gas are distributed over the various groups of single-particle states. Eq. (11.54) is adequate for all applications, i.e. one always sums over neighbouring levels anyway. However it is convenient and easy to eliminate the groups of single-particle states altogether. N_i/G_i is just the mean occupation number of a single-particle state within the ith group. If s labels a single-particle state within this group, we rewrite Eq. (11.54) as

$$\bar{n}_s = \frac{1}{e^{\alpha+\beta\varepsilon_s} \pm 1} \qquad \left\{ \begin{array}{l} \text{FD} \\ \text{BE} \end{array} \right. \qquad (11.55)$$

where $\bar{n}_s$ is the mean occupation number of the sth single-particle state, and where we replaced the energy E_i of the ith group of states by the exact single-particle energy ε_s. Eq. (11.55) represents our final result and in this form it no longer depends on a division into groups of states. Eq. (11.55) is an extremely important equation. It gives the mean occupation number $\bar{n}_s$ of the sth single-particle state for a gas of $\bar{N}$ non-interacting fermions or bosons in thermal equilibrium at temperature $T(=1/\beta k)$ in an enclosure of volume V. (Remember the single-particle states depend on V.) The properties of a perfect quantal gas which follow from Eq. (11.55) will be discussed in sections 11.4 to 11.6.

We must next discuss the meaning of the parameters α or μ, introduced in Eqs. (11.46) and (11.47). We can rewrite the subsidiary condition (11.37), which states that we are dealing with a gas of $\bar{N}$ particles, in terms of the mean single-particle occupation numbers $\bar{n}_s$:

$$\sum_s \bar{n}_s = \bar{N} \qquad (11.56)$$

where the summation is now over single-particle states, of course. With Eq. (11.55) we can rewrite this relation

$$\sum_s \frac{1}{e^{\alpha+\beta\varepsilon_s} \pm 1} = \bar{N} \, . \qquad \left\{ \begin{array}{l} \text{FD} \\ \text{BE} \end{array} \right. \qquad (11.57)$$

We see from this equation that the parameter α must be chosen so that the mean occupation numbers $\bar{n}_s$ add up to the correct total number of particles $\bar{N}$.

The parameters α or μ also admit a thermodynamic interpretation which we now derive. In equilibrium the Helmholtz free energy of the gas is a function of T, V and $\bar{N}$. We shall calculate

$$\left(\frac{\partial F(T, V, \bar{N})}{\partial \bar{N}} \right)_{T, V}$$

from Eq. (11.40), using the occupation numbers $N_1, N_2, \ldots$ which correspond to equilibrium. From Eqs. (11.40) and (11.46a) we obtain

$$\left(\frac{\partial F(T, V, \bar{N})}{\partial \bar{N}}\right)_{T, V} = \sum_i \left[E_i - kT \frac{\partial \ln \Omega_i(N_i)}{\partial N_i}\right] \frac{\partial N_i}{\partial \bar{N}}$$

$$= \mu \sum_i \frac{\partial N_i}{\partial \bar{N}} = \mu \frac{\partial}{\partial \bar{N}} \sum_i N_i = \mu \ , \qquad (11.58)$$

since $\sum_i N_i = \bar{N}$. This equation thus gives a purely thermodynamic definition of the quantity μ:

$$\mu \equiv \mu(T, V, \bar{N}) = \left(\frac{\partial F(T, V, \bar{N})}{\partial \bar{N}}\right)_{T, V} . \qquad (11.59)$$

μ is called the *chemical potential* of the system of $\bar{N}$ particles in a volume V in thermal equilibrium at temperature T.* With $F = -kT \ln Z$ and $\alpha = -\mu/kT$, the last relation can be written in terms of the partition function Z of the gas:

$$\alpha = \left(\frac{\partial \ln Z(T, V, \bar{N})}{\partial \bar{N}}\right)_{T, V} . \qquad (11.60)$$

From the above results one can also obtain the partition function Z of the gas by a straightforward but somewhat tedious calculation. We shall omit all details and only indicate the derivation. We calculate the partition function from

$$\ln Z = \frac{1}{k} S - \beta E \ . \qquad (11.61)$$

In this expression we substitute for E from

$$E = \sum_i N_i E_i \ , \qquad (11.38)$$

and for $S = k \ln \Omega$ from

$$\Omega = \prod_i \Omega_i(N_i) \ , \qquad (11.39)$$

using Eq. (11.52) for $\ln \Omega_i$. With the equilibrium values (11.54) for the occupation numbers N_i, Eq. (11.61) then gives

---

*The chemical potential was introduced earlier in section 8.1 where it was also related to other thermodynamic functions. The present chapter does NOT presuppose a knowledge of Chapter 8.

$$\ln Z = \alpha \bar{N} \pm \sum_i G_i \ln \left[1 \pm e^{-(\alpha + \beta E_i)} \right] \quad . \qquad \left\{ \begin{array}{l} \text{FD} \\ \text{BE} \end{array} \right. \qquad (11.62)$$

This expression can be written as a sum over single-particle states:

$$\ln Z = \alpha \bar{N} \pm \sum_s \ln \left[1 \pm e^{-(\alpha + \beta \varepsilon_s)} \right] \quad . \qquad \left\{ \begin{array}{l} \text{FD} \\ \text{BE} \end{array} \right. \qquad (11.63)$$

Our treatment in this section consisted of minimizing the Helmholtz free energy $F(N_1, N_2, \ldots)$ with respect to the occupation numbers $N_1, N_2, \ldots$, subject to the subsidiary condition $\sum_i N_i = \bar{N}$. We allowed for the latter condition directly by eliminating one of the occupation numbers N_j. An alternative quite general procedure, convenient in more complicated cases, is that of the Lagrange multipliers. We illustrate it for the present problem. We define a new quantity

$$\phi \equiv F(N_1, N_2, \ldots) - \lambda \left(\sum_i N_i - \bar{N} \right)$$

where a new parameter λ has been introduced. We minimize ϕ with respect to λ and the occupation numbers $N_1, N_2, \ldots$, *all* of which are treated as independent variables. This leads to the conditions

$$\frac{\partial \phi}{\partial \lambda} = \left(N - \sum_i N_i \right) = 0$$

$$\frac{\partial \phi}{\partial N_i} = \frac{\partial F}{\partial N_i} - \lambda = 0 \ , \qquad i = 1, 2, \ldots \ .$$

The first of these conditions ensures that the subsidiary condition (11.37) holds. The second set of conditions leads, from Eq. (11.40) for F, to

$$\frac{\partial F}{\partial N_i} = \left[E_i - kT \frac{\partial \ln \Omega_i(N_i)}{\partial N_i} \right] = \lambda \ , \qquad i = 1, 2, \ldots \ .$$

If one uses the method of Lagrange multipliers as the basic approach one must investigate separately the meaning of the Lagrange undetermined multiplier λ. However, we can appeal to our earlier results. Comparison of the last equation with Eq. (11.46a) shows that λ is identically equal to the chemical potential, which we called μ before.

The method of Lagrange multipliers is frequently applied to an isolated system. We now want to maximize the entropy $S(N_1, N_2, \ldots)$, the total particle number $\bar{N}$ and the total energy $\bar{E}$ both being constant. Instead we can introduce two Lagrange undetermined multipliers α and β, and maximize

$$\frac{1}{k} S(N_1, N_2, \ldots) - \alpha \left(\sum_i N_i - \bar{N} \right) - \beta \left(\sum_i N_i E_i - \bar{E} \right) \tag{11.64}$$

treating α, β and all occupation numbers N_i as independent variables. The physical meaning of α and β must again be established separately.* α and β turn out to be the quantities which we denoted by these symbols before: $(-\alpha kT) = \mu$ is the chemical potential and $\beta \equiv 1/kT$ is the temperature parameter. In the Lagrange multiplier method one introduces the undetermined multipliers as auxiliary mathematical quantities whose physical meaning one must afterwards separately establish. It is interesting to compare Eq. (11.64) with Eq. (11.40) which we minimized earlier. The latter equation already contains the correct temperature parameter. It was introduced from the very beginning through the temperature of the heat bath.

★ 11.4 THE CLASSICAL LIMIT

We shall now show that under the appropriate conditions the quantum statistics of this chapter reduce to the classical or Maxwell–Boltzmann statistics (MB statistics for short) of Chapter 7.

From section 7.3, Eq. (7.31), we have the following sufficient condition for the validity of MB statistics:

$$\bar{n}_i \ll 1 \qquad \text{for all } i , \tag{11.65}$$

i.e. the mean occupation numbers of all single-particle states must be small compared to unity. We see from Eq. (11.26) [Eq. (11.55) with Eq. (11.47) for α substituted in it][†] that this condition will be satisfied provided

$$e^{-\beta\mu} \gg 1 . \tag{11.66}$$

Here we have again chosen the zero of the energy scale so that

$$0 = \varepsilon_1 \leqslant \varepsilon_2 \leqslant \varepsilon_3 \leqslant \ldots . \tag{11.67}$$

It follows from Eq. (11.67) that $\exp[\beta(\varepsilon_i - \mu)] \geqslant \exp(-\beta\mu)$, so that on account of the inequality (11.66), Eq. (11.26) [Eq. (11.55)] becomes

$$\bar{n}_i = e^{\beta(\mu - \varepsilon_i)} , \qquad \text{for all } i , \tag{11.68}$$

i.e. *if Eq. (11.66) holds* FD *and* BE *statistics both reduce to* MB *statistics*. The physical reason for this condition is obvious. If condition (11.66) holds,

---

*See, for example, Schrödinger,[15] Chapter 2, or ter Haar,[17] section 4.4.
†Since sections 11.2 and 11.3 were *alternative* treatments of the perfect quantal gas we shall give cross-references to both sections, those in square brackets referring to section 11.3.

then $\bar{n}_i \ll 1$ for all i. And if $\bar{n}_i \ll 1$ for all i, it does not matter how we weight states of the gas with some $n_i \geq 2$. Under the given conditions such states do *not* occur, for practical purposes. So it does not matter that MB statistics underweights them compared with BE statistics (see section 7.1) while FD statistics forbids them altogether.

Summing the MB distribution (11.68) over all single-particle states, we obtain the number of particles in the system:

$$\bar{N} = e^{\beta\mu} \sum_i e^{-\beta\varepsilon_i} \equiv e^{\beta\mu} Z_1(T, V) \qquad (11.69)$$

where $Z_1(T, V)$ is the partition function for one particle. We shall rederive this result from thermodynamics in section 11.8, where the equation of state of a perfect classical gas will also be obtained. From Eq. (11.69) one readily obtains the chemical potential μ, allowing one to eliminate it altogether from calculations for a perfect classical gas. For example, the correctly normalized MB distribution (11.68) becomes

$$\bar{n}_i = \frac{\bar{N}}{Z_1(T, V)} e^{-\beta\varepsilon_i} . \qquad (11.70)$$

Eq. (11.69) for MB statistics is to be contrasted with Eq. (11.27) [Eq. (11.57)] for a perfect quantal gas. The latter equation is an implicit relation for the chemical potential μ which does not usually allow one to eliminate μ.

One can easily show that the criterion (11.66) for MB statistics to hold follows from our earlier validity criterion (7.33). If the latter criterion is satisfied, we can calculate $e^{-\beta\mu}$ using the equations of classical statistics, i.e. from Eqs. (11.69) and (7.20), giving

$$e^{-\beta\mu} = \frac{Z_1(T, V)}{\bar{N}} = \frac{V}{\bar{N}} \left(\frac{2\pi mkT}{h^2} \right)^{3/2} Z_{int}(T) . \qquad (11.71)$$

With our choice $\varepsilon_1 = 0$, Eq. (11.67), it follows from Eq. (7.12) that $\varepsilon_1^{int} = 0$ and that $Z_{int}(T) \geq 1$ (see Eq. (7.15)). Hence Eq. (11.66) certainly holds provided

$$\frac{V}{\bar{N}} \left(\frac{2\pi mkT}{h^2} \right)^{3/2} \gg 1 \qquad (11.72)$$

which is our earlier criterion for the validity of classical statistics, Eq. (7.33).

★ 11.5   THE FREE ELECTRON MODEL OF METALS

As a simple application of FD statistics we shall briefly consider the free electron model of metals.* According to this model the valence electrons, which are only weakly bound, become detached in a metal and roam around freely. If a potential difference, i.e. an electric field, is applied to the metal these electrons produce a current. They are therefore called conduction electrons. In the free electron model these conduction electrons are treated as a perfect gas obeying Fermi–Dirac statistics. Their interaction with the positively charged atomic ions is neglected as is also the interaction of the conduction electrons with each other. Nevertheless this model is very successful in explaining qualitatively many features of metals. The conduction electrons in a metal do move like free particles, with a large mean free path (this can be as large as a centimetre under suitable conditions), little disturbed by the presence of the positive ions and undergoing few collisions with other conduction electrons.

The reasons for this behaviour are quantum-mechanical. The ions of a perfect lattice at $T = 0\,\text{K}$ allow the electrons considered as waves to propagate freely. (For example, the electrical conductivity is infinite under these conditions.) It is only lattice imperfections, such as impurities or lattice vibrations, which disturb the propagation of the electron waves. The effect of the atomic ions is to produce a positively charged background which partially screens the conduction electrons from each other. Even these residual electron–electron interactions, i.e. the electron–electron collisions, are comparatively unimportant. This is a consequence of the Pauli exclusion principle. Consider the collision of two electrons initially in the single-particle states r and s. A collision of these electrons, in which they are scattered into the single-particle states r' and s', can only occur if the states r' and s' are unoccupied. Otherwise this scattering process is forbidden by the Pauli principle. We shall see that the energy distribution of the conduction electrons, given by the FD distribution, is such that most final states r' and s' which are energetically accessible in a collision are already occupied. Hence such scattering events cannot occur. In this way one can partly justify the free electron model. Of course, it is essential to treat the electrons as a *quantal* gas obeying FD statistics. This leads to an understanding of the heat capacity (section ˉ11.5.2) and of the paramagnetic susceptibility (see problem 11.1) of the conduction electrons. This is in contrast to the classical electron gas (i.e. an ideal electron gas obeying Maxwell–Boltzmann statistics) which was incapable of explaining these phenomena.

---

*For a fuller discussion see, for example, Blakemore,[24] Hall[27] or Kittel.[29]

11.5.1   The Fermi–Dirac energy distribution

To study the properties of a perfect electron gas, we take as our starting point the FD distribution (11.26) [or Eq. (11.55) with Eq. (11.47) for α substituted in it],* i.e.

$$\bar{n}_i = \frac{1}{e^{\beta(\varepsilon_i - \mu)} + 1} \; . \tag{11.73}$$

This equation gives the mean occupation number $\bar{n}_i$ of the ith single-particle state. The density of states in momentum space is given by the same expression (7.17) which we had before (the derivation of this expression in Appendix B was from the Schrödinger equation which describes electrons correctly) *with one important modification*. We must multiply Eq. (7.17) by a factor 2 to allow for the spin of the electron of $\hbar/2$. Because of this spin, the state of an electron is not fully specified by its momentum **p**. We must also specify its spin state. Because the electron has spin $\hbar/2$ there are two possible independent spin states which one can take as spin parallel and antiparallel to some axis. We are not interested in a detailed description of these states.[†] We only need the density of states, i.e. all we need to know is that there are two distinct 'internal' states of motion of the electron. Eq. (7.17), for the number of states with momentum **p** whose magnitude lies in the interval p to $p + dp$, becomes

$$f(p)\,dp = \frac{V8\pi p^2\,dp}{h^3} \tag{11.74}$$

where V is the volume of the electron gas, i.e. in our free electron model of a metal, the volume of the piece of metal. (Note h and not $\hbar$ in this and the following formulae.)

We convert Eq. (11.74) into an energy density of states by substituting

$$\varepsilon = \frac{p^2}{2m} \tag{11.75}$$

(m = mass of electron) in it. The number of states with energy in the interval ε to $\varepsilon + d\varepsilon$ then becomes

$$f(\varepsilon)\,d\varepsilon = \frac{4\pi V}{h^3}(2m)^{3/2}\varepsilon^{1/2}\,d\varepsilon \; . \tag{11.76}$$

---

*Since sections 11.2 and 11.3 were *alternative* treatments of the perfect quantal gas we shall give cross-references to both sections, those in square brackets referring to section 11.3.
[†]For discussion of spin see, for example, Eisberg,[39] Fano and Fano,[40] or Willmott.[48]

Correspondingly we write Eq. (11.73) as

$$\bar{n}(\varepsilon) = \frac{1}{e^{\beta(\varepsilon-\mu)}+1} \ . \tag{11.73a}$$

Combining Eqs. (11.76) and (11.73a), we obtain

$$dN(\varepsilon) = \bar{n}(\varepsilon)f(\varepsilon)d\varepsilon = \left[\frac{4\pi V}{h^3}(2m)^{3/2}\right]\varepsilon^{1/2}d\varepsilon\,\frac{1}{e^{\beta(\varepsilon-\mu)}+1} \ . \tag{11.77}$$

Eq. (11.77) gives the FD energy distribution: $dN(\varepsilon)$ is the number of electrons with energy in the interval ε to $\varepsilon+d\varepsilon$. The total number of electrons in the electron gas is obtained by integration of Eq. (11.77) from $\varepsilon=0$ to $\varepsilon=\infty$ (we call it N again, rather than $\bar{N}$ as in the last three sections)

$$N = \left[\frac{4\pi V}{h^3}(2m)^{3/2}\right]\int_0^\infty \frac{\varepsilon^{1/2}d\varepsilon}{e^{\beta(\varepsilon-\mu)}+1} \ . \tag{11.78}$$

This is just Eq. (11.27) [Eq. (11.57)] adapted to the present situation. Eq. (11.78) determines the chemical potential μ for a given number of electrons (N), a given volume (V) and a given temperature $T(=1/\beta k)$:

$$\mu = \mu(T, V, N) \ . \tag{11.79}$$

For a given piece of metal, its volume V is given. We shall also assume that N is fixed. It is determined by the atomic properties of the metal. In the case of the alkalis or the noble metals there is one conduction electron per atom. For simplicity we shall therefore omit V and N from Eq. (11.79) and write

$$\mu = \mu(T) \ . \tag{11.80}$$

The *Fermi energy* ε_F is defined as the value of the chemical potential at the absolute zero temperature

$$\varepsilon_F \equiv \mu(0) \ . \tag{11.81}$$

The physical significance of ε_F will appear in a moment.

We shall first show that ε_F is positive. We put $z=\beta\varepsilon$ as new integration variable in Eq. (11.78):

$$N = \frac{4\pi V}{h^3}(2mk)^{3/2}[\,T^{3/2}e^{\beta\mu}\,]\int_0^\infty \frac{z^{1/2}dz}{e^z+e^{\beta\mu}} \ . \tag{11.82}$$

We shall assume that $\varepsilon_F < 0$ and show that this leads to nonsense! If $\varepsilon_F < 0$, then $\beta\varepsilon_F \to -\infty$ as $T \to 0$, and

$$\lim_{T \to 0} e^{\beta\mu} = \lim_{T \to 0} e^{\beta\varepsilon_F} = 0 \; ;$$

hence, the term in square parentheses in Eq. (11.82) tends to zero. But the integral in (11.82) tends to the finite limit

$$\int_0^\infty e^{-z} z^{1/2} dz$$

as $T \to 0$, hence the right-hand side of (11.82) goes to zero as $T \to 0$. But the left-hand side is a positive number ($\sim 10^{23}$) independent of temperature. Hence our original assumption is not tenable and we must have

$$\varepsilon_F > 0 \; . \tag{11.83}$$

Now we are in business! Fig. 11.4 shows the FD mean occupation number $\bar{n}(\varepsilon)$, Eq. (11.73a), at zero temperature. We see from this equation that with $\beta \to \infty$, $\exp[\beta(\varepsilon - \varepsilon_F)]$ tends to $\exp(\mp\infty)$ for $\varepsilon < \varepsilon_F$ and $\varepsilon > \varepsilon_F$ respectively, i.e. $\bar{n}(\varepsilon)$ tends to 1 or 0 for $\varepsilon < \varepsilon_F$ and $\varepsilon > \varepsilon_F$ respectively. This is shown in Fig. 11.4 and has a simple meaning. At $T = 0$ K the system is in its state of lowest energy. Because of the Pauli principle not all N electrons can crowd into the single-particle state of lowest energy ε_1. We can have only one

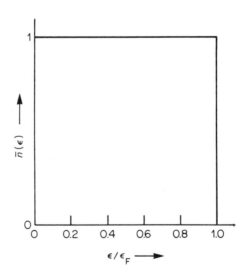

Fig. 11.4. The Fermi–Dirac mean occupation number $\bar{n}(\varepsilon)$, Eq. (11.73a), at zero temperature: $T = 0$ K. ε_F is the Fermi energy.

particle per state. Hence with the single-particle states ordered in increasing energy

$$\varepsilon_1 \leqslant \varepsilon_2 \leqslant \ldots, \qquad (11.84)$$

the first N (lowest) single-particle states are filled. The states above these first N states are empty. At the same time, the Fermi energy ε_F receives a physical meaning. It is the energy of the topmost occupied level at $T = 0$ K. For a perfect gas of bosons, $\bar{n}(\varepsilon)$ will of course not have the striking appearance of Fig. 11.4.

Eq. (11.73a) gives the mean occupation number of a perfect FD gas. Its energy distribution is given by Eq. (11.77). Fig. 11.5 shows this distribution for zero temperature. The sharp cut-off at $\varepsilon = \varepsilon_F$ reflects the behaviour of $\bar{n}(\varepsilon)$; the parabolic rise (with vertical tangent at $\varepsilon = 0$) reflects the density-of-states factor $\varepsilon^{1/2}$.

From the interpretation of ε_F as the topmost energy level occupied at $T = 0$ K, we can at once find the Fermi energy ε_F. At $T = 0$ K, $\bar{n}(\varepsilon)$, Eq. (11.73a), has the form shown in Fig. 11.4. Hence Eq. (11.78) becomes, at $T = 0$ K,

$$N = \left[\frac{4\pi V}{h^3} (2m)^{3/2} \right] \int_0^{\varepsilon_F} \varepsilon^{1/2} \mathrm{d}\varepsilon = \left[\frac{4\pi V}{h^3} (2m)^{3/2} \right] \tfrac{2}{3} (\varepsilon_F)^{3/2}, \qquad (11.85)$$

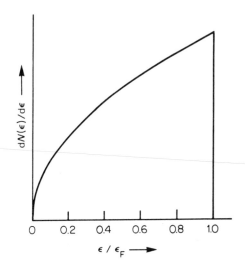

Fig. 11.5. The Fermi–Dirac energy distribution, Eq. (11.77), at zero temperature: $T = 0$ K. $\mathrm{d}N(\varepsilon)$ is the number of electrons with energy in the interval ε to $\varepsilon + \mathrm{d}\varepsilon$. ε_F is the Fermi energy.

so that

$$\varepsilon_F = \frac{h^2}{2m} \left(\frac{3}{8\pi} \frac{N}{V} \right)^{2/3} . \qquad (11.86)$$

We see that the Fermi energy depends on the mass m of the particles (in our case the electron mass) and on their mean concentration N/V. For monovalent metals V/N is the atomic volume, (i.e. the volume per atom of the solid). The *Fermi temperature* T_F is defined by

$$\varepsilon_F \equiv kT_F . \qquad (11.87)$$

We estimate the order of magnitude of ε_F and T_F from Eqs. (11.86) and (11.87). With $N/V \approx 5 \times 10^{22}\,\text{cm}^{-3}$ (which would correspond to an atomic spacing $(V/N)^{1/3} = 2.7\,\text{Å}$) and one conduction electron per atom one obtains $\varepsilon_F \approx 4.5\,\text{eV}$, $T_F \approx 5 \times 10^4\,\text{K}$. Table 11.1 gives the Fermi energy ε_F, the Fermi temperature T_F and the corresponding electron velocity $v_F = (2\varepsilon_F/m)^{1/2}$.

Table 11.1. The Fermi energy, temperature and velocity for the electron gas model.[a]

Element	$N/V\,\text{cm}^{-3}$	$\varepsilon_F\,\text{eV}$	$T_F\,\text{K}$	$v_F\,\text{cm/sec}$
Li	4.70×10^{22}	4.72	5.48×10^4	1.29×10^8
K	1.40	2.12	2.46	0.86
Cu	8.45	7.00	8.12	1.57
Au	5.90	5.51	6.39	1.39

[a] Source: Kittel,[29] p. 134.

We have so far considered a Fermi gas at zero temperature. Such a gas is called *completely degenerate*. We must now look at temperatures different from zero. The significance of the Fermi temperature T_F is that at temperatures T very small compared with T_F, the mean occupation number $\bar{n}(\varepsilon)$ and the energy distribution $dN(\varepsilon)/d\varepsilon$ are only very little different from their behaviour at $T=0\,\text{K}$, shown in Figs. 11.4 and 11.5. Such an electron gas is called extremely degenerate. Since the Fermi temperature T_F is of the order 10^4 to $10^5\,\text{K}$ for metals, it follows that the conduction electrons are in an extremely degenerate state under all usual conditions. Figs. 11.6 and 11.7 show the FD mean occupation number $\bar{n}(\varepsilon)$, Eq. (11.73a), and the FD energy spectrum (11.77) for a temperature T such that $kT = 0.2\,\mu$, plotted against ε/μ. Also shown as dashed lines are the corresponding curves for $T=0\,\text{K}$ (i.e. those of Figs. 11.4 and 11.5). We see that as the temperature is raised from $T=0\,\text{K}$, electrons are excited from single-particle states with energy $\varepsilon < \mu$ to single-particle-states with energy $\varepsilon > \mu$. We see from Fig. 11.6 that most of the transfer occurs for electrons with energy near $\varepsilon = \mu$.

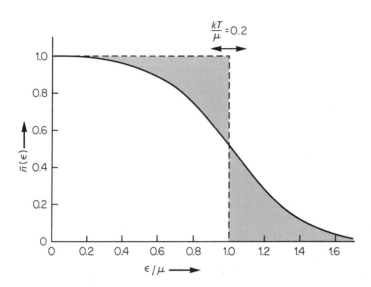

Fig. 11.6. The Fermi–Dirac mean occupation number $\bar{n}(\varepsilon)$, Eq. (11.73a), at a temperature $kT = 0.2\,\mu$.

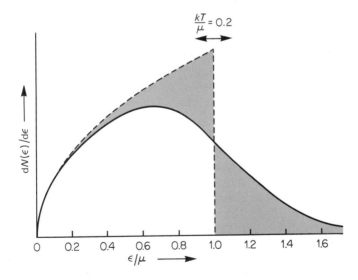

Fig. 11.7. The Fermi–Dirac energy distribution, Eq. (11.77), at temperature $kT = 0.2\,\mu$. $dN(\varepsilon)$ is the number of electrons with energy in the interval ε to $\varepsilon + d\varepsilon$.

Figs. 11.6 and 11.7 contain a 'hidden' temperature dependence since μ itself depends on the temperature. However for $T \ll T_F$ this T-dependence is very weak being given by (see problem 11.2)

$$\mu(T) \approx \varepsilon_F \left[1 - \frac{\pi^2}{12} \left(\frac{T}{T_F} \right)^2 \right] \quad , \qquad (T \ll T_F) \ . \qquad (11.88)$$

For the case shown in Figs. 11.6 and 11.7, $T/T_F = 0.2$, so the change in the chemical potential is about 4 per cent. With $T_F = 5 \times 10^4$ K, say, this corresponds to a temperature $T \approx 10^4$ K. From Fig. 11.7 one can see how the FD energy distribution begins to go over into an energy distribution reminiscent of MB statistics. The latter is shown in Fig. 7.5.

The free electron model of metals is successful for those properties which depend primarily on the energy distribution of the conduction electrons. One of these, their heat capacity, will be considered in the next subsection. Another, their paramagnetic susceptibility, is treated in problem 11.1.

11.5.2 The electronic heat capacity of metals

The free electron model of metals was originally developed using MB statistics by Drude (1900) and Lorentz (1905). Drude was able to derive the Wiedemann–Franz law which gives the temperature dependence of the ratio of thermal to electrical conductivities in metals. One knows now that this success of Drude's theory was fortuitous. However, from its inception Drude's theory faced another serious problem. Treated as a MB gas, the free electrons should contribute an amount $\frac{3}{2}k$ per electron to the heat capacity of metals in addition to the heat capacity of the lattice vibrations. At room temperature, the lattice vibrations lead to Dulong and Petit's law: the heat capacity of a solid is $3R$ per gram-atom. For monovalent metals (i.e. with one conduction electron per atom) the heat capacity should then be $9R/2$ per gram-atom. But metals obey Dulong and Petit's law very well (see Table 6.1, where a few examples are quoted) and do not form an exception. This difficulty is removed if, as we have been doing, the conduction electrons are described according to FD statistics. This was first done by Sommerfeld (1928) and it is his theory we are describing in this section.

We can see the effect on the conduction electrons of raising the temperature from 0 K to T K from Figs. 11.6 or 11.7. For such a temperature change, we expect the energy of an electron to increase by about kT. But for most electrons, with the exception of those whose energy is close to ε_F, the states to which they would be excited are already occupied. So because of the Pauli principle they can't be excited into these states. It is only a small fraction of electrons with energy near ε_F which can be excited, namely those which lie approximately within an energy interval kT of the Fermi energy ε_F. This number of excited electrons N_{ex} is given by

$$N_{ex} \sim f(\varepsilon_F)kT .$$ (11.89)

Since from Eq. (11.76)

$$f(\varepsilon_F) = \left[\frac{4\pi V}{h^3} (2m)^{3/2} \right] \varepsilon_F^{1/2}$$ (11.90)

and

$$N = \left[\frac{4\pi V}{h^3} (2m)^{3/2} \right] \tfrac{2}{3} \varepsilon_F^{3/2} ,$$ (11.85)

it follows from the last two equations by division that

$$f(\varepsilon_F) = \frac{3N}{2\varepsilon_F} .$$ (11.91)

Hence Eq. (11.89) becomes

$$N_{ex} \sim \frac{3NkT}{2\varepsilon_F} = \frac{3}{2} N \frac{T}{T_F} .$$ (11.92)

Only a small fraction $(\sim \tfrac{3}{2}(T/T_F))$ of the conduction electrons are excited. If $T_F = 3 \times 10^4 \, \text{K}$ and $T = 300 \, \text{K}$ a few per cent are excited.

The excitation energy per electron is of the order of kT. Hence the electronic energy (relative to the ground state at $T = 0 \, \text{K}$) is given by

$$E(T) \sim N_{ex} kT \sim \tfrac{3}{2} Nk \frac{T^2}{T_F}$$ (11.93)

and the heat capacity per electron becomes

$$c_V(T) = \frac{1}{N} \frac{\partial E}{\partial T} \sim 3k \frac{T}{T_F} .$$ (11.94)

A proper calculation (which is somewhat tedious, involving formulas like Eq. (11.88)) gives (see problem 11.3)

$$c_V(T) = \frac{\pi^2}{2} k \frac{T}{T_F} .$$ (11.95)

We see from Eq. (11.95) that at room temperature the electronic heat capacity per free electron is very small compared with the lattice heat capacity of about $3k$ per atom and will give a contribution of only a few per cent to the observed heat capacity of metals.

At very low temperatures the situation is different. We saw in Chapter 6, section 6.1, that at low temperatures the lattice heat capacity is proportional to T^3 whereas the electronic heat capacity (11.95) varies linearly with T. At sufficiently low temperatures the electronic contribution dominates. This behaviour is fully borne out by experiment. We had an example of this in Fig. 6.2.

★ 11.6 BOSE–EINSTEIN CONDENSATION

We shall now study the properties of a perfect gas of bosons of non-zero mass. The Pauli principle does not apply in this case, and the low-temperature properties of such a gas are very different from those of a fermion gas, discussed in the last section. A BE gas displays most remarkable quantal features. At low temperatures a phase change occurs in some respects reminiscent of the condensation of a vapour. This phenomenon is known as Bose–Einstein condensation and was first predicted by Einstein in 1925. BE condensation is of interest for two reasons. Firstly it is a case of a phase transition which admits an exact mathematical treatment. Secondly, as first pointed out by F. London (1938), liquid helium 4 (^{4}He) displays phenomena similar to the BE condensation of a perfect BE gas. In this section we shall briefly discuss the properties of a perfect BE gas and shall then point out the similarities which exist in the ^{4}He problem.*

The properties of a BE gas follow from the Bose–Einstein distribution (11.26) [or Eq. (11.55) with Eq. (11.47) for α substituted in it]:[†]

$$\bar{n}_i = \frac{1}{e^{\beta(\varepsilon_i - \mu)} - 1} \, . \tag{11.96}$$

The total number of particles in the system (we call it N again rather than $\bar{N}$) is obtained by summing Eq. (11.96) over all single-particle states

$$N = \sum_i \frac{1}{e^{\beta(\varepsilon_i - \mu)} - 1} \, . \tag{11.97}$$

We rewrite this equation as an integral. The energy density of states follows

---

*See London,[33] McClintock et al.,[34] Tilley and Tilley,[35] and Wilks.[37]
[†]Since sections 11.2 and 11.3 were *alternative* treatments of the perfect quantal gas we shall give cross-references to both sections, those in square brackets referring to section 11.3.

from Eq. (B.25) in Appendix B, together with $\varepsilon = p^2/2m$ (m being the mass of a particle of the gas):

$$f(\varepsilon)d\varepsilon = \frac{V4\pi p^2}{h^3}\frac{dp}{d\varepsilon}d\varepsilon = \frac{V2\pi(2m)^{3/2}}{h^3}\varepsilon^{1/2}d\varepsilon \ , \tag{11.98}$$

whence Eq. (11.97) becomes

$$\frac{N}{V} = \frac{2\pi(2m)^{3/2}}{h^3}\int_0^\infty \frac{\varepsilon^{1/2}d\varepsilon}{e^{\beta(\varepsilon-\mu)}-1} \ . \tag{11.99}$$

Let us vary the temperature of the gas, keeping N, V and hence the particle density N/V constant. In the last equation the left-hand side, and hence the right-hand side also, are constant. Now the crux of the matter is that for a BE gas the chemical potential μ is always negative: $\mu < 0$.* Hence if the right-hand side of Eq. (11.99) is to remain constant as T is lowered, μ, which is negative, must increase, i.e. $|\mu|$ must decrease. The integral in Eq. (11.99) then defines a *minimum temperature* T_c such that for $T = T_c$ the chemical potential vanishes: $\mu = 0$. This temperature T_c is thus given from Eq. (11.99) by

$$\frac{N}{V} = \frac{2\pi(2m)^{3/2}}{h^3}\int_0^\infty \frac{\varepsilon^{1/2}d\varepsilon}{\exp(\varepsilon/kT_c)-1} \ .$$

If in the last equation we change the variable of integration to $z = \varepsilon/kT_c$ we obtain

$$\frac{N}{V} = \left(\frac{2\pi mkT_c}{h^2}\right)^{3/2}\left(\frac{2}{\sqrt{\pi}}\int_0^\infty \frac{z^{1/2}dz}{e^z-1}\right)$$

$$= \left(\frac{2\pi mkT_c}{h^2}\right)^{3/2} \times 2.61 \ , \tag{11.100}$$

where the last step follows from the evaluation of the integral in curly parentheses.†

Now there is clearly something wrong with this argument which says that there exists a temperature T_c *below which the* BE *gas cannot be cooled at constant density*. In fact, Eq. (11.99) only holds for $T \geqslant T_c$, but for $T < T_c$ this equation must be modified. The trouble stems from the replacement of the discrete sum over single-particle states in Eq. (11.97) by the integral in Eq. (11.99). As the temperature of the gas is lowered sufficiently, particles

---

*For the derivation of this result, see the footnote on p. 268 and also the discussion following Eq. (11.26). With $\varepsilon = p^2/2m$, the ground-state energy is $\varepsilon = 0$, as required.
†See London,[33] for detailed evaluation of this and similar later integrals.

will begin to crowd into the single-particle ground state with zero energy: $\varepsilon_1 = 0$. Just this state is *completely* neglected in Eq. (11.99). Because of the factor $\varepsilon^{1/2}$ in the integrand it is given zero weight. At higher temperatures this does not introduce an error but at very low temperatures we must not simply omit this state since it will contain an appreciable number of particles. In replacing Eq. (11.97) by an integral, we must *explicitly* retain the first term $\bar{n}_1 = (e^{-\beta\mu} - 1)^{-1}$, which gives the number of particles in the ground state, and replace only the remaining terms by an integral.* Instead of Eq. (11.99) we should write

$$N = \frac{1}{e^{-\beta\mu} - 1} + V \frac{2\pi(2m)^{3/2}}{h^3} \int_0^\infty \frac{\varepsilon^{1/2} d\varepsilon}{e^{\beta(\varepsilon - \mu)} - 1} \, . \qquad (11.101)$$

In this equation

$$N_1 \equiv \frac{1}{e^{-\beta\mu} - 1} \qquad (11.102a)$$

is the number of particles in the ground state with zero energy and zero momentum, $\varepsilon = 0$, $p = 0$, while

$$N_{\varepsilon > 0} \equiv V \frac{2\pi(2m)^{3/2}}{h^3} \int_0^\infty \frac{\varepsilon^{1/2} d\varepsilon}{e^{\beta(\varepsilon - \mu)} - 1} \qquad (11.102b)$$

is the number of particles whose energy and momentum are non-zero: $\varepsilon > 0$, $p > 0$.

Above the temperature T_c, the number of particles in the ground state is quite negligible, the first term in Eq. (11.101) can be omitted, and the chemical potential is given by the original equation (11.99). Below the temperature T_c the chemical potential stays *extremely close to zero*. (We know from the argument which led to the condition $\mu < 0$, that μ cannot vanish exactly.) For $T < T_c$, the number of particles with non-zero energy is then given by Eq. (11.102b) with $\mu = 0$. If in the integral in this equation, we introduce a new variable of integration $z = \beta\varepsilon$ and use Eq. (11.100) to evaluate the resulting integral, we obtain

$$N_{\varepsilon > 0} = N \left(\frac{T}{T_c} \right)^{3/2} . \qquad (11.103)$$

This expression gives the fraction of particles $N_{\varepsilon > 0}/N$ in states with energy $\varepsilon > 0$. The remaining particles, i.e. a fraction

---

*It can be shown that only the first term need be treated in this special way, provided N is very large.

$$\frac{N_1}{N} = 1 - \left(\frac{T}{T_c}\right)^{3/2} , \qquad (11.104)$$

are in the ground state with zero energy ($\varepsilon = 0$) and zero momentum ($p = 0$).

This result (11.104) is shown in Fig. 11.8. For $T > T_c$, the number of particles in the ground state is negligible. As T decreases below T_c, the number of such particles increases rapidly. These particles possess zero energy and zero momentum. Because of the latter, they contribute neither to the pressure nor do they possess viscosity (since viscosity is related to transport of momentum). The process of concentrating particles into the zero-energy ground state is known as *Bose–Einstein condensation*. It differs from the condensation of a vapour into a liquid in that no *spatial* separation into phases with different properties occurs in BE condensation. Nevertheless many similar features occur in both kinds of condensation. For example, the pressure of a BE gas at $T < T_c$, just like the pressure of a saturated vapour, depends only on its temperature but not on its volume (see problem 11.4).*

A BE gas at temperatures below T_c is called degenerate. T_c is known as the *degeneracy temperature* or the *condensation temperature*.

We next consider the energy of the BE gas. For $T < T_c$, only the 'normal' particles possess energy. There are $N_{\varepsilon > 0}$ such particles and the energy of

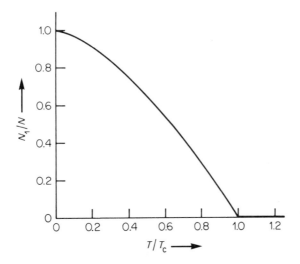

Fig. 11.8. Fraction of particles N_1/N in the zero-energy ground state, as function of the temperature T. T_c is the BE condensation temperature, given by Eq. (11.100).

*Becker,[2] section 54 (particularly, pp. 215–218), shows that there exists a far-reaching analogy between the two cases.

each is of the order of magnitude of kT; hence it follows from Eq. (11.103) that the energy of the BE gas is given by

$$E \sim N_{\varepsilon > 0} kT = Nk \, \frac{T^{5/2}}{T_c^{3/2}} \, , \qquad\qquad T < T_c \, , \qquad (11.105)$$

and hence the molar heat capacity at constant volume by

$$C_V \sim \frac{5}{2} R \left(\frac{T}{T_c} \right)^{3/2} , \qquad\qquad T < T_c \, . \qquad (11.106)$$

The exact calculation from the BE distribution function gives

$$C_V = 1.93 \, R \left(\frac{T}{T_c} \right)^{3/2} , \qquad\qquad T < T_c \, . \qquad (11.107)$$

Eq. (11.107) is easily derived. For $T < T_c$ it follows from an expression analogous to Eq. (11.102b) with $\mu = 0$:

$$E = V \frac{2\pi(2m)^{3/2}}{h^3} \int_0^\infty \frac{\varepsilon^{3/2} d\varepsilon}{e^{\beta\varepsilon} - 1} \, , \qquad\qquad T < T_c \, . \qquad (11.108)$$

If in this expression one changes the variable of integration to $z = \beta\varepsilon$, evaluates the resulting integral and substitutes for V from Eq. (11.100), one obtains

$$E = 0.77 \, Nk \, \frac{T^{5/2}}{T_c^{3/2}} \, , \qquad\qquad T < T_c \, , \qquad (11.109)$$

whence Eq. (11.107) follows directly.

The heat capacity (11.107) exceeds the classical value $\frac{3}{2} R$ at the condensation temperature T_c. At high temperatures the classical result of course holds. But it requires an exact calculation to see how the heat capacity for $T < T_c$, Eq. (11.107), joins on to the heat capacity for $T > T_c$. Such a calculation shows that the heat capacity C_V is continuous at $T = T_c$ but it has a kink there, as shown in Fig. 11.9.

To conclude this section, we discuss briefly the relevance of BE condensation for the low-temperature properties of helium 4 (^{4}He) which has spin zero and therefore satisfies BE statistics.

We note first of all that under its own vapour pressure ^{4}He remains a liquid down to the absolute zero of temperature. Large pressures, about 25 atm, are required to solidify it under these conditions. One can understand this as follows. In the liquid or solid, each helium atom is localized within

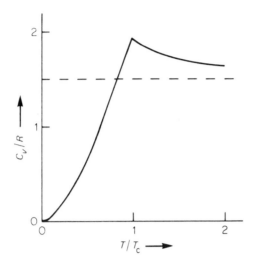

Fig. 11.9. Molar heat capacity at constant volume C_V of a BE gas as a function of the temperature. T_c is the condensation temperature.

a region of linear dimension a, where a is of the order of the mean separation between nearest neighbours. From the uncertainty principle, a particle so confined possesses a momentum of the order of $\hbar/a$ and a kinetic energy of the order of $\hbar^2/2ma^2$, m being the mass of the helium atom. This zero-point energy decreases with increasing separation a. We can think of the zero-point energy as producing a pressure: it tends to increase the interatomic separation, i.e. the molar volume. Helium being an inert gas the only interatomic forces are the extremely weak van der Waals forces. At the comparatively large interatomic spacings of liquid helium these forces are inadequate to overcome the thermal agitation of the zero-point motion, and liquid helium solidifies only under considerable pressures.

The normal boiling point of ^{4}He is 4.2 K. As liquid ^{4}He in contact with its vapour is cooled, its properties change dramatically at a sharply defined temperature of 2.17 K. Above this temperature it behaves like a normal liquid and is known as helium I. Below this temperature it has most remarkable properties, such as flowing rapidly through fine capillaries with apparently no viscosity. This peculiar form of helium is called helium II. Many of the properties of helium II are described well by a two-fluid model. This model assumes that helium II behaves like a mixture of two fluids, the normal fluid and the superfluid, and that there is no viscous interaction between these two fluids. The normal fluid possesses all the usual properties of a fluid. The superfluid has curious properties. It has zero entropy and experiences no resistance to its flow, i.e. it has no viscosity. The fact that in a degenerate

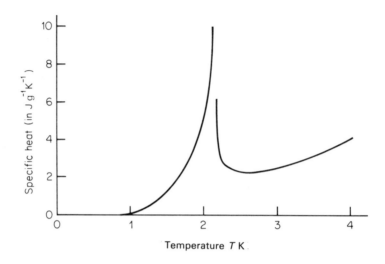

Fig. 11.10. The specific heat of liquid helium 4 in contact with its saturated vapour.

BE gas we were also dealing with two kinds of particle suggests that the two phenomena are related. Indeed there are further similarities which we shall now mention.

The transition from helium I to helium II is accompanied by an anomalous behaviour in the heat capacity. Fig. 11.10 shows the experimental specific heat as a function of temperature. The specific heat curve has a shape similar to the Greek letter lambda. The transition is therefore known as a lambda transition and the transition temperature as the lambda point. The similarity of Figs. 11.9 and 11.10 led F. London to suggest that the lambda transition of ^{4}He is a form of BE condensation. London calculated from Eq. (11.100) the BE condensation temperature T_c for a perfect gas of helium atoms whose density equals the density of helium I at the experimental lambda point. With $V = 27.6 \, \text{cm}^3/\text{mol}$ London obtained $T_c = 3.13 \, \text{K}$ which is not very far from the experimental lambda point of $T_\lambda = 2.17 \, \text{K}$.

We see that although there is a general similarity of the two phenomena the details differ. This also shows up in other respects. For example, the dependence of the transition temperature on pressure is predicted wrongly by the BE model, and more recent measurements of remarkable accuracy (the measurements were to within almost $10^{-6} \, \text{K}$ of the lambda point!) show that the specific heat has a logarithmic infinity at the lambda point.* That there should be such differences is not surprising. The BE theory considers a perfect gas, and in liquid helium the interactions between the atoms are

*See Wilks,[37] Chapter 11 of the 1967 reference, or section 15.3 of the 1970 reference.

certainly not negligible: for example, we are dealing with a liquid and not a gas! All these circumstances strongly suggest that the lambda transition in ⁴He is the analogue in a liquid of the BE condensation phenomenon in a perfect gas. This conclusion is further strengthened by the fact that helium 3 (³He) whose atoms have spin $\frac{1}{2}$ and are fermions does not exhibit a lambda-point transition at a temperature of the order of a few kelvin, predicted by Eq. (11.100) for BE condensation. Helium 3 does display superfluid properties but these occur at temperatures three orders of magnitude lower, i.e. in the millikelvin region, and they have their origin in a quite different mechanism.*

11.7   THERMODYNAMICS OF THE GIBBS DISTRIBUTION

In section 11.1 we considered a system in equilibrium with a heat bath/particle reservoir and derived the Gibbs distribution, i.e. the statistical distribution for a system whose particle number and energy can fluctuate. We now want to obtain the thermodynamic relations for such a system. We take as our starting point the entropy which corresponds to the Gibbs distribution (11.13). According to Eq. (2.33), which gives the entropy for a general probability distribution, we have in the present case

$$S = -k\sum_{Nr} p_{Nr}\ln p_{Nr}$$

$$= -k\sum_{Nr} p_{Nr}[\beta\mu N - \beta E_{Nr} - \ln \mathcal{Z}]$$

$$= \frac{\bar{E}}{T} - \frac{\mu\bar{N}}{T} + k\ln \mathcal{Z} . \tag{11.110}$$

Here $\bar{E}$ and $\bar{N}$ are the mean energy and the mean number of particles of the system in the heat bath/particle reservoir, defined by

$$\bar{E} = \sum_{Nr} p_{Nr}E_{Nr} , \qquad \bar{N} = \sum_{Nr} p_{Nr}N . \tag{11.111}$$

We define the *grand potential* Ω by rewriting Eq. (11.110) as

$$\Omega \equiv \Omega(T, V, \mu) \equiv -kT\ln \mathcal{Z}(T, V, \mu) = \bar{E} - TS - \mu\bar{N} . \tag{11.112}$$

Comparison with Eqs. (2.35) to (2.37), viz.

$$F = -kT\ln Z(T, V, N) = E - TS , \tag{11.113}$$

shows that Eq. (11.112) is the generalization of these equations for the case of variable particle number: the partition function is replaced by the grand

*See, for example, McClintock *et al.*,[34] and Tilley and Tilley.[35]

partition function, the Helmholtz free energy F by the grand potential Ω. The right-hand side of Eq. (11.112) contains mean values. We shall show in section 11.7.1 that, just as in section 2.5, the *fluctuations are usually completely negligible for macroscopic systems*. So the mean values are just the actual, sharply defined values of these quantities for the system. Accordingly, we shall frequently omit the bars from such quantities. Eq. (11.112) then represents a purely thermodynamic definition of the grand potential.

To derive the thermodynamic consequences of Eq. (11.112) we require the generalization of the fundamental thermodynamic relation (4.12) for the case of variable particle number. By differentiation of $S = S(E, V, N)$ we obtain

$$dS = \frac{\partial S}{\partial E} dE + \frac{\partial S}{\partial V} dV + \frac{\partial S}{\partial N} dN = \frac{dE}{T} + \frac{PdV}{T} - \frac{\mu dN}{T} ,$$

i.e.

$$dE = TdS - PdV + \mu dN. \tag{11.114}$$

This is the required generalization of Eq. (4.12). Eq. (11.114) can be expressed in terms of other variables. From $F = E - TS$ and $G = E + PV - TS$ one obtains respectively

$$dF = -SdT - PdV + \mu dN \tag{11.115}$$
$$dG = -SdT + VdP + \mu dN . \tag{11.116}$$

Eqs. (11.114) to (11.116) contain three alternative expressions for the chemical potential

$$\mu = -T\left(\frac{\partial S}{\partial N}\right)_{E,V} = \left(\frac{\partial F}{\partial N}\right)_{T,V} = \left(\frac{\partial G}{\partial N}\right)_{T,P} . \tag{11.117}$$

The first two of these relations were used to define μ in Eqs. (11.10) and (11.59). The third relation for μ can be written

$$G(T, P, N) = \mu N, \tag{11.118}$$

as we shall now show. It follows from the extensive property of the Gibbs free energy G that for any factor λ

$$\lambda G(T, P, N) = G(T, P, \lambda N) . \tag{11.119}$$

Differentiating this equation with respect to λ and putting $\lambda = 1$ after differentiation, we obtain

$$G(T,P,N) = N \left(\frac{\partial G}{\partial N} \right)_{T,P} \; . \tag{11.120}$$

Elimination of $\partial G / \partial N$ between Eqs. (11.117) and (11.120) leads to Eq. (11.118). This last equation allows us to rewrite the grand potential in a particularly simple form. From Eqs. (11.112) and (11.118) we have

$$\Omega = E - TS - \mu N = E - TS - G$$

and since $G = E + PV - TS$ the grand potential Ω reduces to

$$\Omega = -PV \; . \tag{11.121}$$

Lastly, differentiation of Eq. (11.112) leads to

$$d\Omega = dE - TdS - SdT - \mu dN - Nd\mu$$

which, on account of Eq. (11.114) reduces to

$$d\Omega = -SdT - PdV - Nd\mu \; . \tag{11.122}$$

From this differential form follow the thermodynamic relations with T, V and μ as independent variables:

$$S = -\left(\frac{\partial \Omega}{\partial T} \right)_{V,\mu} , \qquad P = -\left(\frac{\partial \Omega}{\partial V} \right)_{T,\mu} , \qquad N = -\left(\frac{\partial \Omega}{\partial \mu} \right)_{T,V} . \tag{11.123}$$

★ 11.7.1   Fluctuations of particle numbers

We shall now calculate the fluctuations in the number of particles contained in a system which is in equilibrium with a particle reservoir, and we shall show that for a macroscopic system these fluctuations are negligible. For this purpose we note that Eq. (11.123) for the particle number, which was derived from thermodynamic reasoning, also follows from statistics. We have from

$$\Omega = -kT \ln \mathcal{Z} \tag{11.124}$$

and Eqs. (11.13)–(11.14):

$$\left(\frac{\partial \Omega}{\partial \mu}\right)_{T,V} = -kT \sum_{Nr} \frac{\exp\left[\beta(\mu N - E_{Nr})\right]}{\mathcal{Z}} \beta N$$

$$= -\sum_{Nr} p_{Nr} N = -\bar{N}. \tag{11.125}$$

To obtain the particle fluctuations we differentiate this equation again:

$$\left(\frac{\partial^2 \Omega}{\partial \mu^2}\right)_{T,V} = -\left(\frac{\partial \bar{N}}{\partial \mu}\right)_{T,V} = -\sum_{Nr} N \left(\frac{\partial p_{Nr}}{\partial \mu}\right)_{T,V}. \tag{11.126}$$

It follows from Eq. (11.13) that

$$\frac{1}{p_{Nr}}\left(\frac{\partial p_{Nr}}{\partial \mu}\right)_{T,V} = \left(\frac{\partial \ln p_{Nr}}{\partial \mu}\right)_{T,V}$$

$$= \beta N - \left(\frac{\partial \ln \mathcal{Z}}{\partial \mu}\right)_{T,V}$$

$$= \beta N - \beta \bar{N}, \tag{11.127}$$

where the last step follows from Eqs. (11.124) and (11.125). Hence Eq. (11.126) becomes, on substitution from (11.127),

$$\left(\frac{\partial^2 \Omega}{\partial \mu^2}\right)_{T,V} = -\sum_{Nr} N p_{Nr} (\beta N - \beta \bar{N})$$

$$= -\beta (\overline{N^2} - \bar{N}^2)$$

$$\equiv -\beta (\Delta N)^2 \tag{11.128}$$

where ΔN is the standard deviation of N. From Eqs. (11.125) and (11.128) we obtain the relative fluctuation in the number of particles in the system

$$\frac{\Delta N}{\bar{N}} = \frac{\left(-kT \frac{\partial^2 \Omega}{\partial \mu^2}\right)^{1/2}}{\left(-\frac{\partial \Omega}{\partial \mu}\right)}. \tag{11.129}$$

We see from Eqs. (11.117) and (11.112), which define the chemical potential μ and the grand potential Ω, that these quantities are intensive and extensive variables respectively. Hence the dependence of $\Delta N/\bar{N}$ on the size of the system is, from Eq. (11.129) given by

$$\frac{\Delta N}{\bar{N}} \propto \frac{1}{\sqrt{\bar{N}}}. \tag{11.130}$$

For a macroscopic system, where $\bar{N} \sim 10^{23}$ and $\Delta N/\bar{N} \sim 10^{-11}$, the relative fluctuations are extremely small. For practical purposes the particle number $\bar{N}$ is fully determined. In section 2.5 we derived a similar result for the relative energy fluctuations of a system with a fixed number of particles in a heat bath. The same result for the energy fluctuations holds in the present case and could be derived independently from Eqs. (11.13) and (11.14).

★ 11.8   THE PERFECT CLASSICAL GAS

We shall now employ the Gibbs formalism, i.e. the grand partition function and the grand potential, to derive the equation of state of a perfect classical gas, previously studied in Chapter 7. We see from Eq. (11.14) that we can write the grand partition function $\mathcal{Z}$ as a weighted sum of partition functions with fixed numbers, $N = 0, 1, \ldots$, of particles:

$$\mathcal{Z}(T, V, \mu) = \sum_N e^{\beta \mu N} Z(T, V, N) \ , \qquad (11.131)$$

where

$$Z(T, V, N) = \sum_r \exp(-\beta E_{Nr}) \qquad (11.132)$$

is the partition function for a system of N particles contained in a volume V and at temperature T. In section 7.1 we showed that for a perfect classical gas $Z(T, V, N)$ is related to the single-particle partition function $Z_1(T, V)$ by

$$Z(T, V, N) = \frac{1}{N!} \ [Z_1(T, V)]^N \ . \qquad (7.11)$$

Hence Eq. (11.131) for the grand partition function becomes

$$\mathcal{Z}(T, V, \mu) = \sum_N \frac{1}{N!} \ [e^{\beta \mu} Z_1(T, V)]^N = \exp[e^{\beta \mu} Z_1(T, V)] \ . \qquad (11.133)$$

Combining Eqs. (11.121), (11.112) and (11.133), we obtain the grand potential

$$\Omega = -PV = -kT \ln \mathcal{Z} = -kT[e^{\beta \mu} Z_1(T, V)] \qquad (11.134)$$

and hence from Eq. (11.123)

$$N = -\left(\frac{\partial \Omega}{\partial \mu}\right)_{T, V} = [e^{\beta \mu} Z_1(T, V)] \ . \qquad (11.135)$$

This result has already been derived above, Eq. (11.69). Eliminating the expression in square brackets between the last two equations, we obtain

$$PV = NkT \ . \qquad (11.136)$$

This derivation of the equation of state is much simpler—once the grand partition function formalism has been mastered—than that used in Chapter 7. In particular we did not have to calculate the single-particle partition function $Z_1(T, V)$.

11.9 CHEMICAL REACTIONS

In section 11.1 we developed a formalism for handling systems whose particle numbers are not fixed. In particular, we considered a system in contact with a particle reservoir so that particles can be exchanged between the system and the reservoir. Other important situations where particle numbers are not fixed arise when the system contains several different species of particles between which reactions occur. Consequently, the particle number of the individual species may change. Examples are chemical reactions, nuclear reactions and the ionization of atoms. (The last of these processes is of fundamental importance in semi-conductor physics.) In all these situations, the methods of this chapter are particularly appropriate. In this section we shall illustrate this for chemical reactions.

So we are now considering a system consisting of several components, i.e. different substances. For example, the system might contain molecular oxygen (O_2), molecular nitrogen (N_2) and nitric oxide (NO), and undergo the chemical reaction

$$N_2 + O_2 \rightleftharpoons 2NO \ . \tag{11.137}$$

In general a multicomponent system will undergo several reactions. For the above system another reaction would be

$$N_2 + 2O_2 \rightleftharpoons 2NO_2 \ .$$

Although the arguments are easily generalized (see ter Haar and Wergeland,[18] Chapter 7) we shall for simplicity assume that only one reaction occurs. In practice one reaction frequently dominates under given conditions of temperature and pressure. Although we are talking of a reaction we are in fact considering systems in equilibrium. In other words one deals with a dynamic equilibrium: Eq. (11.137) and similar equations represent two reactions proceeding in opposite directions. In general, one of these reactions will dominate, producing a change in the composition of the system. In equilibrium, the speeds of the two reactions in opposite directions are such that the amount of each substance present in the system remains constant. The situation is quite analogous to the equilibrium between liquid water and water vapour where water molecules in the liquid are continually escaping from the liquid while others in the vapour, striking the surface of the liquid, are 'recaptured' by the liquid.

We shall write a general chemical reaction as

$$\nu_1 A_1 + \nu_2 A_2 + \ldots + \nu_s A_s = 0 \ . \tag{11.138}$$

Here $A_1, A_2, \ldots$ stands for the chemical formulas of the different components, s of them in all. $\nu_1, \nu_2, \ldots$ gives the relative number of molecules which take part in the reaction: $\nu_i < 0$ for a reaction ingredient (i.e. A_i occurs on the left-hand side of an equation like (11.137)), $\nu_i > 0$ for a reaction product (i.e. A_i occurs on the right-hand side of an equation like (11.137)). For the reaction (11.137), for example, we would have

$$A_1 = N_2 \qquad A_2 = O_2 \qquad A_3 = NO$$

$$\nu_1 = -1 \qquad \nu_2 = -1 \qquad \nu_3 = +2 \ .$$

11.9.1   Conditions for chemical equilibrium

We shall now derive the equilibrium conditions for the reaction (11.138), assuming that the system is kept at a constant temperature T and a constant pressure P. To begin with, we shall also assume that the system consists of a single phase only. For a system at constant temperature and pressure, the equilibrium is characterized by the fact that the Gibbs free energy G of the system is a minimum.* The Gibbs free energy for a given state of the system is now a function of T, P, and $N_1, N_2, \ldots N_s$, the number of molecules of the various components $A_1, A_2, \ldots A_s$:

$$G = G(T, P, N_1, N_2, \ldots N_s) \ , \tag{11.139}$$

and the thermodynamic results of section 11.7 admit immediate generalizations. For example, differentiation of Eq. (11.139) leads to

$$dG = -S dT + V dP + \sum_i \mu_i dN_i \tag{11.140}$$

(summation with respect to i will always be over all components $i = 1, 2, \ldots s$) where μ_i is the chemical potential of the ith component in the system:

$$\mu_i \equiv \mu_i(T, P, N_1, N_2, \ldots N_s) = \left(\frac{\partial G}{\partial N_i} \right)_{T, P, N_j (j \neq i)} \ . \tag{11.141}$$

---

*This important result is derived in section 4.5. Readers who have not studied section 4.5 may like to turn to problem 11.7 and the hints for solving this problem, where a detailed alternative treatment of chemical reactions in an *isolated* system is given.

Eqs. (11.140) and (11.141) are the generalizations of Eqs. (11.116) and (11.117) for a multicomponent system.

For equilibrium at constant temperature and constant pressure, G is a minimum. Hence from Eq. (11.140)

$$dG = \sum_i \mu_i dN_i = 0 \ . \tag{11.142}$$

The particle numbers $N_1, N_2, \ldots$ of the different components cannot change in an arbitrary way, independently of each other. Rather, these changes are related to each other through the chemical reaction (11.138). This requires

$$dN_i \propto \nu_i \ , \qquad i = 1, 2, \ldots s \ , \tag{11.143}$$

so that Eq. (11.142) becomes

$$\sum_i \nu_i \mu_i = 0 \ . \tag{11.144}$$

Eq. (11.144) is the basic equilibrium condition for chemical reactions from which the properties of such a system follow. This equation has a simple interpretation (which makes it easy to remember it). Comparison of Eq. (11.144) with Eq. (11.138) shows that the chemical potentials of the different components must be added in the same proportions in which the components occur in the reactions. For reaction (11.137), for example, Eq. (11.144) becomes

$$\mu_{NO} = \tfrac{1}{2} (\mu_{N_2} + \mu_{O_2}) \ . \tag{11.145}$$

Thus the chemical potential of a compound is made up additively from the chemical potentials of its constituents. From this remark the generalization to several simultaneous reactions is obvious. Suppose in addition to reaction (11.137) there occur the reactions

$$N + N \rightleftharpoons N_2 \ , \qquad O + O \rightleftharpoons O_2 \ , \qquad N + O \rightleftharpoons NO \ . \tag{11.137'}$$

For these we would obtain the equilibrium conditions

$$\mu_{N_2} = 2\mu_N \ , \qquad \mu_{O_2} = 2\mu_O \ , \qquad \mu_{NO} = \mu_N + \mu_O \tag{11.145'}$$

which are consistent with each other and with (11.145).

Eq. (11.144) continues to be the equilibrium condition for the reaction (11.138) even when the system consists of several phases in contact with each other so that the reacting substances can pass between them. This is a consequence of the fact that for a system in equilibrium the chemical potential of a given component has the same value in all phases. (See problem 11.11.)

11.9.2 Law of mass action

We shall now apply the condition (11.144) for chemical equilibrium to a reaction in a gaseous phase, treating the gases as perfect gases. For this case we can write down explicit expressions for the chemical potentials, and we shall find that the equilibrium condition (11.144) reduces to the simple and useful law of mass action. However, the approach we shall use can be generalized and applied to many other situations, giving results reminiscent of those we shall obtain for perfect gases. Two examples are reactions involving gases and pure (i.e. one-component) liquids and solids, and dilute solutions. For these aspects and chemical equilibrium more generally, the reader is referred to books on chemical thermodynamics.*

To obtain the chemical potentials, we shall start from the Helmholtz free energy F which, for a mixture of ideal gases, is given by

$$F \equiv F(T, V, N_1, N_2, \ldots N_s) = \sum_i F_i(T, V, N_i) \qquad (11.146)$$

where F_i is the Helmholtz free energy of the ith component. We have from Eq. (7.11), with an obvious extension of notation, that

$$Z_i(T, V, N_i) = \frac{1}{N_i!} [Z_i(T, V, 1)]^{N_i}. \qquad (11.147)$$

Hence

$$F_i(T, V, N_i) = -kT \ln Z_i(T, V, N_i)$$
$$= -kT\{N_i \ln Z_i(T, V, 1) - N_i(\ln N_i - 1)\}. \qquad (11.148)$$

The chemical potential μ_i of the ith component follows from the generalization of Eq. (11.117) for a multicomponent system (cf. Eq. (11.141):

$$\mu_i = \left(\frac{\partial F}{\partial N_i}\right)_{T, V, N_j(j \neq i)}. \qquad (11.149)$$

From Eqs. (11.149), (11.146) and (11.148) we obtain

$$\mu_i = -kT\{\ln Z_i(T, V, 1) - \ln N_i\}. \qquad (11.150)$$

*See, for example, Denbigh.[4]

The single-particle partition function $Z_i(T, V, 1)$ is, from Eqs. (7.13) and (7.19), given by

$$Z_i(T, V, 1) = V \left(\frac{2\pi m_i kT}{h^2} \right)^{3/2} Z_i^{\text{int}}(T) \tag{11.151}$$

where m_i and Z_i^{int} are the mass and the internal partition function of a molecule of the ith component. We introduce the abbreviation

$$f_i(T) \equiv \left(\frac{2\pi m_i kT}{h^2} \right)^{3/2} Z_i^{\text{int}}(T) = \frac{Z_i(T, V, 1)}{V} \tag{11.152}$$

and define the particle density per unit volume of the ith component

$$c_i = N_i/V \ . \tag{11.153}$$

Combining Eqs. (11.151) to (11.153) gives for the chemical potential of the ith component

$$\mu_i = kT\{\ln c_i - \ln f_i(T)\} \ . \tag{11.154}$$

Substituting Eq. (11.154) into Eq. (11.144), we obtain as the condition for chemical equilibrium for the reaction (11.138)

$$\sum_i \nu_i \ln c_i = \sum_i \nu_i \ln f_i(T) \tag{11.155}$$

or, taking exponentials of this equation,

$$\prod_i (c_i)^{\nu_i} = \prod_i [f_i(T)]^{\nu_i} \equiv K_c(T) \ . \tag{11.156}$$

For example, applied to the reaction (11.137), Eq. (11.156) becomes

$$\frac{(c_{NO})^2}{c_{N_2} c_{O_2}} = K_c(T) \ .$$

Eq. (11.156) states the law of mass action. The left-hand side of this equation contains the equilibrium densities of the gases in the reaction. $K_c(T)$, on the right-hand side, is independent of these densities and is a function of the temperature only. $K_c(T)$ is called the equilibrium constant for the reaction. The law of mass action states that at a given temperature the product of the equilibrium densities, raised to the appropriate powers, is constant. For a mixture of gases at temperature T and initially with *arbitrary* densities, the law of mass action tells us the direction in which the reaction will proceed. If $\nu_1, \nu_2, \ldots \nu_r$ are negative and $\nu_{r+1}, \ldots \nu_s$ positive, we can write the reaction (11.138)

$$\sum_{i=1}^{r} |\nu_i| A_i \rightleftharpoons \sum_{i=r+1}^{s} \nu_i A_i \ . \tag{11.138a}$$

Suppose for the initial densities

$$\prod_i (c_i)^{\nu_i} \equiv \frac{\displaystyle\prod_{i=r+1}^{s} (c_i)^{\nu_i}}{\displaystyle\prod_{i=1}^{r} (c_i)^{|\nu_i|}} < K_c(T) \ , \tag{11.157}$$

i.e. the system is not in chemical equilibrium. Suppose the reaction (11.138a) proceeds from left to right in order to establish equilibrium. Then the densities of the substances on the left-hand side of Eq. (11.138a) decrease and those of substances on the right-hand side increase. This means that all densities in the denominator of the middle expression in (11.157) decrease and all densities in the numerator increase. Hence, if the initial densities satisfy the inequality (11.157), the reaction will proceed from left to right in order to establish the equilibrium densities. The reaction will proceed from right to left, if for the initial densities the inequality sign in (11.157) is reversed.

The law of mass action can be expressed in several other ways. The partial pressure P_i of the ith gas is given by

$$P_i = N_i kT/V = c_i kT \ . \tag{11.158}$$

In terms of P_i, Eq. (11.156) becomes

$$\prod_i (P_i)^{\nu_i} = \prod_i [kT f_i(T)]^{\nu_i} \equiv K_P(T) \tag{11.159}$$

where the P_i are the partial pressures at equilibrium. Introducing the particle concentrations

$$x_i = N_i/N = P_i/P \tag{11.160}$$

($N = \sum_i N_i$ is the total particle number and P is the total pressure of the gas) into Eq. (11.159) leads to

$$\prod_i (x_i)^{\nu_i} = \frac{1}{P^{\Sigma \nu_i}} \prod_i [kT f_i(T)]^{\nu_i} \equiv K_x(T, P) \tag{11.161}$$

where x_i are the particle concentrations at equilibrium.

The interpretation of these alternative forms (11.159) and (11.161) of the law of mass action is similar to that of Eq. (11.156). Note that the equilibrium constant $K_x(T,P)$, unlike $K_c(T)$ and $K_P(T)$, depends on P as well as on T. The three equilibrium constants are seen to be related in a trivial way.

For the determination of the equilibrium constants one requires the internal partition functions of the molecules involved in the reaction, and these are obtained from the molecular energy levels and their degeneracies. In practice, one usually deduces these from spectroscopic data. In principle, they can be calculated from the electronic, vibrational and rotational properties of molecules which were mentioned briefly in section 7.5.

In the expression (11.156) for the equilibrium constant, the internal partition functions Z_i^{int} [which occur in $f_i(T)$] must all be calculated relative to the same zero of energy. This is achieved by writing them in the form

$$Z_i^{int}(T) = e^{-\beta\varepsilon_0(i)}\left\{\sum_\alpha e^{-\beta[\varepsilon_\alpha(i) - \varepsilon_0(i)]}\right\} \tag{11.162}$$

where $\varepsilon_\alpha(i)$ are the energy levels of the ith type of molecule, with $\alpha = 0$ denoting the ground state, and measuring the ground state energies $\varepsilon_0(i)$ relative to a common zero of energy. The term in curly parentheses in Eq. (11.162) now contains only energy differences and is just the internal partition function, for the ith type of molecule, with its ground state energy taken as zero. Substituting Eq. (11.162) into Eq. (11.152) for $f_i(T)$ leads to a factor

$$\exp\left\{-\beta\sum_i \nu_i\varepsilon_0(i)\right\} = \exp\left\{-\beta\left[\sum_{i=r+1}^s \nu_i\varepsilon_0(i) - \sum_{i=1}^r |\nu_i|\varepsilon_0(i)\right]\right\} \tag{11.163}$$

in Eq. (11.156) defining $K_c(T)$. The expression in square brackets in (11.163) is the total energy of the product molecules minus the total energy of the ingredient molecules in the reaction (11.138a), all molecules being in their ground states, i.e.

$$Q_0 \equiv \sum_{i=r+1}^s \nu_i\varepsilon_0(i) - \sum_{i=1}^r |\nu_i|\varepsilon_0(i) \tag{11.164}$$

is just the heat of reaction at absolute zero. This energy is usually of the order of a few electron-volts, so that the exponential factor (11.163) will usually dominate the temperature dependence of the equilibrium constant. (Remember that $kT = 1$ eV means $T \approx 10^4$ K.) Some simple applications of the law of mass action, which illustrate this point, are given in problems 11.8 and 11.9.

11.9.3   Heat of reaction

Next, we shall consider the heat which is absorbed or generated in a chemical reaction in a mixture of perfect gases. We shall assume that the system is kept at a constant temperature T and a constant pressure P. The heat of reaction Q_P of the process (11.138a) is defined as the heat supplied to the system when the reaction proceeds from left to right. For $Q_P > 0$, heat is absorbed in the reaction which is then called endothermic; for $Q_P < 0$, heat is generated and the reaction is called exothermic. For a process at constant pressure, the heat supplied to the system, Q_P, equals the increase in the enthalpy of the system

$$Q_P = \Delta H = \Delta(E + PV) = \Delta E + P\Delta V \ , \tag{11.165}$$

since in such a process work is done by the system against the applied external pressure. (In contrast, in a process at constant volume, the heat supplied to the system, Q_V, equals the increase in the internal energy E of the system: $Q_V = \Delta E$.)

We shall express the enthalpy H in terms of the Gibbs free energy G (and subsequently ΔH in terms of ΔG). We have

$$G = E + PV - TS = H - TS$$

and from Eq. (11.140) we obtain the entropy

$$S = - \left(\frac{\partial G}{\partial T} \right)_{P, N_1, \ldots N_s} .$$

(In the following equations, partial differentiation with respect to T will always be understood to be done with P and all particle numbers $N_1, \ldots N_s$ kept constant.) Combining the last two equations gives

$$H = G - T \frac{\partial G}{\partial T} = - T^2 \frac{\partial}{\partial T} \left(\frac{G}{T} \right) . \tag{11.166}$$

We are considering a process at constant temperature and pressure, i.e. only the particle numbers are changing. Suppose $N_1, N_2, \ldots$ become $N_1', N_2', \ldots$, with

$$N_i' \equiv N_i + \Delta N_i \ , \qquad i = 1, \ldots s \ .$$

The Gibbs free energy then changes by

$$\Delta G = G(T, P, N_1', \ldots N_s') - G(T, P, N_1, \ldots N_s)$$

$$= \sum_i \left(\frac{\partial G}{\partial N_i} \right)_{T, P, N_j (j \neq i)} \Delta N_i = \sum_i \mu_i \Delta N_i \tag{11.167}$$

where we used Eq. (11.141). Substituting Eq. (11.167) in (11.166) gives the corresponding change in the enthalpy

$$\Delta H = - T^2 \frac{\partial}{\partial T}\left(\frac{\Delta G}{T}\right) \ . \tag{11.168}$$

For the reaction (11.138a), $\Delta N_i = \nu_i$ and, from Eq. (11.167),

$$\Delta G = \sum_i \mu_i \nu_i \ . \tag{11.169}$$

In equilibrium G is a minimum, i.e. $\Delta G = 0$, and Eq. (11.169) reduces to the equilibrium condition (11.144). But it is clear from the derivation of Eqs. (11.167) and (11.169) that these equations hold in general, not only in equilibrium.

To calculate ΔH from Eqs. (11.168) and (11.169), we require the chemical potentials as functions of T, P and the particle numbers, since P and $N_1, \ldots N_s$ are held constant in the differentiation in (11.168). For a perfect gas, Eq. (11.154) gives μ_i as a function of $c_i = N_i/V$ and T. We obtain the required form by writing

$$c_i = N_i/V = P_i/kT \ , \tag{11.170}$$

since the partial pressure $P_i = N_i P/(\sum_i N_i)$ is a function of P and $N_1, \ldots N_s$. From Eqs. (11.154) and (11.170) we find

$$\mu_i = kT\{\ln P_i - \ln[kT f_i(T)]\} \ . \tag{11.171}$$

Substituting this expression in Eq. (11.169), we obtain

$$\Delta G = kT\sum_i \nu_i \ln P_i - kT \ln K_P(T) \tag{11.172}$$

where we used the definition (11.159) of the equilibrium constant $K_P(T)$. It follows that

$$\frac{\partial}{\partial T}\left(\frac{\Delta G}{T}\right) = - k\frac{\mathrm{d}}{\mathrm{d}T} \ln K_P(T) \ .$$

Hence we obtain from Eqs. (11.165) and (11.168) the heat of reaction

$$Q_P = kT^2 \frac{\mathrm{d}}{\mathrm{d}T} \ln K_P(T) \tag{11.173}$$

for a single elementary process (11.138a). The molar heat of reaction (i.e. the coefficients ν_i in the reaction equation (11.138a) now represent moles of substances) is obtained from Eq. (11.173) by replacing Boltzmann's constant k by the gas constant R.

Eq. (11.173) is known as van't Hoff's equation. It relates the heat of reaction to the temperature derivative of the equilibrium constant $K_P(T)$ and tells us how changes of temperature affect the equilibrium. Raising the temperature for an endothermic process ($Q_P > 0$) or lowering it for an exothermic one ($Q_P < 0$) increases K_P, so that the equilibrium is shifted in the direction of the reaction from left to right. Conversely, lowering the temperature for an endothermic process or raising it for an exothermic one decreases K_P and the equilibrium is shifted from right to left. We can sum up this state of affairs by saying that if the temperature of the system in equilibrium is raised (lowered) at constant pressure, the reaction proceeds in the direction in which heat is absorbed (released).

11.9.4   Pressure dependence of the reaction equilibrium

The way in which a chemical equilibrium depends on pressure can be analysed in a similar way. Consider the reaction (11.138a) between perfect gases. For this case we obtain from Eq. (11.161)

$$\left(\frac{\partial}{\partial P} \ln K_x(T,P) \right)_T = -\frac{1}{P} \sum_i \nu_i \ . \tag{11.174}$$

The molar volume change for this reaction at constant temperature and pressure follows from the perfect gas law

$$PV = RT \sum_i N_i \qquad (N_i \text{ in moles})$$

whence

$$P\Delta V = RT \sum_i \Delta N_i = RT \sum_i \nu_i \ . \tag{11.175}$$

In particular, there is no volume change if $\Sigma \nu_i = 0$, i.e. if the number of molecules does not change in the reaction. Combining Eqs. (11.174) and (11.175) leads to

$$\left(\frac{\partial}{\partial P} \ln K_x(T,P) \right)_T = -\frac{\Delta V}{RT} \ . \tag{11.176}$$

Suppose $\Delta V > 0$, i.e. the volume increases as the reaction (11.138a) proceeds from left to right at constant temperature and pressure. For an increase in pressure, at constant temperature, it follows from Eq. (11.176) that $K_x(T,P)$ decreases, so that the equilibrium shifts from right to left: the reaction proceeds in the direction which leads to a decrease in volume. If the pressure is decreased, at constant temperature, the reaction proceeds in the direction which leads to an increase in volume. If a reaction occurs without change of volume ($\Delta V = 0$), the equilibrium composition is independent of pressure. The reaction

$$CO + H_2O \rightleftharpoons CO_2 + H_2$$

is an example of this.

PROBLEMS 11

11.1 Due to its spin, the electron possesses a magnetic moment μ_B. Treating the conduction electrons in a metal as a free electron gas, obtain an expression (involving integrals over Fermi–Dirac distribution functions) for the magnetization due to the magnetic moments of the conduction electrons, when placed in a magnetic field. Evaluate this expression at the absolute zero of temperature.

11.2 Derive Eq. (11.88) for the chemical potential of a free electron gas at temperatures T very small compared with the Fermi temperature T_F.

(For $T \ll T_F$ and any function $\phi(\varepsilon)$

$$\int_0^\infty \frac{\phi(\varepsilon)d\varepsilon}{\exp[(\varepsilon - \mu)/kT] + 1} = \int_0^\mu \phi(\varepsilon)d\varepsilon + \frac{\pi^2}{6}(kT)^2 \frac{d\phi(\mu)}{d\mu} . \qquad (11.177)$$

See Landau and Lifshitz,[7] pp. 169–170 for the derivation of this result.)

11.3 Derive Eq. (11.95) for the heat capacity of a free electron gas at low temperatures, i.e. $T \ll T_F$. (Use Eq. (11.177).)

11.4 The molar heat capacity at constant volume of a perfect gas of bosons of mass m at a temperature T below the condensation temperature T_c is given by

$$C_V = 1.93 \, R \left(\frac{T}{T_c}\right)^{3/2} , \qquad T < T_c .$$

For this gas at a temperature $T < T_c$ obtain: (i) the internal energy per mole, (ii) the entropy per mole, (iii) the pressure.

11.5 Obtain the pressure of a free electron gas, of density N/V electrons per unit volume, at the absolute zero of temperature.

11.6 In order to escape from a metal the conduction electrons, whose motion within the metal is described by the free electron model, must overcome a potential barrier of height χ at the surface of the metal, i.e. an electron will escape only if its component of velocity normal to the surface exceeds $(2\chi/m)^{1/2}$. (m = mass of the electron.)

What is the current density of electrons emitted from the surface of the metal when the temperature T is such that $\chi - \mu \gg kT$? (μ = chemical potential of the electron gas.)

11.7 Consider the equilibrium of an isolated single-phase system in which the chemical reaction

$$\nu_1 A_1 + \nu_2 A_2 + \ldots + \nu_s A_s = 0 \qquad (11.138)$$

occurs. Obtain the equilibrium condition for this system by maximizing its entropy.

11.8 Obtain the equilibrium constant $K_c(T)$ for the ionization of atomic hydrogen which occurs at very high temperatures:

$$H \rightleftharpoons e^- + P \ .$$

Assume that the electrons, protons and hydrogen atoms can be treated as perfect classical gases.

11.9 At very high temperatures, molecular hydrogen dissociates into hydrogen atoms:

$$H_2 \rightleftharpoons H + H \ .$$

Obtain the equilibrium constant $K_c(T)$ for this reaction, expressing your answer in terms of the internal partition function of molecular hydrogen.

11.10 Consider the dissociative gaseous reaction

$$N_2O_4 \rightleftharpoons 2NO_2 \ .$$

Let f be the fraction of N_2O_4 molecules dissociated into NO_2 molecules, when the system is in equilibrium. Derive the pressure dependence of f, at a fixed temperature.

11.11 A system of s components, $A_1, A_2, \ldots A_s$, consists of p phases in contact, so that components can pass between them. The Gibbs free energy of the system is given by

$$G = \sum_{\alpha=1}^{p} G^{(\alpha)}(T, P, N_1^{(\alpha)}, \ldots N_s^{(\alpha)}) \qquad (11.178)$$

where the index $\alpha(=1, 2, \ldots .p)$ labels the p phases of the system, $N_i^{(\alpha)}$ is the number of molecules of component i in the αth phase, and $G^{(\alpha)}$ is the Gibbs free energy of this phase. By considering a change of the system at constant temperature and pressure in which dN molecules of component 1 pass from phase 1 to phase 2, all other particle numbers $N_i^{(\alpha)}$ remaining the same, show that in equilibrium the chemical potential of component 1 has the same value in phases 1 and 2. Hence write down the equilibrium conditions for the passage of any component between different phases.

11.12 A semiconductor possesses n donor levels, whose energy is $-\varepsilon_0$. A donor level can be occupied either by a 'spin up' electron or a 'spin down' electron but it cannot be simultaneously occupied by two electrons. Obtain the grand partition function for the electrons in the donor levels and hence find the number of electrons in the donor levels.

11.13 For a system of s components, existing in a single phase, the Gibbs free energy $G(T, P, N_1, \ldots .N_s)$ has the property that

$$G = \sum_{i=1}^{s} \mu_i N_i \qquad (11.179)$$

where μ_i is the chemical potential of the ith component of the system. Derive this relation, which is the generalization of Eq. (11.118) to a multi-component phase. Hence show that

$$\sum_{i=1}^{s} N_i d\mu_i = -SdT + VdP \ . \qquad (11.180)$$

11.14 Consider the chemical reaction (11.138) between perfect gases, the temperature and the volume of the system being kept constant. Show that the molar heat of reaction at constant volume for this process is given by

$$Q_V = RT^2 \frac{\mathrm{d}}{\mathrm{d}T} \ln K_c(T)$$ (11.181)

where $K_c(T)$ is the equilibrium constant (11.156).

11.15 Consider a given region of volume v inside a large volume V of a perfect classical gas at temperature T. Show that the probability of finding N particles in this region of the gas is given by the Poisson distribution

$$p_N = \frac{\bar{N}^N e^{-\bar{N}}}{N!}$$

where $\bar{N}$ is the mean number of particles in the volume v.

APPENDIX

Mathematical results

A.1 STIRLING'S FORMULA

For large values of N, $N!$ can be approximated by Stirling's formula*

$$\ln N! = N \ln N - N + \tfrac{1}{2} \ln N + \tfrac{1}{2} \ln (2\pi) + O\left(\frac{1}{N}\right) \qquad (A.1)$$

where $O(1/N)$ means that a term of order $1/N$ has been neglected. For $N \gg \ln N$, which is the usual situation in statistical physics, the simpler formula

$$\ln N! = N \ln N - N = N \ln \left(\frac{N}{e}\right) \qquad (A.2)$$

is adequate. The origin of this formula is easily seen by writing

$$\ln N! = \sum_{p=1}^{N} \ln p \qquad (A.3)$$

and approximating this sum by the integral (see Fig. A.1, where the area of the rectangles drawn is just the sum (A.3))

$$\int_{1}^{N} \ln p \, dp = N \ln N - N + 1 \ . \qquad (A.4)$$

*For a derivation see, for example, M. L. Boas, *Mathematical Methods in the Physical Sciences*, 2nd edn., Wiley, New York, 1983, pp. 472–473.

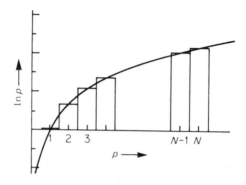

Fig. A.1. Approximation of $\ln(N!)$ by the area under the
curve of $\ln p$.

This agrees with Eq. (A.2) for $N \gg 1$.

Eq. (A.1) is in fact very accurate even for quite small N. Thus for $N = 5$, the value given by Eq. (A.1) and the exact value of ln 5! are both 4.8, to two significant figures. Hence the error in using Eq. (A.2) is $\frac{1}{2}\ln(2\pi N)$ for $N \geqslant 5$. From Table A.1 it can be seen how this error rapidly becomes negligible.

Table A.1

N	$\ln N!$	$N\ln N - N$	$\frac{1}{2}\ln(2\pi N)$
5	4.8	3.0	1.8
25	58.0	55.5	2.5
100	363.7	360.5	3.2

A.2   EVALUATION OF $\int_0^\infty (e^x - 1)^{-1} x^3 \, dx$

The integral

$$I \equiv \int_0^\infty \frac{x^3 \, dx}{(e^x - 1)} \qquad (A.5)$$

can be evaluated by contour integration. We shall give a more elementary derivation, purely formally without rigorous justification.

Since $e^{-x} < 1$ for $0 < x < \infty$, we have

$$\frac{1}{e^x - 1} = \frac{e^{-x}}{1 - e^{-x}} = \sum_{n=1}^\infty e^{-nx} \qquad (A.6)$$

which, on substitution in Eq. (A.5), gives

$$I = \sum_{n=1}^{\infty} \int_0^\infty e^{-nx} x^3 dx = \sum_{n=1}^{\infty} \frac{1}{n^4} \int_0^\infty e^{-y} y^3 dy = 6 \sum_{n=1}^{\infty} \frac{1}{n^4} . \qquad (A.7)$$

The last infinite series is rapidly convergent and is easily evaluated numerically to give $I = 6 \times 1.082$.

To sum the series (A.7) in closed form we start from the Fourier series for the function

$$f(x) = \begin{cases} \dfrac{\pi}{4} , & 0 < x < \pi , \\[2mm] -\dfrac{\pi}{4} , & -\pi < x < 0 , \end{cases} \qquad (A.8)$$

shown in Fig. A.2. Since $f(x)$ is an odd function we expand it in the sine series

$$f(x) = \sum_{n=1}^{\infty} a_n \sin nx, \qquad (A.9)$$

and since

$$\frac{1}{\pi} \int_{-\pi}^{\pi} \sin mx \sin nx \, dx = \begin{cases} 1, & m = n \\ 0, & m \neq n \end{cases}$$

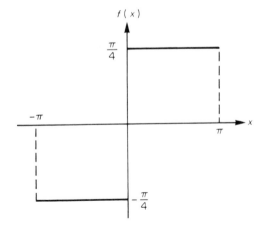

Fig. A.2. The function

$$f(x) = \begin{cases} \pi/4, & 0 < x < \pi \\ -\pi/4, & -\pi < x < 0 . \end{cases}$$

we find that

$$a_n = \begin{cases} 0, & n=2,4,6,\ldots \\ \dfrac{1}{n}, & n=1,3,5,\ldots \end{cases} \tag{A.10}$$

Combining Eqs. (A.8) to (A.10) we can write $f(x)$, for $0<x<\pi$, as

$$\frac{\pi}{4} = \sum_{\text{odd}} \frac{\sin nx}{n} \tag{A.11}$$

where $\sum_{\text{odd}}$ stands for summation over all odd integers only: $n=1,3,5,\ldots$.
We integrate Eq. (A.11) three times in succession from 0 to x obtaining

$$\frac{\pi x}{4} = \sum_{\text{odd}} \frac{1}{n^2}(1-\cos nx), \tag{A.12}$$

$$\frac{\pi x^2}{8} = \sum_{\text{odd}} \left[\frac{x}{n^2} - \frac{\sin nx}{n^3} \right]$$

and

$$\frac{\pi x^3}{24} = \sum_{\text{odd}} \left[\frac{x^2}{2n^2} - \frac{1}{n^4}(1-\cos nx) \right]. \tag{A.13}$$

If we put $x=\pi/2$ in Eqs. (A.12) and (A.13), we obtain

$$\sum_{\text{odd}} \frac{1}{n^2} = \frac{\pi^2}{8} \tag{A.14}$$

and

$$\sum_{\text{odd}} \frac{1}{n^4} = \frac{\pi^2}{8} \sum_{\text{odd}} \frac{1}{n^2} - \frac{\pi^4}{192} = \frac{\pi^4}{96}, \tag{A.15}$$

where the last step follows on using Eq. (A.14). Now

$$\sum_{\text{even}} \frac{1}{n^4} = \frac{1}{16} \sum_{n=1}^{\infty} \frac{1}{n^4}, \tag{A.16}$$

where $\sum_{\text{even}}$ denotes a sum over all even integers: $n=2,4,6,\ldots$, so that

$$\sum_{n=1}^{\infty} \frac{1}{n^4} = \left\{ \sum_{\text{even}} + \sum_{\text{odd}} \right\} \frac{1}{n^4} = \frac{1}{16} \sum_{n=1}^{\infty} \frac{1}{n^4} + \frac{\pi^4}{96}; \tag{A.17}$$

whence

$$\sum_{n=1}^{\infty} \frac{1}{n^4} = \frac{\pi^4}{90}. \tag{A.18}$$

From Eqs. (A.5), (A.7) and (A.18) we finally obtain

$$I \equiv \int_0^\infty \frac{x^3 \, dx}{(e^x - 1)} = \frac{\pi^4}{15} \ . \tag{A.19}$$

The integral

$$I' \equiv \int_0^\infty \frac{x^4 e^x \, dx}{(e^x - 1)^2} \tag{A.20}$$

is related to I, Eq. (A.19), by integration by parts. Since

$$d\left(\frac{1}{e^x - 1}\right) = -\frac{e^x}{(e^x - 1)^2} \ ,$$

it follows that

$$I' = -\int_0^\infty x^4 d\left(\frac{1}{e^x - 1}\right)$$

$$= 4\int_0^\infty \frac{x^3 \, dx}{e^x - 1} - \left[\frac{x^4}{e^x - 1}\right]_0^\infty = 4I \ , \tag{A.21}$$

so that, from Eqs. (A.19) to (A.21),

$$I' \equiv \int_0^\infty \frac{x^4 e^x \, dx}{(e^x - 1)^2} = \frac{4}{15} \pi^4 \ . \tag{A.22}$$

A.3   SOME KINETIC THEORY INTEGRALS

The integrals

$$I_n(a) \equiv \int_0^\infty dx \, x^n e^{-ax^2} \ , \qquad (a > 0) \ , \tag{A.23}$$

are found by calculating I_0 and I_1 directly, and differentiating these successively with respect to a.
  (i)

$$I_0(a) = \frac{1}{\sqrt{a}} \int_0^\infty du \, e^{-u^2} = \frac{1}{2\sqrt{a}} \int_{-\infty}^\infty du \, e^{-u^2} \ . \tag{A.24}$$

Now

$$\left(\int_{-\infty}^{\infty} du\, e^{-u^2} \right)^2 = \int_{-\infty}^{\infty} dx\, e^{-x^2} \int_{-\infty}^{\infty} dy\, e^{-y^2}$$

$$= \int_{-\infty}^{\infty} \int_{-\infty}^{\infty} e^{-(x^2+y^2)}\, dx\, dy \; , \qquad (A.25)$$

and changing this double integral over the whole of the (x, y) plane to polar coordinates, the element of area becomes $r\, dr\, d\theta$ and the double integral (A.25) becomes

$$2\pi \int_0^{\infty} e^{-r^2} r\, dr = \pi \int_0^{\infty} e^{-u}\, du = \pi \; . \qquad (A.26)$$

Hence from Eq. (A.24)

$$I_0(a) = \frac{1}{2} \sqrt{\frac{\pi}{a}} \; . \qquad (A.27)$$

(ii)

$$I_1(a) = \int_0^{\infty} dx\, x\, e^{-ax^2} = \frac{1}{2a} \int_0^{\infty} du\, e^{-u} = \frac{1}{2a} \; . \qquad (A.28)$$

(iii) Differentiation of Eq. (A.23) with respect to a gives

$$\frac{dI_n(a)}{da} = \int_0^{\infty} dx\, x^n e^{-ax^2}(-x^2) = -I_{n+2}(a) \; . \qquad (A.29)$$

Applying this recurrence relation repeatedly to Eqs. (A.27) and (A.28) leads to

$$I_m(a) = \frac{1.3 \ldots (m-1)}{(2a)^{m/2}} \frac{1}{2} \left(\frac{\pi}{a} \right)^{1/2} , \qquad m = 2, 4, 6, \ldots \qquad (A.30a)$$

$$I_m(a) = \frac{2.4 \ldots (m-1)}{(2a)^{(m+1)/2}} , \qquad m = 3, 5, 7, \ldots . \qquad (A.30b)$$

For example

$$\left. \begin{array}{cc} I_0(a) = \dfrac{1}{2} \left(\dfrac{\pi}{a} \right)^{1/2} & I_1(a) = \dfrac{1}{2a} \\[3mm] I_2(a) = \dfrac{1}{4a} \left(\dfrac{\pi}{a} \right)^{1/2} & I_3(a) = \dfrac{1}{2a^2} \\[3mm] I_4(a) = \dfrac{3}{8a^2} \left(\dfrac{\pi}{a} \right)^{1/2} & I_5(a) = \dfrac{1}{a^3} \end{array} \right\} \qquad (A.31)$$

and so on.

Many kinetic theory integrals are conveniently expressed in terms of the gamma function

$$\Gamma(x) = \int_0^\infty e^{-t} t^{x-1} \, dt, \qquad x > 0 \ . \tag{A.32}$$

Integrating by parts one easily proves that

$$\Gamma(x) = (x-1)\Gamma(x-1) \ . \tag{A.33}$$

It follows from Eq. (A.32) directly that

$$\Gamma(1) = 1, \qquad \Gamma(\tfrac{1}{2}) = \sqrt{\pi} \ . \tag{A.34}$$

Hence if x is a positive integer it reduces to the factorial function

$$\Gamma(x) = (x-1)! \tag{A.35}$$

By change of variables one derives

$$\int_0^\infty e^{-\beta t} t^{x-1} \, dt = \frac{\Gamma(x)}{\beta^x} \ , \qquad x > 0 \ . \tag{A.36}$$

APPENDIX

The density of states

The problem of this appendix occurs in this book in several contexts and is a common one in other parts of physics, for example in problems of collisions of particles: electrons with atoms, π mesons with nuclei, etc. We stated in Chapter 2 that we can always force the energy levels of a system to be a discrete set rather than a continuum by putting the system in a box and that, furthermore, if the box is large enough the surface effects introduced by the box cannot affect the physical properties of the system. This means that the discrete states of our system must lie very close together, they must lie very dense—it is from this picture that the term density of states originates. The discrete very densely lying states will for practical purposes be indistinguishable from a continuum of states. Having deliberately introduced discrete states to facilitate unambiguous counting of states—this, classical statistical mechanics could not achieve—we want, for actual manipulation, to represent sums over these states by integrals. The problem is almost identical with that of finding the normal modes, i.e. the harmonics, of a vibrating string. We shall want the normal modes of the three-dimensional wave equation. This is the problem to be solved in this appendix. In section B.1 we shall consider the problem in general; in sections B.2 to B.4 we shall apply the results to particular cases.

B.1 THE GENERAL CASE

We consider the three-dimensional wave equation

$$\nabla^2\phi(\mathbf{r}) + k^2\phi(\mathbf{r}) = 0 \qquad\qquad (B.1)$$

which describes standing or progressive waves,* depending on the boundary conditions. k is the magnitude of the three-dimensional wave vector $\mathbf{k}$ describing the waves. It is related to the wavelength λ by

$$k = \frac{2\pi}{\lambda} \cdot \qquad (B.2)$$

The basic equation relating phase velocity v, frequency ν and λ is

$$v = \lambda\nu \ ; \qquad (B.3)$$

in terms of the circular frequency

$$\omega = 2\pi\nu \qquad (B.4)$$

and k, it becomes

$$v = \frac{\omega}{k} \cdot \qquad (B.5)$$

For a dispersive medium the phase velocity v depends on the frequency of the wave.

We consider waves, described by the wave equation (B.1), inside an enclosure which is a cube of side L, as shown in Fig. B.1. We impose the boundary conditions

$$\phi(\mathbf{r}) = 0 \qquad \text{on all faces of the cube,} \qquad (B.6)$$

so that we shall obtain *standing waves*. This problem is the three-dimensional generalization of waves on a string with fixed end points, and the solutions obviously are

$$\left.\begin{array}{r} \phi_{n_1 n_2 n_3}(\mathbf{r}) = \text{const.} \sin\left(\frac{n_1 \pi}{L}x\right) \sin\left(\frac{n_2 \pi}{L}y\right) \sin\left(\frac{n_3 \pi}{L}z\right) \\ n_1, n_2, n_3 = 1, 2, \ldots \end{array}\right\} \qquad (B.7)$$

with

$$k^2 = \frac{\pi^2}{L^2} (n_1^2 + n_2^2 + n_3^2) \ . \qquad (B.8)$$

*For a general discussion of waves see F. G. Smith and J. H. Thomson, *Optics*, Wiley, London, 1971, or H. J. J. Braddick, *Vibrations, Waves and Diffraction*, McGraw-Hill, London, 1965, or F. S. Crawford, *Waves*, McGraw-Hill, New York, 1968.

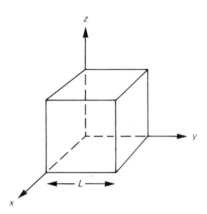

Fig. B.1. Cubic enclosure.

These solutions, and only these, vanish for $x=0$ and $x=L$, and similarly for the other axes. The integers n_1, n_2 and n_3 are restricted to non-zero positive integers. $n_i=0$ would give no solution at all: $\phi \equiv 0$; one or more n_i's negative would produce no *new* solutions but would reproduce one of the solutions (B.7), at most with a change of sign $(\sin(-x) = -\sin x)$. The magnitude k, given by Eq. (B.8), corresponds to a wave vector $\mathbf{k}$ with components

$$\mathbf{k} = \left(\frac{\pi}{L} n_1, \frac{\pi}{L} n_2, \frac{\pi}{L} n_3 \right), \qquad n_1, n_2, n_3 = 1, 2, \ldots . \qquad \text{(B.9)}$$

We can plot these wave vectors in a three-dimensional space with the components k_x, k_y, k_z as cartesian axes, the so-called $\mathbf{k}$-space, shown in Fig. B.2. We see that not all values of $\mathbf{k}$ are permissible (i.e. we can't have any vector in Fig. B.2). In general these won't satisfy the boundary conditions (B.6). There are only certain *allowed* values of $\mathbf{k}$, given by Eq. (B.9). In $\mathbf{k}$-space these form a cubic point lattice with the spacing between points being π/L. The volume per point in $\mathbf{k}$-space is thus $(\pi/L)^3$, i.e. the density of lattice points (allowed $\mathbf{k}$-vectors) in $\mathbf{k}$-space is $(L/\pi)^3$ per unit volume of $\mathbf{k}$-space. (Remember, from Eq. (B.2), that $\mathbf{k}$ has the dimension of (length)$^{-1}$. Hence a density of points in $\mathbf{k}$-space has the dimension (length)^3.)

We now want the number of such normal modes of standing waves with wave vectors whose magnitude lies in the interval k to $k+dk$. This number is equal to the number of lattice points in $\mathbf{k}$-space lying between two spherical shells (centred at the origin) of radii k and $k+dk$ in the positive octant (i.e. $k_x > 0, k_y > 0, k_z > 0$: this corresponds to $n_i > 0$). Fig. B.3 shows the spherical surface of radius k in $\mathbf{k}$-space. The volume of the region lying between the

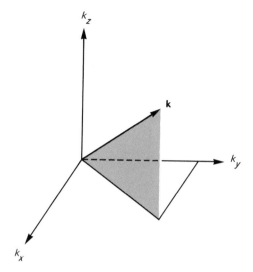

Fig. B.2. **k**-space.

radii k and $k+dk$ in the positive octant is $\frac{1}{8}4\pi k^2 dk$, so *the number of modes of standing waves with a wave vector whose magnitude lies in the range k to $k+dk$ is*

$$f(k)dk = \tfrac{1}{8}(4\pi k^2 dk)\Big/\left(\frac{\pi}{L}\right)^3 = \frac{Vk^2 dk}{2\pi^2} \ . \qquad (B.10)$$

Here we have introduced the volume $V=L^3$ of the cubic enclosure.

Eq. (B.10) is our basic result for the *density of states* from which everything else follows. An alternative way of deriving it, is to say that the number of allowed **k**-vectors of magnitude less than or equal to k is

$$\Gamma(k) = \frac{1}{8}\left(\frac{4\pi}{3}k^3\right)\Big/\left(\frac{\pi}{L}\right)^3 = \frac{Vk^3}{6\pi^2} \ . \qquad (B.11)$$

Hence differentiation of Eq. (B.11) with respect to k gives the number lying within dk, Eq. (B.10).

Eqs. (B.10) and (B.11) can be rewritten in various ways. In particular, the number of normal modes with frequencies in the range ω to $\omega + d\omega$ is, from Eq. (B.10),

$$f(\omega)d\omega = \frac{Vk^2}{2\pi^2}\frac{dk}{d\omega}\,d\omega. \qquad (B.12)$$

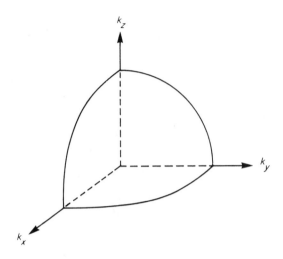

Fig. B.3. The positive octant of the sphere of radius k in **k**-space.

Note that f in Eqs. (B.10) and (B.12) is *not* the same *function*. It denotes in each case the distribution function of the quantity which occurs as the argument of f, i.e. $f(x)\,dx$ is the number of states in which a certain quantity x lies in the interval x to $x + dx$. In Eq. (B.10) x is the magnitude of the wave vector **k**; in Eq. (B.12) it is the circular frequency. Other cases of this convention occur below, e.g. Eqs. (B.15), (B.21) and (B.25).

The group velocity v_g of the waves, whose density of states is given by Eq. (B.12), is defined by

$$v_g = \frac{d\omega}{dk} \tag{B.13}$$

so that, on account of Eq. (B.5), we get

$$f(\omega)\,d\omega = \frac{V\omega^2\,d\omega}{2\pi^2 v^2 v_g} . \tag{B.14}$$

For a non-dispersive medium $v_g = v$ (we shall only be concerned with this case) and Eq. (B.14) becomes

$$f(\omega)\,d\omega = \frac{V\omega^2\,d\omega}{2\pi^2 v^3} . \tag{B.15}$$

There are three comments relevant to our main result (B.10).

Firstly, a certain error has been made in dividing the volume of the spherical shell by that of a cube of side π/L. The lattice points are not arranged in just this neat way *along* the spherical surface. It is easy to show that this error is of the order of $1/Lk$.

The simplest way to see this is to consider the error in Eq. (B.11). This error is of the order of the number of points lying in the surface area of the octant, which is $\frac{1}{8}(4\pi k^2)/(\pi/L)^2$ i.e. we should replace Eq. (B.11) by

$$\Gamma(k) = \frac{Vk^3}{6\pi^2} + O\left(\frac{L^2k^2}{2\pi}\right)$$

$$= \frac{Vk^3}{6\pi^2}\left[1 + O\left(\frac{1}{Lk}\right)\right] . \tag{B.16}$$

For the error to be negligible we thus require

$$Lk \gg 1 \tag{B.17a}$$

which we can also write

$$\lambda/L \ll 1 . \tag{B.17b}$$

(If the enclosure is not a cube, L is replaced by $V^{1/3}$ in Eqs. (B.16) and (B.17).) This condition is practically always very well satisfied, as the reader should verify by considering some particular cases (visible light in an enclosure; an ordinary gas: say molecular hydrogen at all but the most extremely lowest temperatures where a lot of other things would have gone wrong).

Secondly our derivation of the density of states (B.10) was for an enclosure the shape of a cube. One can quite generally show that this result depends *only on the volume* of the enclosure and is *independent of its shape*. This should not surprise the reader. The frequency ranges of musical instruments are largely, though not entirely, determined by their sizes.*

Thirdly, it is possible to derive similar results for more general differential equations, for example for

$$\nabla^2\phi(\mathbf{r}) + [k^2 - U(\mathbf{r})]\phi(\mathbf{r}) = 0,$$

which is the Schrödinger equation of a particle *in a potential* whereas Eq. (B.1), which we treated, represents a *free* particle.

---

*For the proofs of these results, which involve some very beautiful mathematics, we refer the interested reader to R. Courant and D. Hilbert, *Methods of Mathematical Physics*, Interscience, New York, 1953, Vol. 1, Chapter 6.

The result (B.10) is also largely independent of the detailed form of the boundary conditions imposed at the surface of the enclosure.

We consider an alternative boundary condition to Eq. (B.6), which is frequently of use, particularly in collision problems. (We shall further comment on its significance below.) For a cubic enclosure of side L we now impose *periodic* boundary conditions

$$\left. \begin{array}{l} \phi(0,y,z)=\phi(L,y,z) \\ \phi(x,0,z)=\phi(x,L,z) \\ \phi(x,y,0)=\phi(x,y,L) \end{array} \right\} \quad . \tag{B.18}$$

Eq. (B.1) now has the solutions

$$\phi_{n_1 n_2 n_3}(\mathbf{r}) = \text{const.} \exp(i\mathbf{k}\cdot\mathbf{r}) \tag{B.19}$$

with the wave vectors $\mathbf{k}$ restricted to one of the values

$$\mathbf{k}=\left(\frac{2\pi}{L}n_1, \frac{2\pi}{L}n_2, \frac{2\pi}{L}n_3\right), \qquad n_1, n_2, n_3 = 0, \pm 1, \pm 2, \ldots \ . \tag{B.20}$$

Note that the n_i now can be positive or negative. Eq. (B.19) represents a plane progressive wave with wave vector $\mathbf{k}$, given by Eq. (B.20).

We calculate the density of states as before. Instead of the positive octant of a sphere in momentum space, we now require the whole sphere (n_1, n_2, n_3 can each be positive or negative) but the spacing of lattice points in $\mathbf{k}$-space is now $2\pi/L$. Hence we obtain, instead of Eq. (B.10),

$$f(k)\,dk = (4\pi k^2 \,dk)\Big/\left(\frac{2\pi}{L}\right)^3 = \frac{Vk^2\,dk}{2\pi^2}$$

which is *identical* with Eq. (B.10).

We can refine this last result. We can ask for the number of modes of travelling (i.e. progressive) waves with wave vector pointing into a three-dimensional volume element $d^3\mathbf{k}$ at the point $\mathbf{k}$ of $\mathbf{k}$-space, i.e. waves travelling in a given direction. This number of modes is given by

$$f(\mathbf{k})\,d\mathbf{k} = d^3\mathbf{k}\Big/\left(\frac{2\pi}{L}\right)^3 = \frac{V}{(2\pi)^3}\,d^3\mathbf{k} \ . \tag{B.21}$$

If for $d^3\mathbf{k}$ we choose the volume element in $\mathbf{k}$-space which lies between concentric spheres of radii k and $k+dk$, subtending a solid angle $d\Omega$ at the origin (Fig. B.4), then $d^3\mathbf{k} = k^2\,dk\,d\Omega$ and Eq. (B.21) becomes

$$f(\mathbf{k})k^2\,dk\,d\Omega = \frac{V}{(2\pi)^3}\,k^2\,dk\,d\Omega \ . \tag{B.22}$$

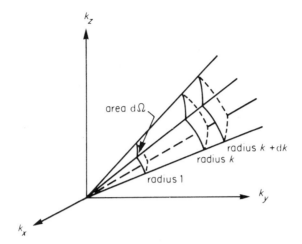

Fig. B.4. Volume element $d^3\mathbf{k} = k^2\,dk\,d\Omega$ in $\mathbf{k}$-space.

Integrating over all directions of $\mathbf{k}$, we get

$$f(k)\,dk = \frac{V}{(2\pi)^3}\, 4\pi k^2\,dk = \frac{Vk^2\,dk}{2\pi^2}\;. \qquad (\text{B.23})$$

The last expression agrees with Eq. (B.10) as it should. But the basic result, as we now see, is Eq. (B.21). This is reflected in the middle expression (B.23) where the factor 4π is the solid angle of the complete sphere in $\mathbf{k}$-space.

The normal modes (B.19) represent plane progressive waves with wave vector $\mathbf{k}$. We see that as V increases the allowed wave vectors (B.20) lie closer and closer together (their spacing is $2\pi/L$). Their density (B.21) increases proportionally to V. For sufficiently large V (in practice all V) the allowed $\mathbf{k}$ fill a continuum: all wave vectors $\mathbf{k}$ are allowed.

B.2   THE SCHRÖDINGER EQUATION

The Schrödinger equation for a particle of mass m, not subject to any forces except that it is confined to a given volume V, is just Eq. (B.1), together with the boundary condition (B.6). The energy E and momentum $\mathbf{p}$ of the particle are related to its wave vector $\mathbf{k}$ by

$$E = \frac{1}{2m}\, p^2\;, \qquad \mathbf{p} = \hbar\mathbf{k}\;. \qquad (\text{B.24})$$

For a particle in an enclosure of volume V, the density of states with momentum $\mathbf{p}$ of magnitude in the range p to $p + \mathrm{d}p$, follows from Eqs. (B.10) and (B.24):

$$f(p)\,\mathrm{d}p = \frac{V4\pi p^2\,\mathrm{d}p}{h^3} \ , \tag{B.25}$$

where $h \equiv 2\pi\hbar$ is Planck's constant.

From Eq. (B.21) we similarly obtain

$$f(\mathbf{p})\,\mathrm{d}\mathbf{p} = \frac{1}{h^3}\,V\mathrm{d}^3\mathbf{p} \ . \tag{B.26}$$

This is very similar to Eq. (7.28). Indeed since the density of states is proportional to the volume, one can infer the latter equation from Eq. (B.26).

B.3 ELECTROMAGNETIC WAVES

The electric field $\mathscr{E}(\mathbf{r}, t)$ of an electromagnetic wave in vacuum satisfies the wave equation

$$\frac{1}{c^2}\frac{\partial^2}{\partial t^2}\,\mathscr{E}(\mathbf{r}, t) = \nabla^2\mathscr{E}(\mathbf{r}, t) \ . \tag{B.27}$$

c is the velocity of light in vaccum which is constant and independent of frequency, i.e. there is no dispersion. With

$$\mathscr{E}(\mathbf{r}, t) = \mathscr{E}(\mathbf{r})\mathrm{e}^{-i\omega t} \ , \tag{B.28}$$

Eq. (B.27) reduces to

$$(\nabla^2 + k^2)\mathscr{E}(\mathbf{r}) = 0 \tag{B.29}$$

where, from (B.5),

$$\omega = kc \ . \tag{B.30}$$

Although Eq. (B.29) is now an equation for a vector field it possesses only *two* independent solutions for a given wave vector $\mathbf{k}$ and *not three*. This is a consequence of the Maxwell equation

$$\mathrm{div}\,\mathscr{E}(\mathbf{r}) = 0 \tag{B.31}$$

which $\mathscr{E}(\mathbf{r})$ must satisfy in free space. If for simplicity we use periodic boundary conditions (B.18), the real electric field of the normal modes can be represented by

$$\mathscr{E}(\mathbf{r}) = \mathscr{E}_0 \cos(\mathbf{k} \cdot \mathbf{r}) \qquad (B.32)$$

where $\mathbf{k}$ is one of the allowed vectors (B.20) and $\mathscr{E}_0$ is a constant vector. Condition (B.31) at once reduces to

$$\mathbf{k} \cdot \mathscr{E}(\mathbf{r}) = 0 \;, \qquad (B.33)$$

i.e. the electric field is perpendicular to the direction of propagation defined by $\mathbf{k}$; it is transverse. Hence for each $\mathbf{k}$, we can choose only two mutually perpendicular directions of polarization, both normal to $\mathbf{k}$. From Eq. (B.15) we then obtain for the number of normal modes of electromagnetic waves in a cavity of volume V

$$f(\omega)\,d\omega = \frac{V\omega^2\,d\omega}{\pi^2 c^3} \;. \qquad (B.34)$$

B.4   ELASTIC WAVES IN A CONTINUOUS SOLID

The propagation of elastic waves in crystalline solids, allowing for their atomic structure, is a hard problem. In the long-wavelength limit (i.e. wavelength large compared to the interatomic spacing) one can describe the solid as a homogeneous continuous elastic medium. Elastic waves are then propagated with a constant frequency-independent velocity, the velocity of sound, which is determined by the elastic stiffness constants of the medium. (See the end of this appendix for a demonstration of this.) The velocities will in general depend on the direction of propagation of the waves. We shall assume the solid elastically isotropic. In that case the velocities of propagation are independent of the direction of propagation. But for each wave vector $\mathbf{k}$ there are still three independent modes: one can have longitudinal (compression) waves and transverse (shear) waves. In the longitudinal waves, the displacements occur in the direction of propagation. In the transverse modes, the displacements are in two mutually orthogonal directions perpendicular to the direction of propagation. The longitudinal and transverse waves will have different velocities v_L and v_T respectively. From Eq. (B.15) we therefore obtain, allowing for the three directions of polarization,

$$f(\omega)\,d\omega = V\frac{\omega^2\,d\omega}{2\pi^2}\left[\frac{1}{v_L^3} + \frac{2}{v_T^3}\right] \;. \qquad (B.35)$$

If the solid is not isotropic then one must in the first place consider waves

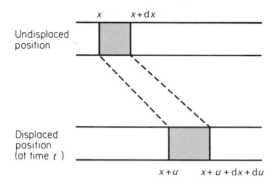

Fig. B.5. Longitudinal deformation of a homogeneous elastic line.

propagating in a given direction and average over all orientations. The result is again of the same form, with $1/v_L^3$ and $1/v_T^3$ being averages of these quantities over all directions of propagation in the solid.

Lastly, we want to show how the propagation velocity is related to the elastic stiffness constants. We shall consider the simplest case of purely longitudinal waves in a one-dimensional problem, a so-called homogeneous elastic line. Consider the displacement of an element of solid of length dx by $u \equiv u(x,t)$, Fig. B.5. The strain, i.e. the fractional change in length, is given by

$$e = \frac{\partial u}{\partial x} \, . \tag{B.36}$$

The elastic stiffness constant c is defined by

$$F = ec \tag{B.37}$$

where F is the force producing the deformation. (Do not confuse c with the velocity of light!) The net force acting on the element dx of the solid is

$$c \left\{ \frac{\partial}{\partial x} \left[u + \frac{\partial u}{\partial x} \, dx \right] - \frac{\partial u}{\partial x} \right\} = c \, \frac{\partial^2 u}{\partial x^2} \, dx \, . \tag{B.38}$$

This force must equal the mass of the element ρdx (where ρ is the mass per unit length) times the acceleration $\partial^2 u/\partial t^2$:

$$\rho \, \frac{\partial^2 u}{\partial t^2} = c \, \frac{\partial^2 u}{\partial x^2} \, . \tag{B.39}$$

The wave velocity resulting from this equation is

$$v = \left(\frac{c}{\rho}\right)^{1/2},$$

(B.40)

independent of frequency ω. For more general three-dimensional problems, the results are of the same form, with ρ the density, and c the stiffness constant appropriate to the kind of wave considered.

APPENDIX

Magnetic systems*

In our treatment of paramagnetism in section 2.2 and in Chapter 3, some results were quoted or made plausible only. We remind the reader of the points involved. In discussing magnetic work in section 1.4, we obtained two alternative formulations of the theory, depending on whether the mutual field energy (resulting from the superposition of the applied magnetic field and the field due to the magnetized sample) is included as part of the system or not. In section 1.4 we stated that the exclusive formulation (where the mutual field energy is not counted as part of the system) is more appropriate in statistical mechanics. It is this exclusive formulation which we used in treating paramagnetism in Chapter 3 but we did not justify its use. We shall now do so and then fill some gaps left in our earlier arguments. In particular, we must derive the definition of the temperature T for a magnetic system which we gave in Eq. (3.20), namely

$$\frac{1}{T} = \left(\frac{\partial S}{\partial E} \right)_{\mathscr{B}, N} \quad . \tag{C.1}$$

We started the discussion of paramagnetism with the assertion that when an external magnetic field $\mathscr{B}$ is applied to a magnetic dipole, with permanent moment μ, the dipole acquires an additional energy $-\mu \cdot \mathscr{B}$. (We called this an interaction energy, being due to the interaction of the dipole with the applied field.) We shall now derive this result, and from the derivation

*This appendix presupposes a knowledge of section 1.4.

it will be clear that the energy $-\boldsymbol{\mu}\cdot\mathscr{B}$ *excludes the mutual field energy.*

We shall again consider a paramagnetic substance in a uniform applied field $\mathscr{B}$ and make the same simplifying assumptions as in Chapter 3.* In a paramagnetic substance, the magnetic dipole moments stem from the individual atoms (or molecules) of the substance. In a given quantum state $r(=1,2,\ldots)$, each atom possesses a definite dipole moment $\boldsymbol{\mu}_r$ which results from the orbital electron currents and the spin orientations of the electrons in the atom in the state r. If an atom in the state r is placed in an external magnetic field $\mathscr{B}$, the energy E_r of the atom will depend on the applied field: $E_r = E_r(\mathscr{B})$. (This dependence of the energy levels of an atom on the applied field $\mathscr{B}$ leads to the Zeeman effect.) If the applied field is changed, work is done on the atom, thereby changing its energy level. For an infinitesimal change this work is given by

$$\text{d} W = \text{d} E_r(\mathscr{B})\ . \tag{C.2}$$

In writing down Eq. (C.2), we are assuming that the magnetic field is being changed infinitely slowly, so that the atom remains in the quantum state r while the field is changing, and does not make a quantum jump to some other state. In other words, we are assuming that the magnetic field is changed quasistatically, and we are appealing to Ehrenfest's principle which we employed in a similar situation in section 4.1. In section 1.4 we defined the energy of a magnetic dipole in an applied field in two ways, depending on whether the mutual field energy is included or not. Eq. (C.2) holds for either definition, $\text{d} W$ being the work corresponding to the choice made. We now define $E_r(\mathscr{B})$ to be the energy of the atom in the state r in the applied field $\mathscr{B}$, *excluding the mutual field energy.* In this case we know from section 1.4, Eq. (1.45), that the work done on the atom in changing the field by $\text{d}\mathscr{B}$ is given by

$$\text{d} W = -\boldsymbol{\mu}_r\cdot\text{d}\mathscr{B}\ . \tag{C.3}$$

Combining Eqs. (C.2) and (C.3), we obtain

$$\text{d} E_r(\mathscr{B}) = -\boldsymbol{\mu}_r\cdot\text{d}\mathscr{B}\ . \tag{C.4}$$

If we are neglecting diamagnetic effects, the magnetic moment $\boldsymbol{\mu}_r$ of the atom in the state r is independent of the applied field $\mathscr{B}$. (This is precisely what one means by a permanent dipole moment.) Hence we can write Eq. (C.4)

---

*The results have wider validity. See the footnote on p. 25. I am indebted to Albert Hillel for the arguments I am using here.

$$dE_r(\mathscr{B}) = -d(\boldsymbol{\mu}_r \cdot \mathscr{B})$$

which, on integration, leads to

$$E_r(\mathscr{B}) = E_r(0) - \boldsymbol{\mu}_r \cdot \mathscr{B}. \tag{C.5}$$

Eq. (C.5) is our final result. $E_r(\mathscr{B})$ is the energy of the atom in the state r and situated in the applied magnetic field $\mathscr{B}$, *exclusive of the mutual field energy*. We see that the dependence of $E_r(\mathscr{B})$ on $\mathscr{B}$ is given by the interaction energy $-\boldsymbol{\mu}_r \cdot \mathscr{B}$. (The reader should not confuse this interaction energy with the mutual field energy of the dipole in the applied field. In fact, the latter is given by $+\boldsymbol{\mu}_r \cdot \mathscr{B}$, as we know from Eq. (1.44).)

In applying Eq. (C.5) to paramagnetism, we note that, except at very high temperatures, the electronic degrees of freedom of an atom are not excited. Hence only the electronic ground state of the atom contributes to paramagnetism, and the index $r(=1,2,\ldots)$ in Eq. (C.5) labels the different orientations of the magnetic moment $\boldsymbol{\mu}$ of the atom in its ground state relative to the direction of $\mathscr{B}$. For $\mathscr{B}=0$, the energy of the atom is, of course, independent of its orientation so that $E_r(0)$ is independent of r and $E_r(0)$ is a constant which can always be omitted. (This corresponds to a particular choice for the zero of the energy scale.) For a spin $\frac{1}{2}$ atomic state there are just two possible orientations of $\boldsymbol{\mu}$: parallel and antiparallel to $\mathscr{B}$, with interaction energy $-\mu\mathscr{B}$ and $+\mu\mathscr{B}$ respectively. This result formed the starting point for our discussion of paramagnetism in section 2.2 and in Chapter 3.

It is now easy to justify Eq. (C.1) for the temperature of a magnetic system. In introducing the absolute temperature in section 2.3, we considered two systems in thermal contact, with each system constrained so that it could do no work. For a fluid (or a solid subject to hydrostatic forces only) this constraint means that its volume is kept constant, and this led to the definition of temperature

$$\frac{1}{T} = \left(\frac{\partial S}{\partial E}\right)_{V,N}. \tag{2.9}$$

If for a magnetic system we do not count the mutual field energy as part of the system, then the magnetic work is given by

$$đW = -\mathscr{M} \cdot d\mathscr{B}, \tag{1.45}$$

and for no work to be done the applied field $\mathscr{B}$ must be kept constant. Thus for a magnetic system, the temperature definition (2.9) is replaced by Eq. (C.1).

We saw in the above discussion that the microscopic energy levels E_r of the system are functions of the applied field $\mathscr{B}$. In fact, the applied field $\mathscr{B}$ is an external parameter of the system, in the sense in which this term was used in sections 2.5 and 4.1. Changing $\mathscr{B}$ involves magnetic work; changes of the system in which $\mathscr{B}$ is kept constant involve no magnetic work; i.e. they correspond to heat transfer provided we can neglect other forms of work, such as work against hydrostatic pressures. Proceeding as in section 4.1—or indeed from the results earlier in this appendix—one derives the fundamental thermodynamic relation for magnetic systems

$$dE = TdS - \mathscr{M} \cdot d\mathscr{B} \tag{4.73}$$

where forms of work other than magnetic work have been omitted.

In the above analysis we have throughout excluded the mutual field energy from the system. We briefly comment on the alternative treatment where the mutual field energy is counted as a part of the system. The magnetic work is now given by

$$đW_1 = \mathscr{B} \cdot d\mathscr{M} \tag{1.46}$$

and changes in which no magnetic work is done correspond to keeping the magnetic moment $\mathscr{M}$ constant. This is a very awkward constraint. Experimentally, it requires any heat transfer to be accompanied by a compensating change in $\mathscr{B}$ such that $\mathscr{M}$ stays constant. In contrast, the applied field $\mathscr{B}$ is a splendid experimental parameter since it is easily controlled. Theoretically too, $\mathscr{B}$ is a much more convenient independent variable than $\mathscr{M}$. The microscopic energy levels of the system are not functions of $\mathscr{M}$. Thus the magnetic moment $\mathscr{M}$ is not an external parameter of the system; rather, it is a property of the sample in a given macrostate, and calculations are much more difficult if $\mathscr{M}$, and not $\mathscr{B}$, is kept constant. For these reasons, the exclusive formulation of the theory, in which the mutual field energy is not counted as a part of the system, is both more natural and simpler to use.

Hints for solving problems

PROBLEMS 1

1.1 From the first law

$$dE + P\,dV = \mathrm{d}Q = 0 \ .$$

For a perfect gas $dE = C_V dT$, $P = RT/V$ and $R = C_P - C_V$. On substituting these into the first law and integrating, the result follows.

1.2 In an adiabatic expansion into a vacuum the energy of the gas is constant. Hence its final temperature T_2 is given by

$$T_2 = T_1 + \frac{2}{3R}a\left(\frac{1}{V_2} - \frac{1}{V_1}\right) \ .$$

Since $V_2 > V_1$, there is a temperature drop. Work is done against the cohesive forces in the gas. This is the Joule effect. It is quite small; e.g. for nitrogen $a = 1.39$ (litre/mol)^2 atm. Doubling the volume of 22.4 litres of N_2 gives $T_2 - T_1 = -0.25$ K.

1.3 The work done *on* the gas

$$W = -\int_{V_1}^{V_2} P\,dV = \frac{1}{\gamma - 1}(P_2 V_2 - P_1 V_1) \ ,$$

where P_1 and P_2 are the initial and final pressures of the gas.

This result could also have been obtained directly from the energy of the gas since during the process $Q=0$. Hence from the first law

$$W = E_2 - E_1 = C_V(T_2 - T_1) \ ,$$

which reduces to the earlier expression for W.

1.4 From

$$dH = dE + P\,dV + V\,dP = \text{đ}Q + V\,dP$$

one obtains

$$C_P = \left(\frac{\text{đ}Q}{\partial T}\right)_P = \left(\frac{\partial H}{\partial T}\right)_P \ .$$

1.5 From Fig. D.1

$$W = -\int_1^2 P\,dV - \int_3^4 P\,dV = (P_2 - P_1)(V_2 - V_1) \ .$$

For the complete cycle $\Delta E = 0$, hence $Q = -W$.

1.6
$$W = -Q = R(T_2 - T_1)\ln\frac{V_2}{V_1} \ .$$

1.7
$$\Delta E = Q - P\Delta V = 3.75 \times 10^4 \ \text{J/mol.}$$

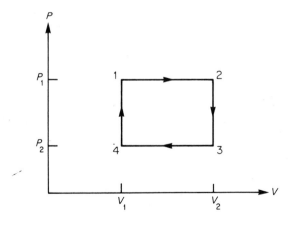

Fig. D.1.

PROBLEMS 2

2.1 In the macrostate considered we have:
(i) n out of N lattice sites vacant: there are

$$\binom{N}{n} = \frac{N!}{n!(N-n)!}$$

ways of choosing these sites;

(ii) n out of N interstitial sites are occupied: there are $\binom{N}{n}$ ways of choosing these occupied sites.

Hence the statistical weight of this macrostate is

$$\Omega(n) = \left[\binom{N}{n}\right]^2 ,$$

and its energy is $E = n\varepsilon$. For $n \gg 1$, $N \gg 1$, one proceeds as in section 2.4 to obtain

$$\frac{n}{N} = \exp\left(-\frac{\varepsilon}{2kT}\right) .$$

This equation gives the required temperature dependence of n/N. At $T = 0$ K, $n = 0$: all atoms at lattice sites: $E = 0$ (the ground state) and $\Omega = 1$, $S = 0$ (the state of complete order). For $kT \gg \varepsilon$, $n/N \approx \frac{1}{2}$: equal populations of lattice and interstitial sites, i.e. random distribution of N atoms over $2N$ sites. This leads to the entropy

$$k\ln\binom{2N}{N} = 2Nk\ln 2 ,$$

which is in agreement with $S = k\ln\Omega(\frac{1}{2}N)$.

This kind of imperfection of a crystal — an atom (or ion) moving from a lattice site to an interstitial site — is known as a Frenkel defect.

For $\varepsilon = 1$ eV (a typical value) and $T = 300$ K, one has $n/N \approx e^{-20} = 10^{-8.7}$.

2.2 The calculation is much simplified if we take the zero of the energy scale at $E_1 = 0$. With $\varepsilon = E_2 - E_1 = E_2$ and $\beta = 1/kT$ one obtains

$$\bar{E} = \frac{N\varepsilon}{e^{\beta\varepsilon} + 1}$$

and

$$C = \frac{\partial \bar{E}}{\partial T} = \frac{Nkz^2 e^z}{(e^z + 1)^2} , \qquad (z \equiv \beta \varepsilon) .$$

The heat capacity C as a function of the temperature possesses a peak which is typical of such a two-level system. This kind of behaviour of the heat capacity is known as a Schottky anomaly. An example of this occurs in Chapter 3, Fig. 3.5. (With $z = 2x$, the above equation for C becomes identical with Eq. (3.12).)

2.3 Take the zero of the energy scale at $E_1 = 0$. From the partition function one obtains the energy and hence the heat capacity

$$C = 2k \frac{x^2 e^x}{(e^x + 1)^2} , \qquad (x \equiv \beta \varepsilon) .$$

2.4 With $F = E - TS$, Eq. (2.37), $\partial F / \partial n = 0$ is easily calculated from section 2.4 which gives $E(n)$ and $S(n)$, and in fact dS/dn which is what we want here.

2.5 The population of the rth excited state is proportional to the Boltzmann factor $\exp(-\beta E_r)$. Hence

$$\frac{\text{population of first excited state}}{\text{population of ground state}} = e^{-\beta \hbar \omega} = 0.03 .$$

The higher vibrationally excited states are even less populated, as should now be obvious to the reader. This is the reason why vibrational excitation does not contribute to the specific heat of gases at ordinary temperatures. (We shall discuss this problem in Chapter 7.)

2.6 3.1×10^{-10}.

2.7 With the origin at the left-hand end of the chain, let n_+ and n_- be the numbers of links pointing in the positive x direction and doubling back respectively. Then

$$L = a(n_+ - n_-) \qquad n = n_+ + n_-$$

whence

$$n_\pm = (na \pm L) \frac{1}{2a} .$$

A macrostate is specified by L, which then determines n_+ and n_-. The statistical weight of this macrostate is

$$\Omega = \frac{n!}{n_+! n_-!} .$$

The entropy follows from $S = k \ln \Omega$ using Stirling's formula:

$$S = kn \left\{ \ln 2 - \frac{1}{2} \left(1 + \frac{L}{na} \right) \ln \left(1 + \frac{L}{na} \right) - \frac{1}{2} \left(1 - \frac{L}{na} \right) \ln \left(1 - \frac{L}{na} \right) \right\}.$$

Analogously to the pressure $P = T(\partial S/\partial V)_E$, we now have for the *tension* τ in the *one-dimensional* case $\tau = -T(\partial S/\partial L)_E$. In the present problem all configurations have the same energy, and S does not depend on E. Hence

$$\tau = \frac{kT}{2a} \left\{ \ln \left(1 + \frac{L}{na} \right) - \ln \left(1 - \frac{L}{na} \right) \right\}.$$

For $L \ll na$ this reduces to $\tau \approx (kT/na^2)L$.

2.8 We showed in section 2.5 that the probability distribution (2.32) is extremely sharply peaked for a macroscopic system (we showed that the relative fluctuation $\Delta E/\bar{E}$ is of order $1/\sqrt{N}$) so that the mean energy $\bar{E}$ of the system agrees with the energy $\tilde{E}$ at which the probability distribution has its maximum. This makes Eq. (2.39) plausible but we must see what has been neglected.

For this purpose let

$$\phi(E) \equiv f(E) e^{-\beta E}.$$

We shall expand $\ln \phi(E)$ in a Taylor series about $E = \tilde{E}$, the energy at which $\phi(E)$ has its maximum:

$$\ln \phi(E) = \ln \phi(\tilde{E}) + \frac{1}{2}(E - \tilde{E})^2 \left(\frac{\partial^2 \ln \phi(E)}{\partial E^2} \right)_{E = \tilde{E}}.$$

The condition $[\partial \ln \phi(E)/\partial E]_{E=\tilde{E}} = 0$ can be written

$$\frac{\partial \ln f(\tilde{E})}{\partial \tilde{E}} = \frac{1}{kT}. \tag{D.1}$$

This equation defines $\tilde{E}$ as a function of T (i.e. each heat bath temperature T determines an energy $\tilde{E}$ at which the Boltzmann distribution (2.32) has its maximum). Hence

$$\left(\frac{\partial^2 \ln \phi(E)}{\partial E^2} \right)_{E=\tilde{E}} = \frac{\partial^2 \ln f(\tilde{E})}{\partial \tilde{E}^2} = -\frac{1}{kT^2 C}$$

where $C \equiv \partial \tilde{E}/\partial T$. ($C$ is just the heat capacity of the macroscopic system.) With the Taylor expansion

$$\ln \phi(E) = \ln \phi(\tilde{E}) - \frac{1}{2kT^2 C}(E - \tilde{E})^2$$

the partition function (2.38) becomes

$$Z = \int \phi(E)\,dE$$

$$= \phi(\tilde{E}) \int dE \exp\left\{ -\frac{1}{2kT^2C}(E - \tilde{E})^2 \right\} .$$

In this integral we can extend the range of integration to $-\infty \leqslant E \leqslant \infty$ since the Gaussian exponential decreases *very* rapidly for E appreciably different from $\tilde{E}$. The value of the integral then follows from Eq. (A.27) of Appendix A, and we obtain

$$Z = \phi(\tilde{E})(2\pi kT^2 C)^{1/2}$$

and

$$\ln Z = \ln f(\tilde{E}) - \beta\tilde{E} + \tfrac{1}{2}\ln (2\pi kT^2 C) . \tag{D.2}$$

Now $\tilde{E}$ and $C = \partial\tilde{E}/\partial T$ are proportional to the number of particles N in the system. Similarly $\ln f(\tilde{E})$ is proportional to N, as follows for example from Eq. (D.1). (See also the remarks following Eq. (2.32).) Hence of the three terms on the right-hand side of Eq. (D.2) the first two terms are of order N, and the third is of order $\ln N$. For a macroscopic system $(N \sim 10^{23})$ this third term is completely negligible and Eq. (2.39) has been justified: we neglect a term of order $\ln N$ compared with terms of order N.

From $F = -kT \ln Z$ and Eq. (2.39) we obtain

$$F = \tilde{E} - T\{k\ln f(\tilde{E})\} .$$

Comparison with $F = E - TS$, Eq. (2.37), shows that $\tilde{E}$ is the energy of the system. This is as expected since the fluctuations are negligible. We must also interpret $k\ln f(\tilde{E})$ as the entropy of the system. Since the system has a well-defined energy this is consistent with Boltzmann's relation $S = k\ln\Omega$. In the present case $f(\tilde{E})$ is the number of microstates of the system, per unit energy interval, with energy near $\tilde{E}$, i.e. it is the statistical weight of the macrostate with energy $\tilde{E}$.

PROBLEMS 3

3.1

$$\mathcal{M} = N\mu \frac{2\sinh x}{1 + 2\cosh x} , \qquad x \equiv \mu\mathcal{B}/kT .$$

(i) $x \ll 1$: $\mathcal{M} = 2N\mu^2 \mathcal{B}/3kT$ (Curie's law).

(ii) $x \gg 1$: $\mathcal{M} = N\mu$ (complete alignment).

3.2 The result for the magnetic moment $\mathcal{M}$ is at once obtained by the method of section 3.1.

If $\mathcal{M} \ll \mu N$, the equation for $\mathcal{M}$ reduces to

$$\chi \equiv \frac{\mathcal{M}}{\mathcal{H}N} = \frac{\mu_0 \mu^2}{k(T-\theta)} \quad .$$

This is the Curie–Weiss law which describes the behaviour of a ferromagnetic material at high temperatures, i.e. above the Curie temperature θ. θ is typically of the order of 1000 K.

For $\mathcal{H} = 0$, the general result for $\mathcal{M}$ reduces to

$$\frac{\mathcal{M}}{\mu N} = \tanh\left(\frac{\theta}{T}\frac{\mathcal{M}}{\mu N}\right) \quad . \tag{D.3}$$

This transcendental equation is best solved graphically as the intersection of the curves $y = \tanh x$ and $y = (T/\theta)x$, as indicated in Fig. D.2(a). For a temperature T, such that $T/\theta < 1$, a solution exists at the point A. (The solution corresponding to the intersection at the origin can be shown to correspond to an unstable state.) Thus for any temperature $T < \theta$ we obtain a spontaneous magnetization. Again, this is typical of a ferromagnetic material. The temperature dependence of the spontaneous magnetic moment $\mathcal{M}$ below the Curie temperature θ is sketched in Fig. D.2(b).

3.4 $\mathcal{B} \sim kT/\mu = 14$ tesla ($= 140$ kG).

PROBLEMS 4

4.1 The entropy changes in the two cases are 68.4 J/K (one heat bath) and 35.6 J/K (two heat baths). Using two heat baths produces a smaller increase in entropy. By raising the temperature of the water by infinitesimal steps (i.e. using an infinite number of heat baths) one can achieve zero net entropy change, i.e. reversible heating of the water.

4.2 Final temperature $T_F = \frac{1}{2}(T_A + T_B)$. Hence the entropy change of the system is, from Eq. (4.32),

$$\Delta S = Nc_P \int_{T_A}^{T_F} \frac{dT}{T} + Nc_P \int_{T_B}^{T_F} \frac{dT}{T} \quad ,$$

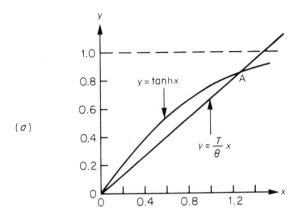

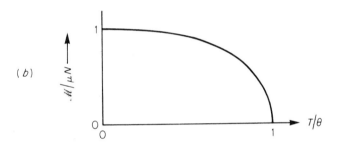

Fig. D.2. The Weiss model of spontaneous magnetization.
(a) Graphical solution of Eq. (D.3).
(b) Temperature dependence of the magnetic moment $\mathcal{M}$ below the Curie temperature θ.

where c_P is the heat capacity per atom at constant pressure. Hence

$$\Delta S = N c_P \ln \frac{(T_A + T_B)^2}{4 T_A T_B} \ .$$

(For a monatomic perfect gas $c_P = \frac{5}{2} k$.)
  $\Delta S \geqslant 0$ follows since

$$(T_A + T_B)^2 \geqslant 4 T_A T_B \ .$$

4.4 For a change for which initial and final states have the same temperature and pressure

$$\Delta G = \Delta E + P \Delta V - T \Delta S \ .$$

Similarly for such a change

$$\Delta H = \Delta E + P\Delta V \ .$$

Hence

$$\Delta G = \Delta H - T\Delta S = -1.05 \times 10^5 \text{ J/mol.}$$

4.5 We break the process up into 3 stages:

> liquid at 1 atm and 110°C
> →liquid at 1.4 atm and 110°C
> →vapour at 1.4 atm and 110°C
> →vapour at 1 atm and 110°C .

The corresponding entropy changes are $\Delta S_1 = 0$,

$$\Delta S_2 = 4 \times 10^4 / 383.15 = 104 \text{ J/K} \ ,$$

and
$$\Delta S_3 = \int \frac{P\,dV}{T} = R\ln 1.4 = 2.8 \text{ J/K} \ .$$

(In calculating ΔS_3 we used the fact that for a perfect gas $E = $ const. for an isothermal process.) For the complete process

$$\Delta S = 107 \text{ J/K} \ .$$

If during the third stage the vapour is not a perfect gas, then from Eq. (4.51) and the expression for $(\partial V/\partial T)_P$:

$$\Delta S_3 = -\int_{P_1}^{P_2} \left(\frac{\partial V}{\partial T}\right)_P dP = R\ln \frac{P_1}{P_2} + 0.46(P_1 - P_2)$$

($P_1 = 1.4$ atm, $P_2 = 1$ atm.) Hence the correction due to non-ideal gas behaviour is

$$0.46 \times 0.4 \text{ atm cm}^3/\text{K} = 0.019 \text{ J/K}$$

which is negligible.

4.6 Let L be the required heat of transformation. Using the third law (see section 4.8) and the data in the question we can write the molar entropy of α-phosphine at 49.43 K as

$$11.22 + L/30.29 + 20.10 = 34.03 \ ,$$

whence $L = 82.1$ J/mol. (See Adkins[1], pp. 248–249, for details of this interesting system.)

4.7 The maximum useful work is, from Eqs. (4.62) and (4.61), given by

$$(-W_u)_{\text{max}} = -(\Delta E + P_0\Delta V - T_0\Delta S) \ .$$

For the change considered: $\Delta V = 0$, $\Delta E = C_V (T_0 - T)$, and $\Delta S = C_V \ln (T_0/T)$. C_V is the molar heat capacity at constant volume. Hence

$$(-W_u)_{max} = C_V(T - T_0) + C_V T_0 \ln \frac{T_0}{T} .$$

4.8

$$(-W_u)_{max} = RT_0 \ln \frac{P_1}{P_2} + RT_0 P_0 \left(\frac{1}{P_1} - \frac{1}{P_2} \right) .$$

PROBLEMS 5

5.1 Consider the heat engine (in the question) and a Carnot engine coupled together, with the heat engine driving the Carnot engine. Both engines operate between the same two heat reservoirs at temperatures T_1 and T_2 ($< T_1$). Then

$$\eta' = 1 - Q_2'/Q_1' , \qquad \eta = 1 - Q_2/Q_1 , \qquad \eta' > \eta$$

(where primed quantities refer to the heat engine, unprimed to the Carnot engine). If we arrange that $Q_2 = Q_2'$ then $Q_1' > Q_1$. Hence the entropy change of the isolated system during one cycle is

$$\Delta S = \frac{1}{T_1} (Q_1 - Q_1') < 0 ,$$

which violates the second law.

Put more directly: during one cycle an amount of heat $Q_1' - Q_1$ is extracted from the hotter heat reservoir and is *completely* converted into work without any compensating change (rejection of heat to a colder body).

The reader should draw a diagram, analogous to Fig. 5.3, of the two engines connected together to see how the energy balance goes.

5.2 With the notation of Fig. 5.6 and section 5.2:

$$Q_1 = Nk\Theta_1 \ln \frac{V_B}{V_A} , \qquad Q_2 = Nk\Theta_2 \ln \frac{V_C}{V_D} .$$

Since AD and BC (Fig. 5.6) are adiabatics: $V_B/V_A = V_C/V_D$, whence the required result for the efficiency $\eta \equiv (Q_1 - Q_2)/Q_1$ follows.

5.3 With the notation of Fig. 5.8:

heat supplied during stage 3: $Q_1 = C_V(T_C - T_B)$
heat extracted during stage 5: $Q_2 = C_V(T_D - T_A)$.

Hence

$$\eta \equiv \frac{Q_1 - Q_2}{Q_1} = 1 - \frac{T_D - T_A}{T_C - T_B} ;$$

using the fact that curves AB and CD are adiabatics, Eq. (5.9) for the efficiency follows.

5.4 $(C_P - C_V)/3R = 0.074$.

5.6

$$\Delta S = \int_{P_1}^{P_2} \left(\frac{\partial S}{\partial P} \right)_H dP .$$

From

$$T dS = dE + P dV = dH - V dP$$

one has

$$\left(\frac{\partial S}{\partial P} \right)_H = - \frac{V}{T} ,$$

whence

$$\Delta S = R \ln \frac{P_1}{P_2} (>0) .$$

5.7 $0.44°C/atm$.

5.8 The initial pressure P_1, of the gas which maximizes the temperature drop is read off from Fig. 5.15. The pressure P_1 corresponds to the point of intersection of the horizontal straight line $T = T_1$ (the initial temperature) with the inversion curve.

This condition also follows from the mathematics. If we consider the temperature drop $\Delta T \equiv T_2 - T_1$ a function of the intitial pressure P_1, the condition for a maximum, i.e. $\partial(\Delta T)/\partial P_1 = 0$, applied to Eq. (5.45), gives at once

$$\left(\frac{\partial T}{\partial P} \right)_H \bigg|_{P = P_1} = 0 .$$

(Here we used

$$\frac{d}{da} \int_a^b f(x) dx = - f(a) .)$$

5.9 (i) From van der Waals' equation

$$dP = \frac{R dT}{V - b} - dV \left\{ \frac{RT}{(V - b)^2} - \frac{2a}{V^3} \right\} \qquad (D.4)$$

whence

$$\alpha \equiv \frac{1}{V}\left(\frac{\partial V}{\partial T}\right)_P = \left(1 - \frac{b}{V}\right)\left\{T - \frac{2a}{RV}\left(1 - \frac{b}{V}\right)^2\right\}^{-1} \tag{D.5}$$

(ii) From Eq. (5.15), van der Waals' equation for P and Eq. (D.4) for $(\partial P/\partial T)_V$

$$\left(\frac{\partial E}{\partial V}\right)_T = \frac{a}{V^2}\,, \tag{D.6}$$

i.e. for a real gas $(a \neq 0)$ the energy depends on the volume: because of the intermolecular potential it depends on the mean separation of the molecules.

(iii) From (D.6) directly $(\partial C_V/\partial V)_T = 0$.

(iv) From Eqs. (5.16a), (D.5) and (D.6):

$$C_P - C_V = \left(P + \frac{a}{V^2}\right)\alpha V = RT\left\{T - \frac{2a}{RV}\left(1 - \frac{b}{V}\right)^2\right\}^{-1} \,.$$

For $a = b = 0$ this reduces to $C_P - C_V = R$, as expected.

(v) From Eqs. (5.11), (D.6) and van der Waals' equation, as well as the definition of C_V,

$$dS = C_V\frac{dT}{T} + \frac{RdV}{V - b} \,.$$

(vi) From Eqs. (5.42c) and (D.5)

$$\left(\frac{\partial T}{\partial P}\right)_H = \frac{VT}{C_P}\left(\alpha - \frac{1}{T}\right)\,,$$

which vanishes for a perfect gas (from the definition of α).

(vii) From the last equation one has for the inversion curve $\alpha = 1/T$. Substituting in this equation for α from Eq. (D.5), one obtains as the equation of the inversion curve, as a function of V and T,

$$\frac{2a}{RT}\left(1 - \frac{b}{V}\right)^2 = b \,.$$

It is more convenient to express the inversion curve as a function of P and T. For this purpose we solve the last equation for V and substitute the resulting expression for V in van der Waals' equation:

$$P = -\frac{3RT}{2b} + \frac{2}{b}\left(\frac{2RTa}{b}\right)^{1/2} - \frac{a}{b^2} \,.$$

5.10 $0.9 \times 10^{-3}\,K.$

5.11 From the Maxwell relation in the question, we have

$$S(\mathcal{M},T) = S(\mathcal{M}=0,T) + \int_0^{\mathcal{M}} \left(\frac{\partial S}{\partial \mathcal{M}}\right)_T d\mathcal{M}$$

$$= S(\mathcal{M}=0,T) - \mu_0 \int_0^{\mathcal{M}} \left(\frac{\partial \mathcal{H}}{\partial T}\right)_{\mathcal{M}} d\mathcal{M} .$$

From

$$\left(\frac{\partial \mathcal{H}}{\partial T}\right)_{\mathcal{M}} \left(\frac{\partial T}{\partial \mathcal{M}}\right)_{\mathcal{H}} \left(\frac{\partial \mathcal{M}}{\partial \mathcal{H}}\right)_T = -1$$

it follows that

$$\left(\frac{\partial \mathcal{H}}{\partial T}\right)_{\mathcal{M}} = - \left(\frac{\partial \mathcal{M}}{\partial T}\right)_{\mathcal{H}} \Big/ \left(\frac{\partial \mathcal{M}}{\partial \mathcal{H}}\right)_T .$$

For $\mathcal{M} = f(\mathcal{H}/T)$ this gives

$$\left(\frac{\partial \mathcal{H}}{\partial T}\right)_{\mathcal{M}} = \frac{\mathcal{H}}{T} = \text{function of } \mathcal{M} \text{ only} \equiv \phi(\mathcal{M}).$$

Hence

$$S(\mathcal{M},T) = S(\mathcal{M}=0,T) - \mu_0 \int_0^{\mathcal{M}} \phi(\mathcal{M}) d\mathcal{M}, \qquad (D.7)$$

and from Eq. (D.7) we obtain the heat capacity at constant magnetization $\mathcal{M}$:

$$C_{\mathcal{M}}(T) = T\left(\frac{\partial S(\mathcal{M},T)}{\partial T}\right)_{\mathcal{M}} = \frac{d}{dT} S(\mathcal{M}=0,T) + 0 = C_{\mathcal{M}=0}(T).$$

PROBLEMS 6

6.1 For a particle of mass M performing simple harmonic motion with amplitude $\frac{1}{10}a$ and frequency ν, the energy is given by

$$E = \frac{1}{2}M(2\pi\nu)^2 \left(\frac{a}{10}\right)^2 .$$

Also from the theorem of equipartition of energy $E = kT_m$. Hence

$$\nu \propto \frac{1}{a} \left(\frac{T_m}{M}\right)^{1/2}$$

follows.

From the Einstein temperatures for Al and Pb $\nu(\text{Al})/\nu(\text{Pb}) = 3.6$. This is what one should expect since: (i) lead has a low melting point compared with aluminium; (ii) lead has a large atomic weight compared with aluminium; (iii) the Pb atoms being large, lead will have a larger lattice spacing than aluminium.

The data for lead and aluminium substituted in the above formula give $\nu(\text{Al})/\nu(\text{Pb}) = 4.3$.

6.2 From $F = E - TS$ and section 6.2

$$S = 3Nk \left\{ \frac{x}{e^x - 1} - \ln(1 - e^{-x}) \right\} \qquad x \equiv \frac{\Theta_E}{T} \equiv \frac{\hbar\omega_E}{kT} \; .$$

Low temperature limit: $x \gg 1$:

$$S = 3Nkx e^{-x} \; . \tag{D.8}$$

This follows also easily from

$$S(T) = \int_0^T C_V(T') \frac{dT'}{T'}$$

using the low temperature approximation (6.12) for C_V.

High temperature limit: $x \ll 1$:

$$S = 3Nk \ln T - 3Nk \ln \Theta_E + 3Nk. \tag{D.9}$$

6.3 We shall find S from $F = E - TS$. We calculate F by superposition of the Helmholtz free energy for one oscillator, Eq. (6.6), using the Debye density of states (6.23);

$$F = \frac{9}{8} Nk\omega_D + \frac{9NkT}{x_D^3} \int_0^{x_D} dt \, t^2 \ln(1 - e^{-t}) \; , \qquad x_D \equiv \frac{\Theta_D}{T} \equiv \frac{\hbar\omega_D}{kT} \; .$$

By integrating by parts we can rewrite this as

$$F = \frac{9}{8} Nk\omega_D + 3NkT \ln(1 - e^{-x_D}) - \frac{3NkT}{x_D^3} \int_0^{x_D} \frac{t^3 dt}{e^t - 1} \; .$$

Combining this equation with the Debye energy (6.24b):

$$S = \frac{12Nk}{x_D^3} \int_0^{x_D} \frac{t^3 dt}{e^t - 1} - 3Nk \ln(1 - e^{-x_D}) \; . \tag{D.10}$$

Low temperature limit: $x_D \gg 1$:

$$S = \frac{4}{5} \pi^4 Nk \left(\frac{T}{\Theta_D}\right)^3 . \tag{D.11}$$

High temperature limit: $x_D \ll 1$: In the integrand in Eq. (D.10) we expand $e^t = 1 + t + \ldots$, whence

$$S = 3Nk \ln T - 3Nk \ln \Theta_D + 4Nk . \tag{D.12}$$

Comparing these results with those of the last problem we see that the high temperature limits agree, except for the absolute value of the entropy. The low temperature limits show the same characteristic difference (T^3 or exponential dependence) as exists for the heat capacities. Eq. (D.12)—and the corresponding result for the Einstein theory—of course are consistent with Dulong and Petit's law.

6.4 This problem deals with a two-dimensional analogue of the Debye theory.

The distribution of normal modes of surface waves is found by the method of Appendix B applied to two dimensions. Analogously to Eq. (B.10) one obtains the number of modes of surface waves, for a surface of area A, with wave vector whose magnitude lies within the range k to $k + dk$:

$$f(k)dk = \frac{Ak\,dk}{2\pi} .$$

For the number of modes with frequencies within $(\nu, \nu + d\nu)$ one then obtains

$$f(\nu)d\nu = \frac{4\pi}{3}\left(\frac{\rho}{2\pi\sigma}\right)^{2/3} A\nu^{1/3} d\nu .$$

The surface energy $E(T)$ is given by

$$E(T) = \int_0^{\nu_0} f(\nu)d\nu \left\{\frac{1}{2}h\nu + \frac{h\nu}{\exp(\beta h\nu) - 1}\right\}$$

where ν_0 is a cut-off frequency determined by the number of degrees of freedom of the system. Hence

$$E(T) = E_0 + \frac{4\pi}{3}\left(\frac{\rho}{2\pi\sigma}\right)^{2/3} Ah \left(\frac{kT}{h}\right)^{7/3} I$$

where E_0 is the zero-point energy and

$$I \equiv \int_0^{x_0} \frac{x^{4/3}\, dx}{e^x - 1} \,,$$

with $x_0 \equiv h\nu_0/kT$.

As usual, low temperatures are defined by $x_0 \gg 1$. We may then replace the upper limit in the last integral by ∞, so that

$$E(T) = E_0 + 1.68 \times \frac{4\pi}{3} \left(\frac{\rho}{2\pi\sigma} \right)^{2/3} A \frac{(kT)^{7/3}}{h^{4/3}} \,.$$

To estimate the range of validity of this formula we require the cut-off frequency ν_0. With n atoms per unit area of surface and each atom having one degree of freedom,* we have

$$nA = \int_0^{\nu_0} f(\nu)\, d\nu = \pi A \left(\frac{\rho}{2\pi\sigma} \right)^{2/3} \nu_0^{4/3} \,,$$

and hence one obtains the cut-off temperature $\Theta_0 \equiv h\nu_0/k$. Taking

$$n = (0.145 \times \tfrac{1}{4} \times 6 \times 10^{23})^{2/3},$$

one obtains from the other data in the question $\Theta_0 = 11.7\,\mathrm{K}$, i.e. for $T \ll 12\,\mathrm{K}$ the above formula for the surface energy holds.

PROBLEMS 7

7.1

$$Z_{\mathrm{rot}} = \sum_{r=0}^{\infty} (2r+1)\exp\left\{ -\frac{\hbar^2}{2IkT}\, r(r+1) \right\} \,.$$

Low temperature limit: $kT \ll \hbar^2/2I$

$$\ln Z_{\mathrm{rot}} = \ln(1 + 3\,e^{-\hbar^2/IkT}) = 3\,e^{-\hbar^2/IkT}$$

$$C_{\mathrm{rot}} = 3R \left(\frac{\hbar^2}{IkT} \right)^2 e^{-\hbar^2/IkT} \,.$$

High temperature limit: $kT \gg \hbar^2/2I$.

---

*All one can safely say is that the number of degrees of freedom per unit area of surface is of order n. But it is very difficult to say what the number of degrees of freedom per surface atom is. This number depends on exactly how the surface atoms move. It will be about 1 or 2.

In this case the partition function can be written as an integral. With $x \equiv r(r+1)$

$$Z_{rot} = \int_0^\infty dx \exp \left\{ -\frac{\hbar^2}{2IkT} x \right\} = \frac{2IkT}{\hbar^2}$$

$$C_{rot} = R.$$

The energy levels ε_r are of the form of rotational energies, i.e.

$$\frac{(\text{angular momentum})^2}{2\,(\text{moment of inertia})}.$$

I is the moment of inertia of the molecule about an axis through the centre of mass and perpendicular to the line joining the two atoms. $\hbar\sqrt{[r(r+1)]}$ are the quantum-mechanically allowed values of the angular momentum of the molecule.

The high temperature limit is the classical result as given by the equipartition theorem. A linear molecule possesses only two rotational degrees of freedom, each contributes $\frac{1}{2}R$ to the molar heat capacity (see problems 7.9 and 7.10).

The distinction between high and low temperature regimes is determined by the moment of inertia I. Small molecules (i.e. light atoms, small internuclear separations) have a high 'rotational temperature' Θ_{rot} given by

$$k\Theta_{rot} \equiv \frac{\hbar^2}{2I}.$$

Deviations from classical behaviour are only observable for small light molecules. Some typical values for Θ_{rot} are

	H_2	HCl	HI	N_2	Cl_2	I_2
Θ_{rot} K	85.4	15.2	9.0	2.86	0.346	0.054

The above expressions for the partition function and the heat capacity need modifying in the case of molecules consisting of identical nuclei, called homonuclear molecules. At low temperatures quantum mechanical effects occur connected with the symmetries of wave functions describing identical particles (see, for example, Hill,[5] section 22.8, or Rushbrooke,[14] Chapter 7). At high temperatures, in the classical regime, the modification is simple. We must divide the partition function by a factor 2. This allows for the fact that a rotation of the molecule through

180 degrees about an axis through the centre of mass and perpendicular to the line joining the nuclei does not produce a new configuration of the molecule in the case of identical nuclei. This is quite analogous to the situation we encountered in section 7.1, Eq. (7.11), where division by $N!$ was necessary to allow for the identity of the particles. The rotational partition function of a homonuclear diatomic molecule therefore becomes

$$Z_{rot} = \frac{T}{2\Theta_{rot}}, \qquad T \gg \Theta_{rot}.$$

The high-temperature limit of the rotational heat capacity is not affected by this change.

7.2

$$Z_{vib} = \sum_{r=0}^{\infty} \exp\left\{-\beta\hbar\omega\left(r+\frac{1}{2}\right)\right\} = \frac{e^{-x/2}}{1-e^{-x}}, \qquad x \equiv \beta\hbar\omega.$$

$$C_{vib} = Rx^2 \frac{e^x}{(e^x-1)^2}.$$

Low and high temperature limits correspond to $x \gg 1$ and $x \ll 1$ respectively and give

$$C_{vib} = R\left(\frac{\Theta_{vib}}{T}\right)^2 \exp\left(-\frac{\Theta_{vib}}{T}\right), \qquad T \ll \Theta_{vib},$$

$$C_{vib} = R, \qquad T \gg \Theta_{vib},$$

where we introduced the vibrational temperature Θ_{vib}, defined by

$$k\Theta_{vib} \equiv \hbar\omega.$$

Readers of Chapter 6 will recognize the above theory as being just the Einstein theory of specific heats (section 6.2.1).

For diatomic molecules Θ_{vib} is usually of the order of a few thousand degrees absolute. So the vibrations do not contribute much to the heat capacity at room temperature. For the same gases which were quoted in the Hints to problem 7.1 one has

	H_2	HCl	HI	N_2	Cl_2	I_2
Θ_{vib} K	6210	4140	3200	3340	810	310

The result for $T \gg \Theta_{vib}$ is the classical equipartition result. The one vibrational degree of freedom contributes $\frac{1}{2}R$ from the kinetic energy and $\frac{1}{2}R$ from the potential energy to the heat capacity. (See problem 7.9.)

We are given that for N_2 $\hbar\omega = 0.3$ eV. Hence at 1000 K $x \equiv \beta\hbar\omega = 3.5$ and $e^x = 33.12$ so the low temperature approximation is reasonable and $C_{vib} = 0.37R = 3.1$ JK^{-1} mol^{-1}.

7.3 From Eq. (7.52)

$$S(\text{vapour}) = 190 \text{ J}K^{-1}\text{mol}^{-1} .$$

From

$$S(\text{liquid}) = S(\text{vapour}) - \frac{L}{T}$$

$$S(\text{liquid}) = 96 \text{ J}K^{-1}\text{mol}^{-1} .$$

This value of S(liquid) is in complete agreement with that found directly from the measured heat capacities of solid and liquid mercury between 0 K and 630 K and from the latent heat of fusion at the normal melting point. This agreement represents experimental verification of the entropy formula (7.51) for a monatomic gas, *and in particular of the constant term in that equation.*

7.4 $C_V = 3R$.

7.6 From Eq. (7.72) and Eqs. (A.33) and (A.36) in Appendix A

$$A' = \frac{\beta^{3N/2}}{\Gamma(3N/2)}$$

$$\bar{E} = A' \frac{\Gamma(3N/2 + 1)}{\beta^{3N/2 + 1}} = \frac{3}{2} NkT$$

$$\overline{E^2} = A' \frac{\Gamma(3N/2 + 2)}{\beta^{3N/2 + 2}} .$$

Eq. (7.75) follows from

$$\left(\frac{\Delta E}{\bar{E}}\right)^2 = \frac{\overline{E^2}}{\bar{E}^2} - 1 .$$

7.7

$$P = \frac{NkT}{V} + \left(\frac{N}{V}\right)^2 \frac{1}{2} \xi \left(\frac{\pi}{\alpha}\right)^{3/2} .$$

7.8 From $E = -\partial \ln Z / \partial\beta$ and $Z = Z_P Q_N$, Eq. (7.88), we obtain

$$E = \frac{3}{2} NkT - \frac{\partial \ln Q_N}{\partial\beta} . \tag{D.13}$$

From Eq. (7.101) for the configurational partition function Q_N and Eq. (7.98)

$$\ln Q_N = \frac{N^2}{2V} \int d^3\mathbf{r}(e^{-\beta u(r)} - 1)$$

and therefore, with $\rho = N/V$, up to terms of order ρ:

$$E = \tfrac{3}{2}NkT + \tfrac{1}{2}N\rho \int d^3\mathbf{r}u(r)e^{-\beta u(r)} . \qquad (D.14)$$

We have the *exact* relation

$$E = \tfrac{3}{2}NkT + E_{\text{pot}}$$

where E_{pot} is the potential energy of interaction between the molecules. Since $u(r)$ is the potential energy between two molecules distance r apart, we obtain for the potental energy of *one particular* molecule, due to the molecules surrounding it,

$$\int_0^\infty dr 4\pi r^2 \rho g(r, \rho, T)u(r) . \qquad (D.15)$$

For the gas as a whole we must sum over all molecules. Each molecule gives a contribution (D.15); for N molecules we must multiply expression (D.15) by $N/2$. The factor $\tfrac{1}{2}$ is necessary as without it the interaction between each pair of molecules would be counted twice. Hence

$$E = \tfrac{3}{2}NkT + \tfrac{1}{2}N\rho \int_0^\infty dr 4\pi r^2 g(r, \rho, T)u(r) . \qquad (D.16)$$

Compare Eqs. (D.14) and (D.16). The latter is exact, the former is only valid for $\rho \ll 1$ (we have neglected terms in ρ^2, ρ^3, . . .). Hence in the limit when ρ tends to zero the two equations agree, so that

$$\operatorname*{Lim}_{\rho \to 0} g(r, \rho, T) = e^{-\beta u(r)} . \qquad (D.17)$$

The form of this radial distribution function can be seen from Fig. 7.9, which shows

$$e^{-\beta u(r)} - 1 = g(r, 0, T) - 1 .$$

The single peak results from the attractive intermolecular forces which enhance the probability of finding a molecule near to another molecule.

The distribution functions in solids and liquids similarly show characteristic features. In a crystalline solid the radial distribution function has regularly spaced peaks, corresponding to the lattice structure of the crystal. In a liquid the atoms are packed fairly tightly and so are arranged in regular patterns over short distances, of the order of a few atomic diameters. Correspondingly, the radial distribution function for liquids shows a few peaks. (See Hill,[28] section 6.3.)

7.9 Go over to centre-of-mass and relative coordinates

$$\mathbf{R} = \frac{M_1 \mathbf{r}_1 + M_2 \mathbf{r}_2}{M_1 + M_2} \qquad \mathbf{r} = \mathbf{r}_2 - \mathbf{r}_1 \ .$$

With (θ, ϕ) being the polar angles of the vector $\mathbf{r}$ and

$$m = M_1 M_2 / (M_1 + M_2)$$

being the reduced mass one obtains for the energy of a molecule

$$\tfrac{1}{2}(M_1 + M_2)\dot{\mathbf{R}}^2 + [\tfrac{1}{2}m\dot{r}^2 + V(r)] + \tfrac{1}{2}mr^2(\dot{\theta}^2 + \dot{\phi}^2 \sin^2\theta) \ . \tag{D.18}$$

The first term corresponds to the translational centre-of-mass motion, the second and third terms — in square parentheses — to the relative vibrational motion of the atoms. For an attractive potential we can expand $V(r)$ into a Taylor series about the equilibrium distance of separation r_0:

$$V(r) = V(r_0) + \tfrac{1}{2}(r - r_0)^2 V''(r_0) + \ \ldots \ , \tag{D.19}$$

and for small amplitudes of oscillation we need only retain the terms up to the quadratic term, which results in simple harmonic vibrations. The last two terms in Eq. (D.18) are the kinetic energy of the *two* independent rotational motions of the *linear* molecule. mr^2 is the moment of inertia. Since r varies, due to the vibrations, the vibrational and rotational motions are coupled. But the vibrational amplitude is in general quite small, about 10 per cent of the equilibrium separation r_0. Hence one may replace the variable moment of inertia by mr_0^2. The vibrational and rotational motions are now independent of each other. We shall not require this approximation in the following discussion.

By comparing Eq. (D.18) with Eq. (7.119) we can at once write the former equation in terms of conjugate momenta. On account of Eq. (7.132) we obtain the Hamiltonian function

$$H = \frac{\mathbf{P}^2}{2M} + \frac{1}{2m} p_r^2 + \frac{1}{2} V''(r_0)(r - r_0)^2$$

$$+ \frac{1}{2mr^2} \left[p_\theta^2 + \frac{1}{\sin^2\theta} p_\phi^2 \right] + V(r_0)$$

where $M = M_1 + M_2$ and $\mathbf{P} = M\dot{\mathbf{R}}$. The Boltzmann distribution is given by

$$\text{const.} \, e^{-\beta H} \, d^3\mathbf{R} \, d^3\mathbf{P} \, dr \, d\theta \, d\phi \, dp_r \, dp_\theta \, dp_\phi \ .$$

The equipartition theorem now gives for the mean energy of a molecule, at temperature T,

$$E = \tfrac{1}{2}kT(3 + 2 + 2) \ ,$$

the three terms in the bracket coming from the three translational degrees of freedom, the one vibrational degree (with contributions from the kinetic and potential terms) and the two rotational degrees.

7.10 Denoting the number of translational, rotational and vibrational degrees of freedom of the molecules by f_t, f_r and f_v, we have $f_t = 3$, $f_r = 2$ or 3 according as to whether the molecule is linear or non-linear, and hence

$$f_v = 3n - (f_t + f_r) = \begin{cases} 3n - 5 & \text{linear molecule} \\ 3n - 6 & \text{nonlinear molecule.} \end{cases}$$

As regards the heat capacity of a gas of these molecules, in addition to the translational part $\tfrac{3}{2}R$, the rotational degrees of freedom are practically always fully excited. (See problem 7.1. For polyatomic molecules the moments of inertia are even larger and the rotational temperatures Θ_{rot} even lower.) The vibrational degrees of freedom result, at not too high temperatures where the harmonic approximation holds, in f_v normal modes, each performing simple harmonic vibrations with a definite frequency, independently of each other. The vibrational degrees of freedom are only rarely fully excited, and one must use the Einstein quantum theory to describe each mode (see problems 7.2 and 7.11).

7.11 In CO_2 the rotations are fully excited at 312 K. Hence the contribution to the heat capacity from translational and rotational motion is $2.50R$.

The vibrational contribution we obtain from Fig. 6.5. For the modes corresponding to $\Theta_{vib} = 954$, 1890 and 3360 K we obtain $0.49R$, $0.09R$ and ≈ 0. Hence $C_V = 3.57R$. The experimental value is $C_V = 3.53R$.

PROBLEMS 8

8.2 Since at constant pressure

$$đQ = dE + P\,dV = dH - V\,dP = dH$$

we can write the Clausius–Clapeyron equation (8.29) in terms of the enthalpy change $\Delta H = T\Delta S$:

$$\frac{dP}{dT} = \frac{\Delta H}{T\Delta V} \ .$$

From the data given, we obtain ΔV and ΔH and hence $\Delta T = -5.7°C$.

8.3 Eq. (8.36) applies also to the sublimation curve. Hence

$$L = 1.4 \times 10^5 \text{ J/mol.}$$

8.4 For a transition from phase 1 to phase 2 at constant temperature T and at constant pressure P: $\Delta S_{12} = \Delta H_{12}/T$. (Here $\Delta x_{12} \equiv x_2(T,P) - x_1(T,P)$ for the quantity x.) Hence ΔH_{12} is the latent heat for the phase change and the mass considered. Hence one obtains the molar latent heat of sublimation $= 2138$ J/mol.

The molar entropy change associated with the liquid–vapour transition at the triple point is $73.4 \text{ J K}^{-1} \text{mol}^{-1}$.

8.5 From the data given, the best one can do is to interpolate the molar latent heat linearly over the temperature range 273.16 to 298.15 K. Integration of Eq. (8.36) then gives the vapour pressure of water at 273.16 K as 4.59 mmHg. The measured value is 4.58 mmHg.

8.6 From the Clausius–Clapeyron equation, using the third law of thermodynamics and the fact that the solid and liquid have different densities, one obtains

$$dP/dT = 0 , \qquad \text{at } T = 0 ,$$

i.e. a horizontal slope.

8.7 If S_{vap} and S_{sol} are the molar entropies of the vapour and solid, and L is the molar latent heat of sublimation then

$$S_{\text{vap}} - S_{\text{sol}} = \frac{L}{T} .$$

The entropy of the vapour, treated as a perfect monatomic gas, is from Eq. (7.51)

$$S_{\text{vap}} = R \left\{ \frac{5}{2} \ln T - \ln P + \ln \frac{(2\pi m)^{3/2} (ek)^{5/2}}{h^3} \right\} .$$

The entropy of the solid for the Einstein theory was considered in problem 6.2. For 'high' temperatures ($T \gg \Theta_E$) it is given by Eq. (D.9):

$$S_{\text{sol}} = R(3\ln T - 3\ln \Theta_E + 3) , \qquad T \gg \Theta_E .$$

Combining these two equations we obtain the final result for the vapour pressure

$$P = \frac{(2\pi m)^{3/2} k^{5/2}}{h^3 e^{1/2}} \Theta_E^{3} \frac{1}{\sqrt{T}} e^{-L/RT} , \qquad T \gg \Theta_E .$$

This equation contains no arbitrary constant of proportionality in contrast to vapour pressure equations obtained by integration of the Clausius–Clapeyron equation (see section 8.5.3, in particular Eqs. (8.36) and (8.37), and problem 8.3; see also the discussion of Eq. (7.54)). Nor did we have to make any assumptions about the temperature dependence of the latent heat, as is necessary when integrating the Clausius–Clapeyron equation (see section 8.5.3 and also problem 8.5). This is reflected in the additional $T^{-1/2}$ dependence of the above result compared with Eq. (8.37).

If more realistically we had described the solid by the Debye theory the only difference would have been to replace the factor $e^{1/2}$ in the denominator of the above equation for P by $e^{3/2}$ and, of course, Θ_E by the Debye temperature Θ_D. (See problem 6.3 and Eq. (D.12).) From the exact expressions for the entropy of a solid derived in problems 6.2 and 6.3, one can obtain vapour pressure formulas valid at all temperatures. (Of course the Einstein model is not very realistic.)

8.8

$$V_c = 2b \ , \qquad T_c = \frac{a}{4Rb} \ , \qquad P_c = \frac{a}{4b^2} \, e^{-2} \ .$$

PROBLEMS 9

9.1 The number of spin $\frac{1}{2}$ particles in the composite particle determines the statistics. Thus ^{12}C obeys BE statistics, etc.

PROBLEMS 10

10.1 $T\lambda_{max} = 0.290 \text{ K cm}, \qquad T = 6000 \text{ K}.$

10.2 For $\lambda = 2 \text{ cm}$ and $T = 6000 \text{ K}$, $\hbar\omega \ll kT$. Hence the simpler long-wavelength form of Planck's law, Eq. (10.21), can be used to calculate the power P. $P = 8.0 \times 10^9 \text{ W}$.

10.3

$$\frac{d}{dU}\left(\frac{1}{T}\right) = -\frac{k}{U^2} \ , \qquad \hbar\omega \ll kT \ ,$$

$$\frac{d}{dU}\left(\frac{1}{T}\right) = \frac{-k}{\hbar\omega U} \ , \qquad \hbar\omega \gg kT \ ;$$

take as interpolation formula

$$\frac{d}{dU}\left(\frac{1}{T}\right) = \frac{-k}{\hbar\omega U + U^2} \ .$$

On the integration one obtains

$$\frac{1}{T} = \frac{k}{\hbar\omega} \ln \frac{1 + U/\hbar\omega}{U} + \text{const.}$$

and the constant of integration must be chosen so that $U \to \infty$ as $T \to \infty$, i.e. const. $= (k/\hbar\omega)\ln(\hbar\omega)$. The last equation then becomes

$$U = \frac{\hbar\omega}{\exp(\hbar\omega/kT) - 1} .$$

This is Planck's law, and Planck originally derived it in this way.

PROBLEMS 11

11.1 In the presence of a magnetic field $\mathscr{B}$ an electron with momentum p will have energy

$$\varepsilon \mp \mu_B \mathscr{B} = \frac{p^2}{2m} \mp \mu_B \mathscr{B}$$

for magnetic moment parallel and antiparallel to $\mathscr{B}$ respectively. The total number of electrons with magnetic moment parallel or antiparallel to $\mathscr{B}$ is

$$N_\pm = \tfrac{1}{2} \int_0^\infty f(\varepsilon) d\varepsilon \, \bar{n}(\varepsilon \mp \mu_B \mathscr{B}) , \qquad (D.20)$$

where $\tfrac{1}{2} f(\varepsilon) d\varepsilon$ is the density of states (11.76) and, from Eq. (11.73a),

$$\bar{n}(\varepsilon \mp \mu_B \mathscr{B}) = \frac{1}{\exp[\beta(\varepsilon \mp \mu_B \mathscr{B} - \mu)] + 1} .$$

The *same* chemical potential μ enters *both* these distributions since the system is in equilibrium.

The magnetization $\mathscr{I}$ of the specimen due to the alignment of electron spins is given by

$$\mathscr{I} = \frac{\mu_B}{V} (N_+ - N_-)$$

where V is the volume of the specimen.

At $T = 0\,\mathrm{K}$

$$\bar{n}(\varepsilon \mp \mu_B \mathscr{B}) = \begin{cases} 1, & \text{if } \varepsilon \mp \mu_B \mathscr{B} < \varepsilon_F \\ 0, & \text{if } \varepsilon \mp \mu_B \mathscr{B} > \varepsilon_F, \end{cases}$$

where $\varepsilon_F \equiv \mu(T = 0)$. Hence Eq. (D.20) becomes

$$N_\pm = \tfrac{1}{2} \int_0^{\varepsilon_F \pm \mu_B \mathscr{B}} f(\varepsilon) d\varepsilon$$

$$= \frac{4\pi V}{3h^3} (2m)^{3/2} (\varepsilon_F \pm \mu_B \mathscr{B})^{3/2}$$

$$= \frac{4\pi V}{3h^3} (2m\varepsilon_F)^{3/2} \left(1 \pm \frac{3}{2} \frac{\mu_B \mathscr{B}}{\varepsilon_F} \right). \qquad (D.21)$$

The last step follows since in practice $\mu_B \mathscr{B} \ll \varepsilon_F$. ($\mu_B = 5.8 \times 10^{-5}$ eV/tesla, ε_F is of the order of a few eV.) With $N = N_+ + N_-$, we have

$$N_\pm = \frac{1}{2} N \left(1 \pm \frac{3}{2} \frac{\mu_B \mathscr{B}}{\varepsilon_F} \right)$$

and

$$\mathscr{I} = \frac{3}{2} \frac{N}{V} \frac{\mu_B^2 \mathscr{B}}{\varepsilon_F} . \qquad (D.22)$$

These results allow a simple interpretation. If in Eq. (D.21) we introduce new variables of integration

$$E = \varepsilon \mp \mu_B \mathscr{B},$$

we obtain

$$N_\pm = \tfrac{1}{2} \int_{\mp \mu_B \mathscr{B}}^{\varepsilon_F} f(E \pm \mu_B \mathscr{B}) dE$$

$$= \frac{2\pi V}{h^3} (2m)^{3/2} \int_{\mp \mu_B \mathscr{B}}^{\varepsilon_F} (E \pm \mu_B \mathscr{B})^{1/2} dE. \qquad (D.23)$$

The two densities of states $f(E \pm \mu_B \mathscr{B})$ which enter this equation are shown in Fig. D.3. The two shaded areas represent $N_\pm$. For $\mathscr{B} = 0$ the two halves of the parabola would join smoothly at $E = 0$. For $\mathscr{B} \neq 0$ the energy levels of electrons with magnetic moments parallel (antiparallel) to $\mathscr{B}$ are displaced downwards (upwards) leading to $N_+ > N_-$.

Comparison of Eq. (D.22) with Eq. (3.8) shows that it is approximately only a fraction T/T_F of the conduction electrons which contribute to the magnetization. Because of the Pauli principle, it is only the electrons with energy within an interval of the order of kT near ε_F which can 'flip' their spins to align themselves with the magnetic field. The same argument occurred in the discussion of the electronic heat capacity: only electrons with energy near ε_F could be excited.

The spin paramagnetism of the conduction electrons was first considered by Pauli in 1927. In addition to this effect, an applied

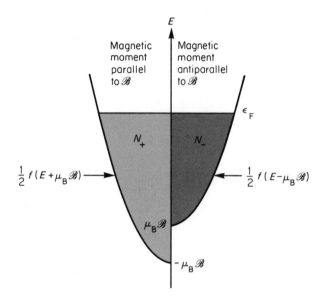

Fig. D.3. Pauli spin paramagnetism at 0 K. $E = \varepsilon \mp \mu_B \mathscr{B}$ is the total energy of an electron with magnetic moment parallel/antiparallel to $\mathscr{B}$ respectively. At $T = 0$ K all levels up to the Fermi energy ε_F are occupied.

magnetic field also alters the spatial motion of the electrons. This can be shown to lead to a diamagnetic effect which produces an additional magnetization equal to one-third of the spin paramagnetism and having the opposite sign. This diamagnetic effect was first calculated by Landau in 1930.

11.2 Applying Eq. (11.177) to Eq. (11.78) one obtains

$$N = \left[\frac{4\pi V}{h^3} (2m)^{3/2} \right] \left\{ \frac{2}{3} \mu^{3/2} + \frac{\pi^2}{12} \frac{(kT)^2}{\mu^{1/2}} \right\}$$

whence, to order $(kT/\varepsilon_F)^2$,

$$\mu = \varepsilon_F \left\{ 1 - \frac{\pi^2}{12} \left(\frac{T}{T_F} \right)^2 \right\} . \tag{D.24}$$

11.3 The energy E of the electron gas is given by

$$E = \int_0^\infty \varepsilon \bar{n}(\varepsilon) f(\varepsilon) \, d\varepsilon$$

(see Eq. (11.77)). Applying Eq. (11.177) to this integral, and remembering that to order $(T/T_F)^2$ one must represent μ by Eq. (D.24), one obtains

$$E = \text{const.} + N \frac{\pi^2}{4} \frac{(kT)^2}{\varepsilon_F}$$

from which Eq. (11.95) follows directly.

11.4

(i) $E = \int C_V dT = 0.772RT \left(\frac{T}{T_c}\right)^{3/2} = 0.128 \frac{(kT)^{5/2} m^{3/2}}{\hbar^3} V$

(ii) $S = \int \frac{C_V}{T} dT = \frac{5E}{3T}$

(iii) $P = - \left(\frac{\partial F}{\partial V}\right)_T = + \frac{2}{3} \left(\frac{\partial E}{\partial V}\right)_T = 0.085 \frac{(kT)^{5/2} m^{3/2}}{\hbar^3}$.

We see that below the condensation temperature T_c the pressure is independent of the volume and depends only on the temperature T.

11.5

$$P = \frac{1}{5} (3\pi^2)^{2/3} \frac{\hbar^2}{m} \left(\frac{N}{V}\right)^{5/3} .$$

N.B. If you did *not* get this answer, this could be because you modelled your answer on problem 11.4 and forgot about the zero-point energy of the electron gas.

11.6 With the outward normal to the metal surface as x-axis, the current density of emitted electrons is

$$j = - \frac{e}{\hbar^3} \int_{p_0}^{\infty} dp_x \frac{p_x}{m} \int_{-\infty}^{\infty} dp_y \int_{-\infty}^{\infty} dp_z \frac{2}{\exp\left[\beta(\varepsilon - \mu)\right] + 1} . \qquad \text{(D.25)}$$

The p_x-integration is restricted to

$$p_x \geqslant p_0 \equiv (2m\chi)^{1/2} ,$$

since only electrons with $p_x \geqslant p_0$ can escape from the metal. It follows from

$$\varepsilon = \frac{1}{2m} (p_x^2 + p_y^2 + p_z^2)$$

that

$$\varepsilon - \mu \geqslant \chi - \mu \equiv \phi .$$

Since $\phi \gg kT$, we can omit the term $(+1)$ compared to the exponential in the denominator of Eq. (D.25), i.e. Eq. (D.25) becomes a Maxwellian distribution. Only electrons from the tail of the FD distribution have sufficient energy to escape from the metal, and the tail has a MB distribution. The resulting integrals in Eq. (D.25) are of the type evaluated in Appendix A. One obtains

$$j = - \frac{4\pi em}{h^3} (kT)^2 e^{-\phi/kT} \ .$$

$\phi = \chi - \mu$ is known as the work function of the metal. Only electrons with energy exceeding the chemical potential μ by at least an amount ϕ can escape from the metal. At not too high temperatures $\mu = \varepsilon_F$, the Fermi energy, and $\phi = \chi - \varepsilon_F$. The work function ϕ, like ε_F, is of the order of a few electron-volts which justifies the assumption $\phi \gg kT$. Although the picture of the electrons moving in a constant potential inside the metal and experiencing a sharp potential step at its surface is an oversimplification, it reproduces the characteristic features of thermionic emission.

11.7 The entropy of the system is a function of the energy, the volume and the particle numbers $N_1, N_2, \ldots N_s$ of the different species of molecules:

$$S = S(E, V, N_1, \ldots N_s) \ .$$

For equilibrium S is a maximum. With E and V fixed this means that

$$dS = \sum_{i=1}^{s} dN_i \left(\frac{\partial S}{\partial N_i} \right)_{E,V,N_j, j \neq i} = 0 \ .$$

The differential coefficient $\partial S/\partial N_i$ is related to the chemical potential μ_i of the ith component in the system (see Eq. 11.117)):

$$\mu_i = - T \left(\frac{\partial S}{\partial N_i} \right) \ .$$

(The variables which are held constant during differentiation are the same as in the previous equation.)

The particle numbers of the different components cannot vary independently of each other but are related through the chemical reaction (11.138). This requires

$$dN_i \propto \nu_i \ .$$

Combining the last three relations we obtain the equilibrium condition

$$\sum_i \nu_i \mu_i = 0 \ ,$$

which was already derived in section 11.9, Eq. (11.144), where a system at constant temperature and pressure was considered.

11.8

$$\frac{c(P)c(e^-)}{c(H)} = K_c(T) \equiv \frac{f_P(T)f_{e^-}(T)}{f_H(T)} \; .$$

To calculate the equilibrium constant K_c we need the functions $f_i(T)$, Eq. (11.152), i.e. the partition functions for the protons, electrons and hydrogen atoms. Care is required to take the *same* zero of the energy scale in calculating the three internal partition functions. We shall take as the zero of the energy scale the state in which a proton and an electron are infinitely far apart and both at rest. The energy of the ground state of the hydrogen atom is then $\varepsilon_0 = -13.6\,\text{eV}$, i.e. the ionization energy of the hydrogen ground state is 13.6 eV. From Eq. (11.152) we then have

$$f_{e^-}(T) = \frac{1}{V} Z_{e^-}(T, V, 1) = \left(\frac{2\pi m_e kT}{h^2}\right)^{3/2} \times 2$$

where the factor 2 comes from the two possible spin states of the electron (i.e. it is the internal partition function for this case). Similarly

$$f_P(T) = \frac{1}{V} Z_P(T, V, 1) = \left(\frac{2\pi m_P kT}{h^2}\right)^{3/2} \times 2 \; ,$$

where the factor 2 comes from the two spin states of the proton. For the hydrogen atom we have

$$f_H(T) = \frac{1}{V} Z_H(T, V, 1) = \left(\frac{2\pi m_H kT}{h^2}\right)^{3/2} Z_H^{\text{int}} \; . \qquad \text{(D.26)}$$

For the internal partition function we need only take the ground state of hydrogen into account, since the first excited state lies 10.2 eV above the ground state so that its population is negligible even at very high temperatures. The ground state contribution to the partition function is

$$Z_H^{\text{int}} = 4e^{-\beta\varepsilon_0} \; .$$

The factor 4 is the degeneracy of this state due to the two possible spin orientations for the proton and two for the electron. The exponential factor $\exp(-\beta\varepsilon_0)$ arises from our choice of the zero of the energy scale at the ionization limit of the hydrogen atom instead of at its ground state. From the above equations we obtain, with $m_H = m_P$,

$$K_c(T) = \left(\frac{2\pi m_e kT}{h^2}\right)^{3/2} e^{\beta\varepsilon_0} \; .$$

11.9

$$\frac{[c(\text{H})]^2}{c(\text{H}_2)} = K_c(T) \equiv \frac{[f_{\text{H}}(T)]^2}{f_{\text{H}_2}(T)} \; . \tag{D.27}$$

From Eq. (11.152)

$$f_{\text{H}_2}(T) = \left(\frac{4\pi m_{\text{H}} kT}{h^2}\right)^{3/2} Z_{\text{H}_2}^{\text{int}} \; . \tag{D.28}$$

Using the results of Chapter 7 one can show that

$$Z_{\text{H}_2}^{\text{int}} = \frac{T}{2\,\Theta_{\text{rot}}} \times \frac{e^{-x/2}}{1-e^{-x}} \times 4\exp(-\beta\varepsilon_{\text{D}}) \; . \tag{D.29}$$

Here the first factor is the rotational partition function for $T \gg \Theta_{\text{rot}}$ (see problem 7.1) and the second factor, with $x \equiv \hbar\omega/kT$, is the vibrational partition function (see problem 7.2). The last factor is the partition function of the electronic and nuclear motion. Only the ground state (i.e. the lowest electronic state) of the H_2 molecule need be taken into account since all other electronic states lie at much higher energies. The factor 4 comes from the 4 possible spin states of the two protons. The spins of the electrons in the H_2 ground state are uniquely determined. The factor $\exp(-\beta\varepsilon_{\text{D}})$ stems from our choice of the common zero of energy in calculating the partition functions of atomic and molecular hydrogen. We choose for this zero the state with the two hydrogen atoms at rest, infinitely far apart. Relative to this state, the H_2 ground state energy is $\varepsilon_0 = -4.48$ eV. In this state the hydrogen molecule possesses the vibrational energy $\frac{1}{2}\hbar\omega = 0.27$ eV. This vibrational energy has already been included in calculating the vibrational partition function [compare Eqs. (6.3) and (6.5)]. Writing

$$\varepsilon_0 = \tfrac{1}{2}\hbar\omega + \varepsilon_{\text{D}} \; ,$$

it follows that with our choice of zero of energy, $Z_{\text{H}_2}^{\text{int}}$ will contain a factor $\exp(-\beta\varepsilon_{\text{D}})$ instead of $\exp(-\beta\varepsilon_0)$, with $\varepsilon_{\text{D}} = -4.75$ eV.

$f_{\text{H}}(T)$ is again given by Eq. (D.26), but with our present choice of the zero of energy

$$Z_{\text{H}}^{\text{int}} = 4 \; . \tag{D.30}$$

Substituting Eqs. (D.26) and (D.28)–(D.30) in (D.27) gives

$$K_c(T) = \left(\frac{4\pi m_{\text{H}} kT}{h^2}\right)^{3/2} \frac{\Theta_{\text{rot}}}{T} \frac{1-e^{-x}}{e^{-x/2}} \exp(\beta\varepsilon_{\text{D}}) \; .$$

If dissociation is the only reaction which need be considered and if the fraction f of H_2 molecules which is dissociated is small, then f can be expressed in terms of the equilibrium constant by

$$f = \frac{c(H)}{2n} \approx \frac{1}{2} \left[\frac{K_c(T)}{n} \right]^{1/2}$$

where $2n$ is the *total* number of hydrogen atoms per unit volume, whether in the form of free hydrogen atoms or as constituents of hydrogen molecules.

11.10 The partial pressures of NO_2 and N_2O_4 are

$$P(NO_2) = \frac{2f}{1+f} P , \qquad P(N_2O_4) = \frac{1-f}{1+f} P$$

where P is the total pressure. From the law of mass action (11.159):

$$\frac{4Pf^2}{1-f^2} = K_P(T) .$$

It follows from this equation that increasing the pressure, at constant temperature, decreases f, i.e. the fraction of molecules dissociated decreases. This is in agreement with our discussion following Eq. (11.176).

11.11 For equilibrium at constant T and P, $G = $ minimum. Hence for

$$- dN_1^{(1)} = + dN_1^{(2)} = dN , \qquad \text{all other } dN_i^{(\alpha)} = 0 ,$$

Eq. (11.178) gives

$$dG = - (\mu_1^{(1)} - \mu_1^{(2)})dN = 0$$

where

$$\mu_i^{(\alpha)} = \left(\frac{\partial G(T,P,N_1^{(\alpha)}, \ldots N_s^{(\alpha)})}{\partial N_i^{(\alpha)}} \right)_{T,P,N_j(j \neq i)}$$

is the chemical potential of the ith component in the αth phase. Hence when the system is in equilibrium $\mu_1^{(1)} = \mu_1^{(2)}$. More generally, we have the equilibrium conditions

$$\mu_i^{(1)} = \mu_i^{(2)} = \ldots = \mu_i^{(p)} , \qquad i = 1, \ldots s ;$$

i.e. in equilibrium, the chemical potential of each component has the same value in all phases.

11.12 For a single donor level we have three possible states: the level is empty (energy zero), the level is occupied by a 'spin up' electron (energy $-\varepsilon_0$), or it is occupied by a 'spin down' electron (energy $-\varepsilon_0$). Each of the n donor levels can, independently of the others, be in any one of these three states. Hence the grand partition function is given by

$$\mathcal{Z} = (1 + 2\,e^{\beta(\mu + \varepsilon_0)})^n \ .$$

The number of electrons in the donor levels is given by

$$n_d = -\frac{\partial \Omega}{\partial \mu}$$

with $\Omega = -kT\ln\mathcal{Z}$; whence

$$n_d = \frac{n}{1 + \frac{1}{2}e^{-\beta(\mu + \varepsilon_0)}} \ .$$

11.13 Since G is an extensive variable

$$\lambda G(T,P,N_1, \ldots .N_s) = G(T,P,\lambda N_1, \ldots .\lambda N_s) \ .$$

Differentiation with respect to λ followed by taking $\lambda = 1$ gives Eq. (11.179). Hence

$$dG = \sum_{i=1}^{s} (\mu_i dN_i + N_i d\mu_i) \ ,$$

and comparing this equation with Eq. (11.140) leads to Eq. (11.180) which is known as the Gibbs–Duhem equation. It shows, in particular, that at constant temperature and pressure the chemical potentials $\mu_1, \ldots .\mu_s$ are not independent variables but must satisfy the differential relation

$$\sum_{i=1}^{s} N_i d\mu_i = 0 \ .$$

11.14 From $E = H - PV$ and

$$\Delta E = Q_V, \qquad \Delta H = Q_P, \qquad \Delta(PV) = \Delta(RT\sum N_i) = RT\sum \nu_i$$

we get

$$Q_V = Q_P - RT\sum \nu_i = RT^2 \frac{d}{dT} \ln K_P(T) - RT\sum \nu_i \ .$$

From Eqs. (11.159) and (11.156)

$$\frac{d}{dT} \ln K_P(T) = \frac{d}{dT} \ln K_c(T) + \frac{1}{T} \sum \nu_i \ ,$$

and combining the last two equations, Eq. (11.181) follows. Alternatively, one can derive Eq. (11.181) directly from

$$Q_V = \Delta E = - T^2 \left(\frac{\partial}{\partial T} \frac{\Delta E}{T} \right)_{T, V, N_1, \ldots N_s} \tag{D.31}$$

which is the analogue of Eq. (11.168) when T and V, rather than T and P, are the independent variables. (Compare also Eq. (4.46).) The further calculation from Eq. (D.31) is similar to that from Eq. (11.168).

11.15 From Eq. (11.13)

$$p_N = \frac{e^{\beta \mu N}}{\mathcal{Z}} \sum_r e^{-\beta E_{Nr}} = \frac{e^{\beta \mu N}}{\mathcal{Z}} Z(T, V, N),$$

and from Eq. (7.11) this becomes

$$p_N = \frac{[e^{\beta \mu} Z_1(T, V)]^N}{\mathcal{Z} \, N!} \ .$$

From Eq. (11.135) (where we must now write $\bar{N}$ rather than N for the mean, since we are considering fluctuations)

$$\bar{N} = e^{\beta \mu} Z_1(T, V)$$

and from Eq. (11.133)

$$\mathcal{Z} = e^{\bar{N}} \ .$$

Hence

$$p_N = \frac{\bar{N}^N e^{-\bar{N}}}{N!} \ .$$

Bibliography

This bibliography lists books referred to in this text or which are particularly suitable for further study. It is in no way comprehensive. References are arranged approximately according to subject, though overlaps necessarily occur.

THERMODYNAMICS, STATISTICAL MECHANICS AND KINETIC THEORY

1. C. J. Adkins, *Equilibrium Thermodynamics*, 3rd edn., Cambridge University Press, Cambridge, 1983. (Good standard approach plus a new version of Carathéodory's treatment of 2nd law.)
2. R. Becker, *Theory of Heat*, 2nd edn., Springer, Berlin, 1967. (Contains a very stimulating and catholic account of statistical mechanics.)
3. H. B. Callen, *Thermodynamics and an Introduction to Thermostatistics*, 2nd edn., Wiley, New York, 1985. (Good account of thermodynamics, based on an axiomatic approach, plus an introduction to statistical physics. Postgraduate level.)
4. K. Denbigh, *The Principles of Chemical Equilibrium*, 4th edn., Cambridge University Press, Cambridge, 1981. (A good book on thermodynamics. The applications cover most topics of interest in chemistry.)
5. T. L. Hill, *An Introduction to Statistical Thermodynamics*, Addison–Wesley, Reading, Mass., 1960. (A good introduction with many applications to physics and chemistry.)

6. C. Kittel and H. Kroemer, *Thermal Physics*, 2nd edn., W. H. Freeman, San Francisco, 1980. (Emphasizes the Gibbs chemical potential approach. Many diverse applications.)

7. L. D. Landau and E. M. Lifshitz, *Statistical Physics*, 3rd edn., part 1 (revised and enlarged by E. M. Lifshitz and L. D. Pitaevskii), Pergamon, Oxford, 1980. (First class postgraduate text.)

8. D. K. C. MacDonald, *Introductory Statistical Mechanics for Physicists*, Wiley, New York, 1963. (Contains much good physics, little formalism. Refreshing style.)

9. A. B. Pippard, *The Elements of Classical Thermodynamics*, Cambridge University Press, Cambridge, 1957. (A superb mature treatment.)

10. M. Planck, *Theory of Heat*, Macmillan, London, 1932.

11. R. D. Present, *Kinetic Theory of Gases*, McGraw-Hill, New York, 1958. (Good elementary treatment.)

12. L. E. Reichl, *A Modern Course in Statistical Physics*, University of Texas Press, Austin, 1980, and Edward Arnold, London, 1980. (Theoretical postgraduate treatment.)

13. F. Reif, *Fundamentals of Statistical and Thermal Physics*, McGraw–Hill, New York, 1965. (A good book, covering many topics, including very readable introductions to transport theory and irreversible thermodynamics.)

14. G. S. Rushbrooke, *Introduction to Statistical Mechanics*, Oxford University Press, Oxford, 1949. (A clear, elementary account oriented towards physical chemistry. No quantum statistics.)

15. E. Schrödinger, *Statistical Thermodynamics*, Cambridge University Press, Cambridge, 1946. (pp. 1–26 undergraduate level. All of it is illuminating and fun, if not particularly useful for applications.)

16. A. Sommerfeld, *Thermodynamics and Statistical Mechanics*, Academic Press, New York, 1956. (Good physics, mature approach.)

17. D. ter Haar, *Elements of Thermostatistics*, Holt, Rinehart and Winston, New York, 1966. (Very good, detailed, theoretical treatment of statistical mechanics. On the whole, postgraduate level. Presupposes thermodynamics.)

18. D. ter Haar and H. Wergeland, *Elements of Thermodynamics*, Addison–Wesley, Reading, Mass., 1966. (A good account of the mathematical skeleton.)

19. J. R. Waldram, *The Theory of Thermodynamics*, Cambridge University Press, Cambridge, 1985. (An individualistic and stimulating account. Advanced undergraduate level.)

20. G. H. Wannier, *Statistical Physics*, Wiley, New York, 1966. (First class; detailed applications to several fields. Postgraduate level.)

21. A. H. Wilson, *Thermodynamics and Statistical Mechanics*, Cambridge University Press, Cambridge, 1966. (Very good theoretical account. Contains many detailed applications. Postgraduate level.)

22. M. W. Zemansky and R. H. Dittman, *Heat and Thermodynamics*, 6th edn., McGraw–Hill, New York, 1981. (Good standard text, containing many detailed applications.)

PROPERTIES OF MATTER, SOLID STATE PHYSICS

23. N. W. Ashcroft and N. D. Mermin, *Solid State Physics*, Holt, Rinehart and Winston, New York, 1976. (Theoretical, postgraduate treatment.)
24. J. S. Blakemore, *Solid State Physics*, 2nd edn., Cambridge University Press, Cambridge, 1985. (Undergraduate level but less advanced than Hall[27] or Kittel.[29])
25. A. H. Cottrell, *The Mechanical Properties of Matter*, Wiley, New York, 1964. (Reprint edition: Krieger, New York, 1981.)
26. B. H. Flowers and E. Mendoza, *Properties of Matter*, (Manchester Physics Series), Wiley, London, 1970.
27. H. E. Hall, *Solid State Physics*, (Manchester Physics Series), Wiley, London, 1974.
28. T. L. Hill, *Lectures on Matter and Equilibrium*, Benjamin, New York, 1966.
29. C. Kittel, *Introduction to Solid State Physics*, 6th edn., Wiley, New York, 1986.

LOW TEMPERATURE PHYSICS

30. J. S. Dugdale, *Entropy and Low Temperature Physics*, Hutchinson, London, 1966. (Good, elementary account.)
31. F. E. Hoare, L. C. Jackson and N. Kurti, *Experimental Cryogenics*, Butterworths, London, 1961.
32. C. T. Lane, *Superfluid Physics*, McGraw–Hill, New York, 1962. (A very readable introduction.)
33. F. London, *Superfluids*, Vol. 2, Wiley, New York, 1954. (Reprint edition: Dover, New York, 1964.) (Theoretical postgraduate treatment.)
34. P. V. E. McClintock, D. J. Meredith and J. K. Wigmore, *Matter at Low Temperatures*, Blackie, Glasgow, 1984. (Concise, up-to-date account at advanced undergraduate/introductory postgraduate level.)
35. D. R. Tilley and J. Tilley, *Superfluidity and Superconductivity*, 2nd edn., Adam Hilger, Bristol, 1986. (Up-to-date postgraduate book.)
36. G. K. White, *Experimental Techniques in Low-Temperature Physics*, 3rd edn., Oxford University Press, Oxford, 1979.
37. J. Wilks, *The Properties of Liquid and Solid Helium*, Oxford University Press, Oxford, 1967. (A detailed discussion of the experimental data. Postgraduate level.) See also: J. Wilks, *An Introduction to Liquid Helium*, Oxford University Press, Oxford, 1970. (An abridged version of the 1967 book.)

38. J. Wilks, *The Third Law of Thermodynamics*, Oxford University Press, Oxford, 1961.

MISCELLANEOUS REFERENCES

39. R. M. Eisberg, *Fundamentals of Modern Physics*, Wiley, New York, 1964.
40. U. Fano and L. Fano, *Basic Physics of Atoms and Molecules*, Wiley, New York, 1959.
41. R. P. Feynman, R. B. Leighton and M. Sands, *The Feynman Lectures on Physics*, Vol. 1, Addison–Wesley, Reading, Mass., 1963.
42. A. P. French and E. F. Taylor, *An Introduction to Quantum Physics*, Thomas Nelson, Middlesex, 1978.
43. H. Goldstein, *Classical Mechanics*, 2nd edn., Addison–Wesley, Reading, Mass., 1980.
44. L. D. Landau and E. M. Lifshitz, *Mechanics*, 3rd edn., Pergamon, Oxford, 1976.
45. F. Mandl, *Quantum Mechanics*, 2nd edn., Butterworths, London, 1957.
46. F. Mandl and G. Shaw, *Quantum Field Theory*, Wiley, Chichester, 1984.
47. A. I. M. Rae, *Quantum Mechanics*, 2nd edn., Adam Hilger, Bristol, 1986.
48. J. C. Willmott, *Atomic Physics*, (Manchester Physics Series), Wiley, London, 1975.

Index

Conversion Factors

1 ångström (Å)	10^{-8} cm
1 newton (N)	10^5 dyn
1 joule (J)	10^7 erg
1 cal	4.1840 J
1 electron-volt (eV)	$\begin{cases} 1.60 \times 10^{-19}\,\text{J} \\ 1.1605 \times 10^4\,\text{K} \\ 23.05\,\text{kcal mol}^{-1} \end{cases}$
1 atmosphere (atm)	$\begin{cases} 760\,\text{mmHg} \\ 1.013 \times 10^5\,\text{N m}^{-2} \\ 1.013 \times 10^6\,\text{dyn cm}^{-2} \end{cases}$

$\nu = 10^{10}$ cycles/s:

associated energy $h\nu$ 4.14×10^{-5} eV

associated temperature $T = h\nu/k$ 0.48 K

The symbols $\approx$ and $\sim$ are used to mean 'approximately equal to' and 'of the order of' respectively.